AHDV

ORDINATION OF PLANT COMMUNITIES

ORDINATION OF PLANT COMMUNITIES

edited by

Robert H. Whittaker

Dr W. Junk Publishers The Hague-Boston-London 1982

ISBN 90 6193 565 2

Cover design Max Velthuijs

PREFACE

A large part of ecological research depends on use of two approaches to synthesizing information about natural communities: classification of communities (or samples representing these) into groups, and ordination (or arrangement) of samples in relation to environmental variables. A book published in 1973, 'Ordination and Classification of Communities,' sought to provide, through contributions by an international panel of authors, a coherent treatise on these methods. The book appeared then as Volume 5 of the *Handbook of Vegetation Science*, for which R. TÜXEN is general editor. The desire to make this work more widely available in a less expensive form is one of the reasons for this second edition separating the articles on ordination and on classification into two volumes.

The other reason is the rapid advancement of understanding in the area of indirect ordination—mathematical techniques that seek to use measurements of samples from natural communities to produce arrangements that reveal environmental relationships of these communities. Such is the rate of change in this area that the last chapter on ordination in the first edition is already, 4 or 5 years after it was written, out of date; and new techniques of indirect ordination that could only be mentioned as possibilities in the first edition are becoming prominent in the field. In preparing the second edition the chapter on evaluation of ordinations has been rewritten, a new chapter on recent developments in continuous multivariate techniques has been included, and references to recent work have been added to other chapters.

This volume is designed to serve several functions for different readers. It offers an introduction to concepts of and approaches to ordination, and to the broader area of gradient analysis—the approach to natural communities through continuous or gradient relationships of environmental variables, species populations, and community characteristics to one another. It provides a survey of research approaches and ordination techniques, with some evaluation of these, for those interested in using the techniques. Ways an investigator can carry out direct ordinations are described (although the mathematical details of indirect techniques must be found in special publications). This is also, as the *Handbook* was designed to be, a scholar's reference work. The rapid advancement in indirect ordination must be expected to continue; the treatment of such techniques cannot be definitive. We trust, however, that this and the companion volume on classification will have durable value as introductions and references for the areas of ecological method that they treat.

Authors and addresses

GRANT COTTAM
Department of Botany, University of Wisconsin, Madison, Wisc. 53706, U.S.A.

PIERRE DAGNELIE
Faculté des Sciences Agronomiques, Statistique, Avenue de la Faculté 59, Gembloux, Belgique.

HUGH G. GAUCH, JR.
Ecology and Systematics, Cornell University, Ithaca, N.Y. 14853, U.S.A.

F. GLENN GOFF
Ecology Section, Oak Ridge National Laboratory, Oak Ridge Tenn. 37830, U.S.A.

DAVID W. GOODALL
CSIRO Division of Land Management, Private Bag, Wembley, W. A. 6014, Australia

ROBERT P. MCINTOSH
Department of Biology, University of Notre Dame, Notre Dame, Ind. 46556, U.S.A.

LÁSZLÓ ORLÓCI
Department of Plant Sciences, University of Western Ontario, London 72, Ont., Canada

IMANUEL NOY-MEIR
Department of Botany, Hebrew University, Jerusalem, Israel

L. N. SOBOLEV
Institute of Geography, USSR Academy of Sciences, Staromonetny 29, Moscow, U.S.S.R.

V. D. UTEKHIN
Institute of Geography, USSR Academy of Sciences, Staromonetny 29, Moscow, U.S.S.R.

ROBERT H. WHITTAKER
Ecology and Systematics, Cornell University, Ithaca, N.Y. 14853, U.S.A.

GEORGE M. WOODWELL
The Ecosystems Center, Marine Biological Laboratory, Woods Hole, Mass. 02543, U.S.A.

CONTENTS

(A companion volume on Classification of Plant Communities comprises articles 12 to 20.)

1 INTRODUCTION

Over the land surface of the world there extends a mantle of vegetation, a living fabric of plant communities that is diverse and subtle in its response to environment, varied in its structure and composition, expressive of the biological productivity of the land, and interesting and esthetically appealing to man. The mantle of vegetation is part of man's life in evolution and in the present; we see the mantle now in many areas subject to accelerating destruction and cannot yet know the implications of that destruction for man's environment and the psychology of a species that evolved in natural landscapes. The destruction may increase the importance of one of science's endeavors – knowledge and understanding of plant communities. It is always of interest and often of value to know the plant communities of an area, a country, or a continent. We wish more than the description of those communities; we seek understanding – of how the different communities relate to one another in the mantle of vegetation and express their environments, of how plant species evolve and distribute themselves in relation to one another in communities, and of how communities develop and are organized and function as living systems.

Especially, for the concerns of this book, we desire to understand the relations of plant communities to one another and the particular environments (biotopes) those communities occupy. Understanding is based on abstraction. From phenomena that are often intricate and obscure, significant relationships are to be detected, embodied in concepts, related to one another, and tested and revised in the systems of abstraction that are science. As much as any other phenomena, plant communities offer complexities that challenge our efforts at abstraction and understanding. In dealing with these complexities students of vegetation seek ways of relating to one another information on three levels – environmental factors, species populations, and community characteristics. We may distinguish two broadly conceived research methods as ways of abstracting and of seeking relations among these (WHITTAKER 1962, 1967).

One, the more traditional and familiar, is classification. When plant communities are classified, a number of samples representing communities are grouped together on the basis of shared characteristics into an abstract unit or *class* of plant communities. Throughout this discussion such units of classification, on any level, by any definition of shared characteristics, may be referred to as *community-types*. When vegetation samples have been classified into a community-type one may then determine for this type the ranges of environmental factors, species compositions, and community characteristics that its samples represent. Through classification we relate kinds of communities to kinds of environments as an approach to understanding the landscape and its vegetation.

The alternative method deals not primarily with classes (which may be treated as discontinuous with one another) but with continuities and gradient relationships. Study of communities in terms of gradients of phenomena on the three levels is termed *gradient analysis* (WHITTAKER 1951, 1967). Samples from plant communities may be arranged in sequence by their positions along a gradient of environment (or of community characteristics). In this sequence of samples (a *transect*) changes in species populations and community characteristics are related to changes in environment. In many cases the samples are arranged in relation to a coordinate system of two or more environmental or community gradients. The arrangement permits us to interrelate samples (and species populations) as parts of an abstract pattern of variation in species composition and other characteristics of communities, and to seek understanding of vegetation in terms of the major directions of variation in that pattern. If the samples are arranged in terms of one or more environmental gradients accepted as given, as a basis of the study, the study is *direct* gradient analysis. Alternatively, measurements of sample similarity or species correlation may be used to derive abstract axes or directions of community variation. If the samples are arranged in terms of such abstract axes (which may or may not correspond to environmental gradients), the study is one of *indirect* gradient analysis. Direct and indirect gradient analysis are not sharply separated from one another; various intermediates and combinations are possible. They represent different perspectives on the solution of vegetation problems, however, and will be separately treated here.

The process of arranging samples (or species) in relation to environmental gradients or axes is termed *ordination* (RAMENSKY 1930, GOODALL 1954). Ordination is an essential technique of gradient analysis but is not (or should not be) its purpose. Classification likewise should not be an objective in itself but a means of ordering information, describing, interrelating, and generalizing about phenomena. In dealing with individual organisms the more technical approach to classification is called *taxonomy*; whereas the broader concern with relationships of organisms, particularly on the basis of evolution, is called *systematics*. In dealing with plant communities broader approaches to description and understanding that are based on classification are often termed *phytosociology* (BRAUN-BLANQUET 1932, 1964). Use of this term has varied between English-language and Continental countries (EGLER 1942, WESTHOFF 1970); and for some Continental students phytosociology represents the whole of vegetation science. Phytosociology may be distinguished, however, from plant geography with its concern with floristics and relations of vegetation structure to environment on a continental or world-wide scale. Within the study of natural communities (biocoenology, symbiology, synecology in its Anglo-American use) we may well regard phytosociology as the approach to vegetation that emphasizes the classificatory perspective and the species composition of communities, the approach we think most characteristic of phytosociology in western Europe. We may then set in approximate parallel three pairs of terms, each including an essential procedure and the approach to understanding of which that procedure is part: taxonomy and systematics, community classification ('syntaxonomy') and phytosociology, ordination and gradient analysis.

The present articles are concerned more with procedures than with application to particular communities; this is an account of ordination and classification, rather than of gradient analysis and phytosociology. Some of the results from gradient analysis must, however, be considered as a basis for understanding both ordination and classification of communities. The articles that follow will proceed from direct gradient analysis and some of its results, through indirect ordination (from the less to the more mathematical forms of this), to the basis of community classification, to reviews of various major approaches to classification. Articles have been arranged by their primary emphasis, but the reader will find that a number of them concern both ordination and classification. Ordination and classification are in no wise antagonistic. Many studies in gradient analysis use a classification to present some of their results, and studies based on classification may approach the interrelations of community-types and species through gradient analysis. These two major approaches are thus complementary and can be combined in various ways for the most effective study and interpretation of plant communities and their relations to environment.

SUMMARY

Two major ways of relating the plant communities of a landscape to one another may be distinguished. Classification groups communities into classes (community-types) that may be characterized and treated as discontinuous with one another. Gradient analysis studies vegetation in terms of gradients of environment, species populations, and community characteristics in relation to one another. The arrangement of samples in relation to environmental gradients as a basis for gradient analysis is termed *ordination*. Ordination may be either direct (arranging samples by positions along one or more environmental gradients accepted as given), or indirect (from comparisons of samples are derived compositional axes, directions of change in community composition, along which the samples are arranged.) The articles (2 to 11) of this book are arranged in sequence from direct gradient analysis, through similarity measurements and indirect ordinations, to evaluations of and recent developments in ordination, while its companion volume deals with major approaches to the classification of communities (in articles 12 to 20).

References for this article are given with those for article 2.

2 DIRECT GRADIENT ANALYSIS

Robert H. Whittaker

Contents page

2 DIRECT GRADIENT ANALYSIS

2.1 Samples

2.1.1 Sample Characteristics

Community samples ('Aufnahmen,' relevés) are the essential working material for gradient analysis. The procedures to be described below can (with some limitations) be applied to any kind of community sample. Thus one can, in principle, use for gradient analysis samples of the vascular plants of forests, lichens on the barks of trees, singing bird pairs of grasslands, marine zooplankton, or soil microarthropods. We shall concern ourselves primarily, however, with samples that include the vascular plants (with or without accompanying data on thallophytes) from land communities. A sample will normally include (i) a list of plant species present in a given study area, the plot or quadrat, (ii) some indications of relative importance of these species (and usually of their growth-forms, heights, and stratal relationships), and (iii) supporting information on environment, soil, community structure, and evidences of disturbance and community change.

There are a wide range of kinds of samples that can be applied to plant communities and used for gradient analysis. Some considerations affecting choice and kind of sample are: (i) The sample should be large enough in area, or should include counts of a sufficient number of plants, to represent effectively the composition of the plant community. If the samples are too large, however, difficulties are encountered in meeting the two following considerations. (ii) The sample should be homogeneous – there should be no trend of change in community composition or structure from one edge of the sample to another. (iii) The sample should be efficient. Since considerable numbers of samples may be needed, the samples should be designed to obtain and record rapidly the kinds of information regarded as most important. (iv) The sample should be appropriate. Among the many kinds of information that might be gathered on a plant community, some are more interesting or significant, appropriate to the character of the community, and informative in relation to time spent and the purposes of gradient analysis than others. These considerations of minimum area, homogeneity, efficiency, and appropriateness of samples are to be discussed in the *Handbook of Vegetation Science*, Part IV. Certain types of samples have been most widely used for gradient analysis of plant communities. The most widely used, and from many standpoints the most efficient, of all types of samples for classification is the relevé of the School of BRAUN-BLANQUET (1932, 1964, ELLENBERG 1956, and article 20 of this book). There is no reason the BRAUN-BLANQUET relevé cannot be used also for gradient analysis (MAAREL 1969, MOORE et al. 1970). In American work in gradient analysis more strongly quantitative samples, with additional time spent to obtain more effective measurements on plant populations, have been preferred. The measurements that express relative abundance, or conspicuousness, or massiveness, or vigor of species in natural communities are termed 'importance values.' Among the various possible importance values the following are of most interest to us: (i) *Density* is the number of individuals of a species per unit ground surface area (or other spatial measurement). (ii) *Coverage* is the percentage of ground-surface (or other substrate) area above which foliage of a given species occurs. (iii) *Basal area* is the area occupied by cross-sections of stems at 1.3 m above the ground (or other plant basal measurements) per unit ground surface area. (iv) *Frequency* is the percentage of small subquadrats within a larger sample plot, in which a given species is present. (v) Species *biomass* is the total mass (usually dry weight of organic matter) present at a given time per unit ground surface area. (vi) Species net *production*, for plants, is total dry weight growth (or its energy equivalent), or organic matter production by photosynthesis minus the plants' own respiration.

These and other measurements are often divided by the total of the same kind of importance values for all species in the sample; the resulting percentages are *relative* importance values. When two or more importance values are combined, the result is a *synthetic* importance value; for example the Wisconsin synthetic importance value (CURTIS & MCINTOSH 1951, GOFF & COTTAM 1967) sums for each species its relative density, relative frequency, and relative basal area. Two types of samples used in the United States for efficient measurement of importance values are:

2.1.2 QUADRAT SAMPLES

For the tenth-hectare strip quadrat (WHITTAKER 1960, WHITTAKER & NIERING 1965) a 50-meter tape is laid out on the ground, and a rectangular plot extending 10 meters on each side of the tape is marked out. All trees rooted within the plot are tallied by species and diameters at breast height (1.3 m). Shrub individuals (and tree seedlings in the shrub stratum) are tallied for the same area or, if very dense, in a narrower inner strip along the tape. Herbs (and tree and shrub seedlings in the herb stratum) are sampled in 25 subquadrats. The subquadrats may be either (i) bounded by eye and a meter stick, on alternate sides of the tape, every other meter, or (ii) marked out with ring-stakes tied into cord (with the cord measured to produce 1×1 or 0.5×2.0 m^2 areas) at random numbers of meters distant from the central tape, on alternate sides of that tape at 2 meter intervals. Occurrence and coverage (visually estimated by dm^2 units) of plant species in these subquadrats are recorded; in some studies densities in the subquadrats are also recorded. Tree and shrub coverages are measured either for the 100 points at corners of the subquadrats, or as intercept coverage along the 50-meter tape, or both. (In open vegetation intercept coverage should be taken along the long sides as well as the center of the quadrat, hence for 150 m in all). Lichen and moss occurrence in the plot should be recorded. The sample may be completed with information on elevation, topographic position, parent material and soil, location, disturbance and developmental trends, and voucher specimens of plant species as needed.

Such samples are effective for obtaining measurements of density, frequency, and coverage of plant species of different sizes in a wide range of vegetation types (WHITTAKER & NIERING 1965). The 0.1 hectare (20×50 m) size has been found suitable for forest, woodland, grassland, and desert samples; but the quadrats have been expanded to 0.2 or 0.5 ha in forests of large trees, and contracted to 0.05 or 0.01 ha in some dense shrub and pygmy forest communities. When procedures are established and plant species are known, samples from most communities (that are not too rich floristically) can be taken in 45 minutes to 1.5 hour by an investigator with an assistant. It is thus possible to take several samples a day for the systematic sampling of a vegetation pattern.

2.1.3 PLOTLESS OR POINT-CENTERED SAMPLES

The Wisconsin School (COTTAM & CURTIS 1949, 1956, CURTIS 1959, and article 7) have developed a quite different set of sample procedures. In a tract of forest judged homogeneous a number of points (often 50 or 100) are located by random or systematic procedures. Each of these points (or the tree nearest the point) is used as sample-point, and distances from these sample-points to trees nearest them, and species and diameters of these trees, are recorded. Undergrowth subquadrats may be located at the same sample-points. Details of the various techniques cannot be described here, but are discussed by GREIG-SMITH (1964). The distance measurements permit calculation of tree densities; from densities together with diameters may be calculated basal areas; and from occurrence at sample-points as subquadrats may be calculated frequencies.

The point-centered systems are usually (though not necessarily for all applications) more time-demanding than the tenth-hectare quadrats. Point-centered samples, with systematic or random location of points through the sample area and undergrowth information and environmental data, generally take several hours. The point-centered samples are more effective in representing the composition of a large forest tract (of the order of 10 ha or more); the tenth-hectare samples are more effective in representing in greater detail community composition within a limited area. Plotless or point-centered samples were used as standard for sampling forest tracts on the more level topography of Wisconsin; the tenth-hectare samples have been used in mountains and other areas where steep environmental gradients make point-centered samples inappropriate because of the large areas needed for such samples. The two sampling approaches have different advantages for different applications; but the tenth-hectare samples are more widely applicable and for many purposes more efficient.

Point-centered samples can be applied to grassland tufts and desert shrubs, but it is not clear that such application is advantageous. Applied to a grassland without trees or shrubs, the tenth-hectare plot becomes a sample of herbs in 25, 1-m^2 subquadrats. A number of gradient analyses of grasslands and desert have been based on rectangular, square, or circular subquadrats of varying numbers and sizes (CURTIS 1955, DIX 1959, PERRING 1959, ITOW 1963, GITTINS 1965, DIX & SMIENS 1967, BEALS 1969b). Further discussions of quantitative sample procedures are given in the books by BROWN (1954), CAIN & OLIVEIRA (1959), PHILLIPS (1959), GREIG-SMITH (1964), GOUNOT (1969), MUELLER-DOMBOIS & ELLENBERG (1974) and in the *Handbook of Vegetation Science*, Part IV.

2.1.4 SAMPLE SETS

Whatever the type of sample used, a study in gradient analysis is based on a sample *set* – a number of samples taken with similar procedures to represent a range of variation in environment and community composition. It is natural to think that samples of the set might best be taken from random locations in a landscape. Formal randomization of sample location has been abandoned by most who have tried it (cf. MOORE et al. 1970), because location of the sample points in the field with surveying techniques is time consuming, because a high proportion of the samples obtained may be heterogeneous or disturbed, and because randomization may be inefficient to represent the range of variation of communities in the landscape. Procedures often used are: (i) Along a well-defined gradient of vegetation and environment, samples are taken at fixed intervals (for example, of elevation up a mountain slope, of compass degrees around a hill, of distance down a desert bajada or from a water body outward through its bordering vegetation). (ii) In a substantially disturbed landscape, samples may be taken from all, or many, of the tracts of vegetation that are sufficiently large, undisturbed, and homogeneous to give useful samples. Such has been a procedure of the Wisconsin school. (iii) In an area of complex environmental variation, such as mountain topography, samples are taken at frequent but unspecified intervals as the investigator encounters new combinations of community composition and environment. The investigator may walk through the mountains taking samples with each change of 50 m elevation, or change of compass direction of exposure, as he goes, excluding samples which are heterogeneous (or, if succession is not being studied, disturbed). Such sampling may be continued until the samples are judged to represent adequately the range of communities and environments that has been observed and is to be studied. If some topographic positions of limited areal extent (e.g. ravines) are poorly represented, samples of these may be specially sought (WHITTAKER 1956, 1960, WHITTAKER & NIERING 1965).

Numbers of samples desired for sets may vary widely. In procedure (i) relatively few samples, of the order of 5 to 20, may serve the purposes of some studies. In procedure (ii) in the Wisconsin school, sample sets have been of the order of 50 to 100 samples. For procedure (iii) the author has used as a standard 50 to 60 samples from each 300-meter or 1000-foot elevation belt on a given parent material in mountains, hence several hundred samples to represent the full range of variation of the mountain vegetation. For phytosociological studies samples are often chosen to represent community types already recognized by the investigator. For gradient analysis it is better to avoid such bias, and to base sample choice on environmental variations and not primarily on community composition according to recognized types. (If samples have been taken to be representative of community-types, they cannot then be used to test the continuity or discontinuity of those types.) All three of the procedures suggested can provide for sample choice that is unbiased in this respect, though not formally random.

2.2 Transects of Single Complex-Gradients

2.2.1 Simple and Composite Transects

Our concern in direct gradient analysis is with the way in which species are distributed and vegetation changes along environmental gradients. These relationships are best observed when a set of vegetation samples are arranged in sequence along an environmental gradient; such a sequence of samples forms a *transect.* In some cases the simplest approach to an environmental gradient – a single series of samples taken at intervals along it – suffices to reveal what the investigator wants to know. Such simple or field transects have been used for study of elevation gradients (Mark 1963, Mark & Sanderson 1962, Scott et al. 1964, Wells & Mark 1966, Beals 1969a), topographic moisture gradients (Horikawa & Okutomi 1955, Okutomi 1958, Mowbray 1966, Mowbray & Oosting 1968), and the vegetation surrounding a swamp or water body (Beschel & Webber 1962).

In many cases, however, such a simple transect provides data that are too limited or too irregular and subject to sample error. It is often desirable to group a larger number of samples, taken from similar environmental conditions at different locations within the study area, into a *composite transect.* Samples may be grouped by sets of five, say, for 30-meter elevation intervals, or for positions along a topographic moisture gradient. Such grouping tends to average out the irregularities affecting individual samples, and to produce a clearer, more easily interpreted picture of species distributions and community change along the gradient. Table I is such a composite transect. Each column in the table is an average of tree-species densities in five, tenth-hectare samples that are similar in composition and represent similar positions along the complex-gradient. The manner in which species populations rise and fall along the gradient in Table I is representative of the way species populations are distributed in a large number of transects, simple and composite, direct and indirect, that have been studied (Whittaker 1967).

2.2.2 Complex-Gradients and Environmental Ordination

Environmental gradients are complex phenomena. Along a given spatial gradient, for example of elevation up a mountain slope,

Table I. A composite transect of the topographic moisture, 1830-2140 m elevation, in the Santa Catalina Mountains, southern Arizona (Whittaker & Niering 1965). Fifty samples are ordinated by weighted average indexes (first column) and grouped in subsets of five per transect step. Topographic positions range from mesic ravines (step 1) to xeric open southwest-facing slopes (step 10). Data are numbers of tree stems over 1 cm dbh in 0.5 ha. The communities are pine-evergreen oak woodlands. See also Fig. 1.

	Species Weight	Transect steps 1	2	3	4	5	6	7	8	9	10	Mean Density /ha	Mean Coverage %	Transect Constancy %
Acer glabrum var. neomexicanum	1	6										1.2	0.1	8
Alnus oblongifolia	1	53										10.6	3.0	10
Salix lasiolepis	1	55										11.0	0.6	10
Abies concolor	1	57	1									11.6	0.4	8
Quercus gambelii	1	32	10									8.4	0.05	6
Juglans major	1	7	6	1								2.8	0.08	16
Pinus strobiformis	1	39	12	2								10.6	0.36	18
Pseudotsuga menziesii	2	44	65	27	1							27.4	1.79	32
Robinia neomexicana	2	8	5	2	1							3.2	0.06	16
Rhamnus californica ssp. ursina	2	3	13	4	1							4.2	0.31	28
Prunus virens	2	6	38	21	6	1						14.4	0.85	28
Pinus ponderosa	2	18	322	236	106	24	3	1				142.0	10.1	52
Quercus rugosa	2	126	93	58	36	22	12	16	16	11	5	79.0	2.5	40
Quercus hypoleucoides	3	2	223	335	526	536	362	124	53.	62	11	446.8	19.1	90
Arbutus arizonica	3		2	8	27	16	8					12.2	0.27	32
Pinus chihuahuana	3		1	4	10	32	36	7				18.0	0.85	24
Juniperus deppeana	4		1	4	4	13	46	51	39	29	25	42.4	1.95	72
Quercus arizonica	4		9	18	47	72	98	100	82	25	19	94.0	3.03	62
Pinus cembroides	4			1	3	13	29	93	161	114	85	99.8	6.0	70
Garrya wrightii	4			.1	4	9	5	13	68	57	10	33.4	17.3	66
Arctostaphylos pringlei	4				6	21	134	54	154	154	34	111.4	9.8	50
Arctostaphylos pungens	4				1	10	17	14	5	12	7	13.2	6.9	50
Quercus emoryi	5						5	8	18	14	76	24.2	0.77	22
Total stems/0.5 ha		456	801	722	779	769	755	481	596	478	272			
Mean weighted average		1.46	2.27	2.54	2.89	3.12	3.43	3.67	3.89	3.85	4.20			
Percentage similarity to step 1.		100	31.8	20.2	10.5	7.02	2.81	4.06	3.42	2.78	1.92			
" " " " 10.		1.92	4.85	8,04	10.1	19.4	27.3	54.2	48.8	56.0	100			
Mean coverage %, trees		59.	96.	80.	76.	60.	44.	41.	26.	25.	25.			
shrubs		0.5	1.6	1.5	3.2	22.	34.	42.	33.	39.	33.			
herbs		6.4	3.9	1.3	0.3	1.6	1.5	0.5	0.8	1.2	2.5			
Mean vascular plant species per sample		28.5	24.8	17.8	16.4	21.0	16.6	13.6	17.4	15.2	17.2			

many environmental factors of precipitation and humidity, temperature and wind velocity, and soil characteristics change together. An environmental gradient may be conceived as a *complex-gradient* comprising gradients of many particular environmental factors, or characteristics of environmental complexes, that change together in space (WHITTAKER 1956, 1967). Gradient analysis may be based on arrangement or *ordination* of samples either (i) in relation to complex-gradients that are accepted as given, in direct gradient analysis, or (ii) along axes of variation in composition of samples, axes derived from study of the samples themselves, in indirect gradient analysis. The latter axes may or may not correspond to complex-gradients. (In either case some gradients or axes may represent disturbance or successional time, rather than spatial variation.) In direct gradient analysis ordination may be based either on environmental measurements, or on indices expressing relative position in a gradient of community composition as this changes in response to an environmental complex-gradient.

For some complex-gradients a convenient environmental measurement is available by which samples may be arranged. Elevation provides an index of sample position along a complex-gradient, even though elevation itself has no bearing on plant physiology. Soil characteristics can also be used as indices of ordination for some studies; for example moisture conditions, acidity, and other qualities of agricultural soils (ELLENBERG 1950, 1952), nutrient levels and acidity along a forest gradient of increasingly intense leaching (WESTMAN 1975), salinity or mechanical analysis of particle sizes along a desert slope or bajada (YANG & LOWE 1956). For the topographic moisture gradient samples may be arranged by topographic positions–from deep ravines or valley bottoms, through smaller canyons and lower sheltered slopes, to open slopes of varying exposures (in the northern hemisphere, from northeast and north, through northwest, east, and west, to southeast, south, and southwest—note the horizontal axis of Fig. 6). Topographic positions are crude indices of position along the complex-gradient; other approaches to topographic moisture gradients are discussed by LOUCKS (1962) and WENTWORTH (1976, section 2.2.4).

2.2.3 COENOCLINES AND COMPOSITIONAL ORDINATION

There are other ways of arranging samples along complex-gradients when no single environmental measurement provides a satisfactory index. Underlying such possibilities is a principle: Community composition expresses the characteristics of the community environment or biotope. (Community composition may express disturbance and time-gradients of succession as well as stable environmental factors.) Corresponding to, and expressing, a complex-gradient is a compositional gradient of species importance values in communities (see Table I). The community gradient may also be termed a *coenocline* (WHITTAKER 1960, 1967, MAAREL 1960, MAAREL & LEERTOUWER 1967). The complex-gradient and coenocline together form a gradient of communities-and-environments, or of ecosystems, an *ecocline* (Clements 1936, WHITTAKER 1960, 1967, 1970a). Since the community gradient expresses the environmental gradient, we may use relative positions of samples in the community-gradient as indices of their relative positions along the complex-gradient. There are two major approaches to ordinating samples by their composition, when an environmental complex-gradient is accepted as given:

2.2.3.1 *Weighted averages*

Fig. 1 gives smoothed curves for the species populations along an environmental gradient (data of Table I). Note that, as indicated by the numbers above the centers or modes for these species, the species can be grouped by relative positions of their centers along the gradient. For the transect of Fig. 1 and Table I species were grouped and numbered by relative position along the topographic

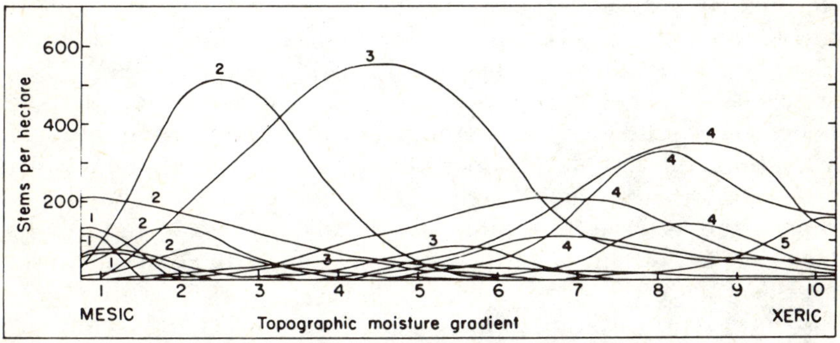

Fig. 1. Species populations and ecological groups in a coenocline. Population distributions are shown for major tree species in the 1830–2140 m elevation belt in the Santa Catalina Mountains, southern Arizona from mesic ravines (left) to xeric southwest slopes (right) (Table I and WHITTAKER & NIERING 1965). Ecological groups for these species in relation to the topographic moisture gradient are indicated and used for the weighted averages of Table I; the groups are: mesic (1), submesic (2), subxeric (3), xeric (4), and more strongly xeric (5).

moisture gradient as: moist or mesic (1), submesic (2), subxeric (3), dry or xeric (4), and one more strongly xeric species (5) (WHITTAKER & NIERING 1965). Such species groupings, based on distributional similarity, are termed *ecological groups* (DUVIGNEAUD 1946, ELLENBERG 1950, 1952, TANGHE 1964, WHITTAKER 1967). The technique of arranging samples by weighted averages was independently discovered by several authors and has been applied to both environmental gradients (ELLENBERG 1948, 1950, 1952, WHITTAKER 1951, 1954a, 1956, 1960, WHITTAKER & NIERING 1965, 1968a, 1968b, CURTIS & MC INTOSH 1951, CURTIS 1955, 1959, ROWE 1956, KNIGHT 1965, WARING & MAJOR 1964, BRAY 1956, 1960, MAYCOCK & CURTIS 1960, DIX & SMIENS 1967, BEALS 1969b, see also MCINTOSH 1967, GOFF & COTTAM 1967, LEWIN 1974) and disturbance gradients (DYKSTERHUIS 1948, 1958, DIX 1959, ITOW 1963).

The ecological groups are derived from field data – either extensive observation of species behavior (ELLENBERG 1950, 1952, CURTIS 1955), or field transects (WHITTAKER 1956), or preliminary composite transects (WHITTAKER 1960, WHITTAKER & NIERING 1965, cf. BROWN & CURTIS 1952, DIX & SMIENS 1967). For analysis of the topographic moisture gradient, for example, a preliminary composite transect may use topographic positions to arrange the samples. Distributions of species in this preliminary transect are used to classify them into ecological groups on which a more effective, weighted-average ordination may be based (WHITTAKER 1960, WHITTAKER & NIERING 1965). If species distributions are used to arrange samples for the study of species distributions, the method involves a clear circularity (WHITTAKER 1956), and this circularity has been part of criticisms of gradient analysis (DAUBENMIRE 1966, LANGFORD & BUELL 1969, GOUNOT 1969). The circularity does not apply to transects of elevation gradients. For topographic gradients and other transects using compositional ordination, the circularity does not invalidate the method when ordination is based on correctly observed gradient relations of species and is used to refine sample arrangements and results that can be obtained also by use of environmental measurements.

Since representation of the ecological groups changes continuously along the coenocline, weighted averages expressing representation of the ecological groups can be used as indices of relative position along the coenocline (and, hence, the complex-gradient). For example, in the first column of Table I, stems of species designated as mesic (249 stems in all) are multiplied by 1 as a weight, submesic stems (205) by 2, subxeric stems (2) by 3, and xeric stems (0) by 4. The sum, $249 \times 1 + 205 \times 2 + 2 \times 3 + 0 \times 4 = 665$, is divided by the unweighted sum (456) to give a quotient, the weighted average 1.46. The 50 individual samples were arranged in sequence by weighted averages, and then grouped in sets of five to produce the composite transect summarized in Table I. The progressive change in weighted averages along the coenocline is indicated in a bottom row of Table I. It is sometimes desirable to apply independent weightings to different strata of a set of samples, and to use the two weighted averages as axes of a chart (Fig. 2). The two weightings provide a check on each other; and samples which are deviant for the set (affected by disturbance or environmental factors other than the complex-gradient under study) may appear above or below the main axis of the chart (WHITTAKER 1960, WHITTAKER & NIERING 1965).

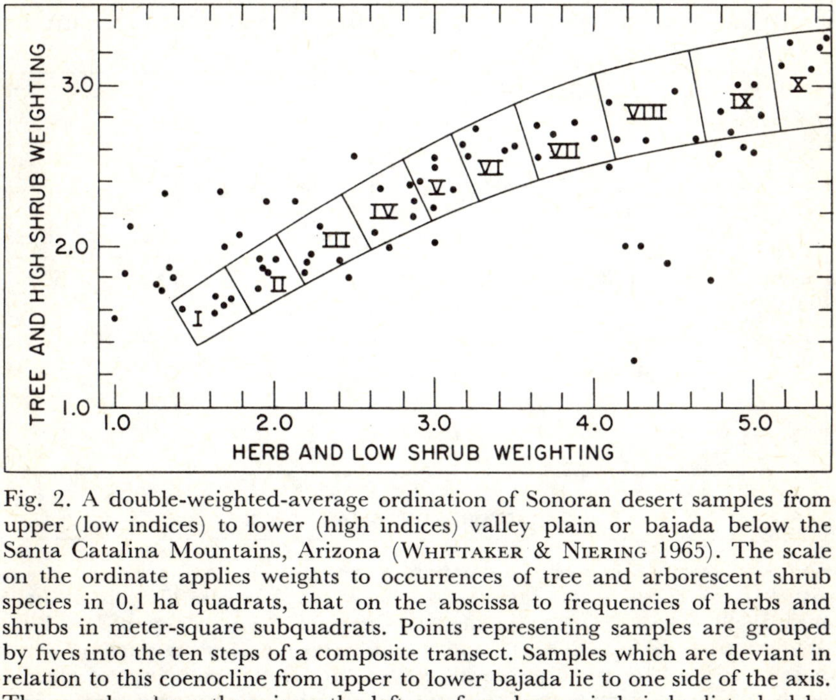

Fig. 2. A double-weighted-average ordination of Sonoran desert samples from upper (low indices) to lower (high indices) valley plain or bajada below the Santa Catalina Mountains, Arizona (WHITTAKER & NIERING 1965). The scale on the ordinate applies weights to occurrences of tree and arborescent shrub species in 0.1 ha quadrats, that on the abscissa to frequencies of herbs and shrubs in meter-square subquadrats. Points representing samples are grouped by fives into the ten steps of a composite transect. Samples which are deviant in relation to this coenocline from upper to lower bajada lie to one side of the axis. The samples above the axis on the left are from less xeric bajada disturbed by grazing, those below the axis on the right are from desert washes.

2.2.3.2 *Sample comparisons*

If we start with samples representing one end of a coenocline, other samples will be increasingly unlike these end-point samples with increasing distance along the gradient. We can thus use quantitative comparison of samples to express distance of samples from one another along a coenocline (compositional distance or ecological distance, WHITTAKER 1952, 1967, article 3). There are a number of ways of measuring sample similarity, discussed in article 5; one of the simplest is percentage similarity in the form, $PS = 100 - 0.5\,\Sigma\,|\,a - b\,| = \Sigma \min\,(a, b)$, in which a and b are, for a given species, the relative importance values for that species in two samples. End-point samples (or averaged groups of samples) representing the opposite extremes of an environmental gradient may be chosen. Using percentage similarities or related measurements we can arrange samples from those most like the first end-point, through various intermediate positions, to those most like the opposite end-point. The manner in which such values change along a gradient is indicated by the percentage similarities for the transect steps, compared with steps 1 and 10, at the bottom of Table I. (The arrangement may be based on the same calculations

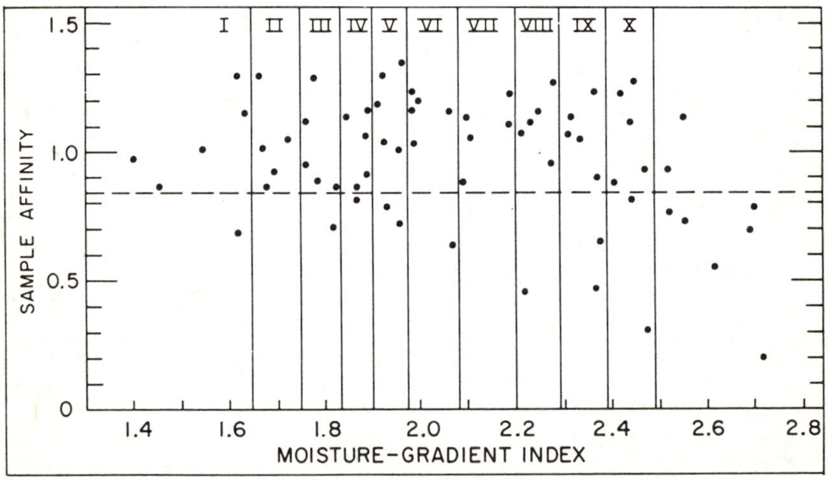

Fig. 3. A sample-similarity ordination technique, applied to low-elevation (610–910 m) forests in the Siskiyou Mountains, southwestern Oregon (WHITTAKER 1960). Three reference points were used, each an average of 5, 0.1–ha samples: most mesic ravines (1), intermediate, open east-facing slopes (2), and most xeric south to southwest-facing slopes (3). These numbers were used as weights. For each sample its similarity values with the three reference points were first summed to give the 'sample affinity' index on the ordinate. The similarity values were then multiplied by the weights, and the sum of weighted values was divided by the unweighted 'sample affinity' to give the moisture-gradient index on the abscissa, by which the samples are arranged in sequence from those most like (1) to those most like (3). Samples with low sample affinity indexes were deviant in some respect (of parent material or disturbance) from the main body of samples for the coenocline. Samples with affinity values above 0.85 were grouped by fives into ten steps for a composite transect of the topographic moisture gradient.

as those used in Wisconsin comparative and other indirect ordinations (BRAY & CURTIS 1957, BEALS 1960, LOUCKS 1962, WHITTAKER 1967, articles 7 and 10). The procedure then differs from indirect ordination only in that the end-point samples have been chosen to represent a complex-gradient accepted as given.) Fig. 3 represents a variant of this ordination technique (WHITTAKER 1960). Again, samples have been ordinated and grouped in sets of five, to form a composite transect.

2.2.4 EFFECTIVENESS OF ORDINATION

Ways in which the relative effectiveness of different ordinations of the same set of samples can be compared are discussed by WHITTAKER (1960, 1967), BRAY (1961), LOUCKS (1962), and article 10. Three approaches to ordination effectiveness are:

(i) It is assumed that increased effectiveness of ordination will be expressed in reduced dispersions of species distributions through the intervals of the transects. The dispersions can be measured as standard deviations if the data are adequate; but it is simpler, and may suffice, to compare the numbers of transect intervals through which species distributions extend. These numbers are averaged for two transects being compared, using the same lists of species that are represented by sufficient measurements to be useful and that occur in several but not all steps of the transect (WHITTAKER 1960, BRAY 1961).

(ii) It is assumed that more effective ordination will be expressed in closer approach to binomial patterns of species importance values along the gradient, and hence greater consistency in decline of importance values away from the peak or modal values. Differences in species importance values in successive transect steps are averaged to give an index that decreases with increasing effectiveness of ordination (WHITTAKER 1960). For a measure that is probably somewhat more sensitive, the importance value differences that decline in sequence away from the peak value are summed, and from this sum are subtracted the amounts by which values that increase away from the peak would have to be reduced to make them consistent with the trends (BRAY 1961).

(iii) It is assumed that more effective ordination will result in closer mean similarities of adjacent samples in the transect. LOUCKS (1962) used a statistical test of sample similarity; but percentage similarity or other sample similarity measurements (articles 5, 10) can be averaged for the sets of samples combined into transect steps, or for adjacent transect steps, to express the ordination's success in placing similar samples next to one another.

Ordinations by composition (weighted averages or sample comparisons) are in many cases more effective than ordination by any single environmental measurement. Such is particularly true of topographic moisture gradients, for which samples are best ordinated by composition; elevation in contrast, may provide as effective an index for ordination as any index based on composition. Between the two major means of compositional ordination – weighted averages and sample comparisons – the advantage is less clear. Weighted averages are probably more effective when the range of community change is large along a complex-gradient that can be accepted as given, hence for many direct gradient analyses. Ordination by sample comparison may be equally effective for shorter ranges of community variation, and may be preferred when the major environmental variables are less easily recognized in the field. Sample comparison may also be preferred when, because the number of samples is limited or because their composition is affected by more than one complex-gradient, it is difficult to assign species to ecological groups. It is quite possible to derive weighted averages from indirect techniques (Bray 1956, 1960, Dix & Butler 1960, Beals 1969b) and the indirect technique of reciprocal averaging (Hill 1973, articles 10 and 11) simultaneously produces species ordinations and weights, and ordinated sample positions.

Tests were used to compare a number of direct and indirect ordinations in the study of Wentworth (1976, article 10.4.1.3). Effectiveness was measured by a weighted average of species dispersions on the ordination axes and by mean per cent similarity of adjacent samples in transects as ordinated; the latter was the more effective measure. Several ordinations were used for two sets of about 70 samples representing elevation belts: (1) direct ordination by topographic aspect (direction of exposure), (ii) scalars combining elevation and aspect (Fig. 4, above) or combining elevation and aspect with corrections for topographic position (Fig. 4, below), (iii) calculations of relative insolation from aspect, position, and slope inclination, (iv) direct weighted averages, (v) polar ordination (Bray & Curtis 1957, article 7), (vi) reciprocal averaging, (vii) Gaussian ordination (article 11), and (viii) principal components analysis (articles 9 and 10). In addition, various similarity measurements and transformations were compared. The scalars gave significant improvement in ordination over topographic position alone and were somewhat better than the insolation calculation (which made no allowance for elevation). Polar ordination gave better results than direct weighted averages; but reciprocal averaging and Gaussian ordination seemed best over-all for arranging samples along a first, major direction of vegetational variation. Reciprocal averaging results were improved by removing samples from canyons and draws, and other extreme sites, from sample sets (10.4.2.3). Principal components analysis seemed the least effective ordination. Data transformations made no obvious difference in ordination effectiveness; but the tests suggested a marginal improvement in ordination using log transformed scores over that using raw scores. The work of Loucks (1962) and Wentworth (1976) shows the effectiveness of scalars for some direct ordination; and the studies of Wentworth (1976) and Peet (1976) suggest the effectiveness of combinations or direct and indirect ordination (e.g. direct for the elevation gradient, indirect by reciprocal averaging or Gaussian ordination within elevation belts).

2.3 **Analysis of Intercommunity Patterns**

Most areas have a complex mantle of vegetation affected by more than one complex-gradient. Means are needed by which effects of these different gradients on community composition can be determined and related to one another. A number of techniques will be described as approaches to intercommunity patterns (that is, landscape patterns involving different kinds of communities and their relations to one another, as distinguished from intra-community patterns involving contagion and species association within a given community).

2.3.1 Crossed Transects

In some cases two complex-gradients are predominant in their effects on communities of a landscape. Such may be the case with the elevation and topographic moisture gradients in moutains, and it may then be reasonable to analyze mountain vegetation in terms of these two gradients (if disturbance and parent material differences do not introduce major complications). Vegetation of the Great Smoky Mountains was analyzed (Whittaker 1956) through four transects of the topographic moisture gradient, holding the elevation gradient constant in each (within 1000-foot elevation belts), and four transects of the elevation gradient, holding 'con-

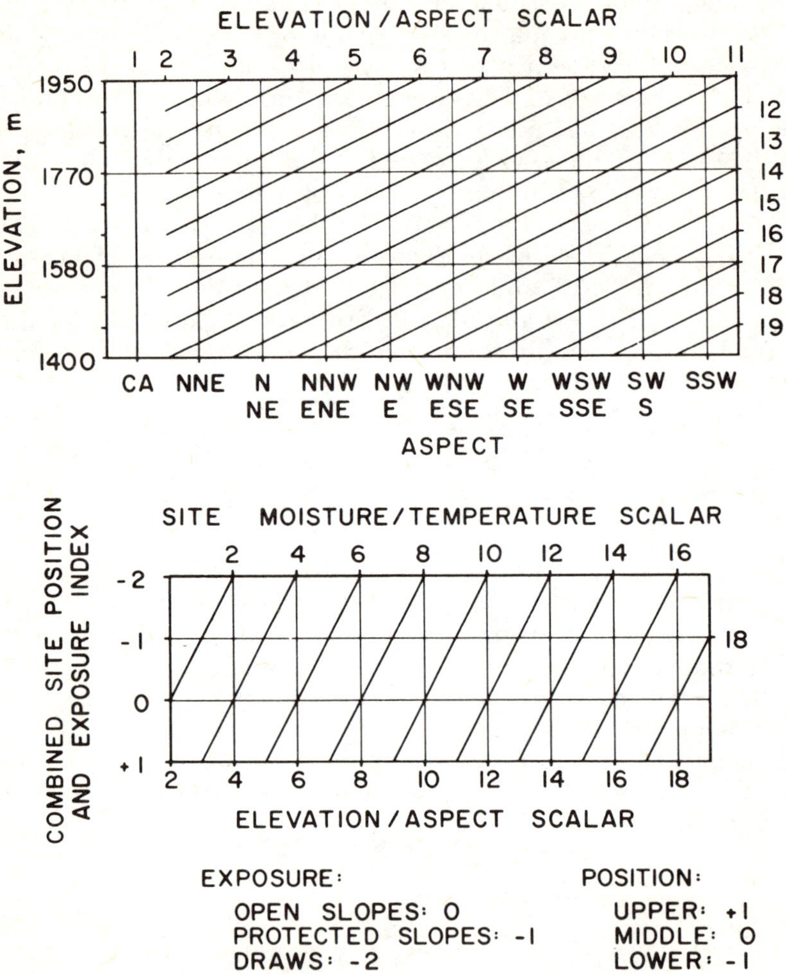

Fig. 4. Ordination scalar, developed from composite topographic transects in the Mule Mountains, Arizona, by WENTWORTH (1976)

Above: an elevation-aspect scalar combining elevation in meters with aspect, or direction of slope exposure. Data on mean rates of change in sample composition with change in elevation and aspect were used to determine the slope of the oblique lines relating these (about 60m/compass unit, but this relationship should be different in other mountain ranges). Canyons (CA) are treated as scalar value 1 (most mesic) regardless of elevation.

Below: a further correction of the elevation-aspect scalar for topographic exposure and position. (For example, if the elevation-aspect scalar is 10, the value remains 10 for an open middle slope, but becomes 11 for an open upper slope, 8 for a draw.)

stant' topographic moisture conditions. (Weighted averages were used to classify the samples into mesic, submesic, subxeric, and xeric groups, and each of these four groups was used for an elevation transect.) The pattern is thus approached through a grid of crossed transects representing effects of the major gradients separately and in different combinations. Where elevation difference is not a concern, direction of exposure and steepness of slope may be used as axes (PERRING 1959, 1961, AYYAD & DIX 1964). It is possible also to use ordinations by composition to treat vegetation in relation to two or more axes. In some cases it is feasible to classify species in the same vegetation into different sets of ecological groups by their responses to different environmental gradients. The same species, as differently classified in relation to different gradients, can then be used for weighted-average ordination of samples in relation to each of these gradients (ELLENBERG 1950, 1952, WARING & MAJOR 1964). Ordination by sample comparisons should be effective for sample sets that are not extensive enough to permit weighted-average treatment or to justify a transect grid. End-point samples may be chosen for two or more complex-gradients, and the samples may be ordinated into composite transects in relation to each of these. The samples may also be arranged in a coordinate system defined by these axes with the techniques of Wisconsin polar ordination (BRAY & CURTIS 1957, article 7) used for, in this case 'direct' gradient analysis.

2.3.2 COENOCLINE COMPARISONS

A related approach uses community-gradients as units of comparison between different landscapes (or elevation belts, or parent materials, or disturbance conditions). Coenoclines can be compared with one another in a variety of ways: (i) Relative importances and distributions along the gradients of species populations. Fig. 4, for example, illustrates a few species populations in their distributions along topographic moisture gradients on three parent materials, under similar climates at low elevations in the Siskiyou Mountains (WHITTAKER 1954b, 1960, cf. WHITTAKER & NIERING 1968a). Striking differences appear in the relative importances and distributions of species on the three parent materials. (ii) Character and composition of stands from the mid-points of the coenoclines. Mid-point samples, if undisturbed, may be used to

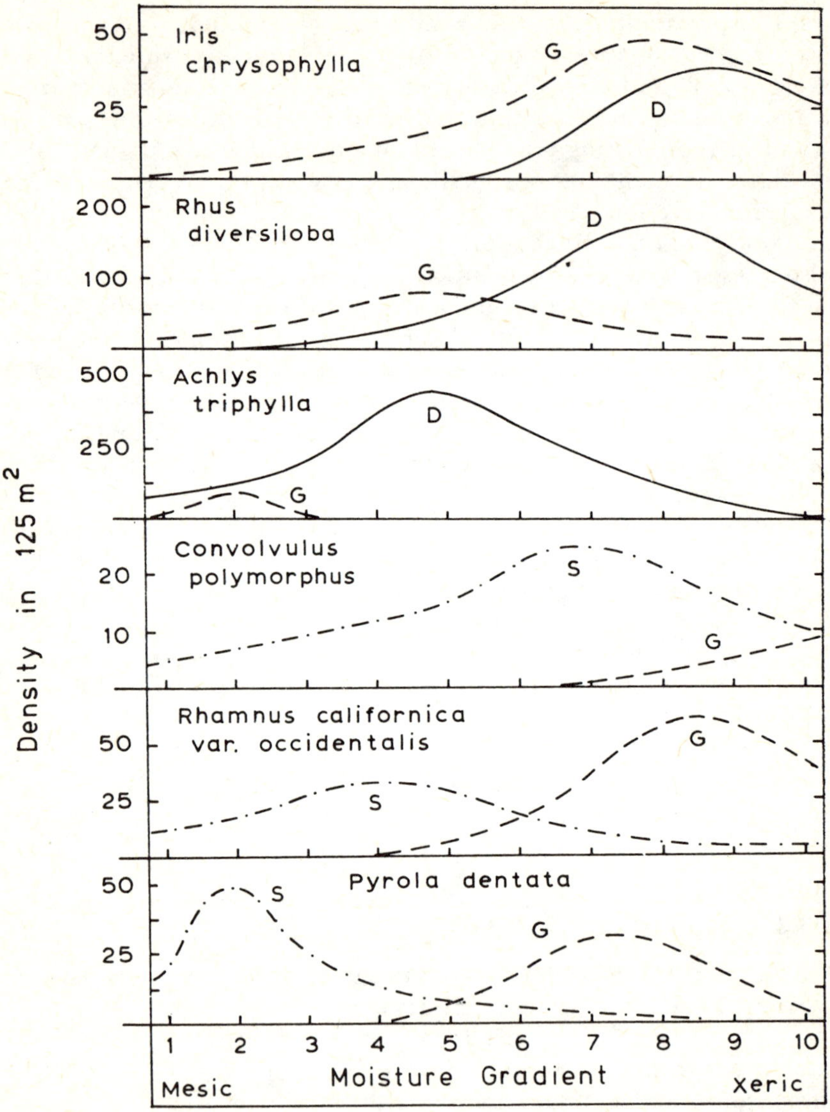

Fig. 5. Population shifts toward the mesic in response to parent materials, at low elevations in the Siskiyou Mountains, southwestern Oregon (WHITTAKER 1960). In each case the population distribution is shifted to the left (more mesic in terms of topographic positions) from acid parent material (D = quartz diorite) to basic parent material (G = olivine gabbro), or from basic parent material to ultrabasic (S = serpentinite and peridotite). The three study areas are near one another in similar climates, but comparable shifts are observed on a given parent material following species from a more to a less humid climate.

represent 'intermediate climax composition' (WHITTAKER 1960); along topographic gradients such mid-point samples may be from open east and west-northwest slopes. (iii) Representation of community-types. As species change in relative importance and distribution from one coenocline to another, so must the community-types they characterize. On a 10-point topographic transect scale a given kind of community may occupy positions 3 to 9 in a more humid climate and appear reduced in distribution and displaced toward the mesic end, occupying positions 1 to 2 or 2 to 4, in a more arid climate. (iv) Floristic composition of the coenoclines as wholes, combining all samples for each of these. To the coenocline floras may be applied such characterizations as RAUNKIAER life-form spectra, representation of areal-types, and over-all species diversity ('gamma diversity') (WHITTAKER 1960, WHITTAKER & NIERING 1965, 1968b). (v) Degree of floristic differentiation of communities along the gradient. Coenoclines differ in the average width of species distributions along the gradient, and consequently in the degree of turnover or replacement of species populations from one end-point sample to the other. Degrees of compositional change along coenoclines have been measured as 'beta diversity' (WHITTAKER 1960, 1965, 1967, 1972, WHITTAKER & NIERING 1965, 1968b, article 3).

2.3.3 MOSAIC CHARTS

Two complex-gradients may be used as axes of a chart, as in Fig. 6. The samples from the landscape are classified into community-types, or transitions between these; and each sample is plotted on the mosaic chart according to its elevation and topographic position. Boundaries are drawn around community-types, thus outlining the ranges of environments these occupy, and showing how they relate to one another and the landscape pattern as a whole. (Because of the crudity of topographic position as an expression of moisture conditions, the samples for different types may overlap broadly, and mean positions must be sought for the boundaries.) Fig. 6 gives two such mosaic charts, by which major characteristics of vegetation patterns may be compared for acidic (diorite) and ultrabasic (serpentinite and peridotite) soils within the same mountain range in southern Oregon (WHITTAKER 1960, cf. WHITTAKER 1951, 1956, 1967, WHITTAKER & NIERING 1965, 1968b, WARING & MAJOR 1964).

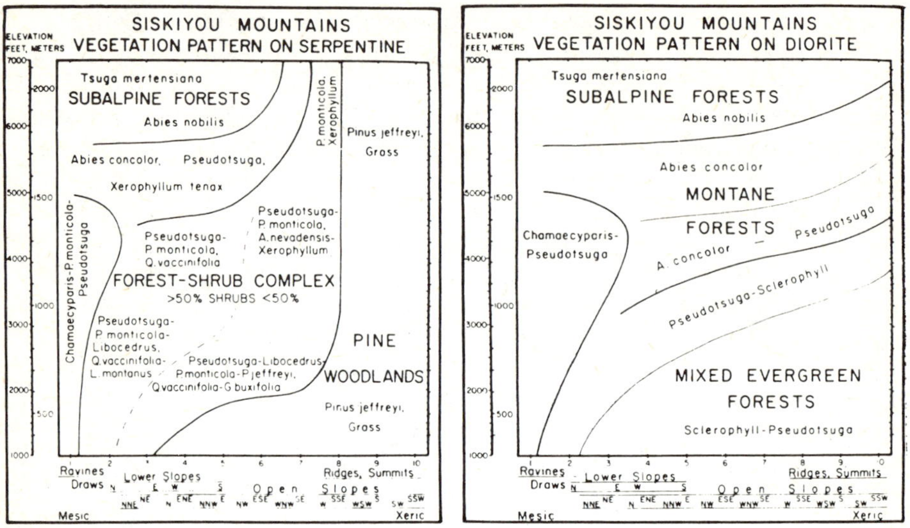

Fig. 6. Mosaic diagrams for vegetation on quartz diorite (right) and serpentines (left) in the Siskiyou Mountains, southwestern Oregon (WHITTAKER 1960, cf. WHITTAKER 1956, WHITTAKER & NIERING 1965, 1968b). Vegetation samples were classified (into dominance-types) and plotted by elevation and topographic position on the chart. Boundaries for community-types were drawn at average positions of transitions between types. Vegetation on each parent material is considered a complex pattern of continuously intergrading communities (WHITTAKER 1960).

2.3.4 NOMOGRAMS

Further data can be plotted at the points on the mosaic chart that represent samples. (It is usually advantageous to group samples in sets of 5 or 10, average the composition for a group, and plot the data at the mean positions on the chart for the groups.) When importance values for a species have been plotted at points on the mosaic chart, isodems may be drawn to show the distribution of the population. Such population charts typically appear as elliptical hills (Fig. 7, upper left); but genetically complex species may show more complex patterns, as in the other three charts of Fig. 7. Community-level characteristics may be plotted and outlined in the same manner, as illustrated in Fig. 8 (WHITTAKER 1952, 1956, WHITTAKER & NIERING 1965).

2.4 Conclusion on Techniques

The diagrams represent ways in which techniques serve the purposes of gradient analysis. A first purpose is ordering of infor-

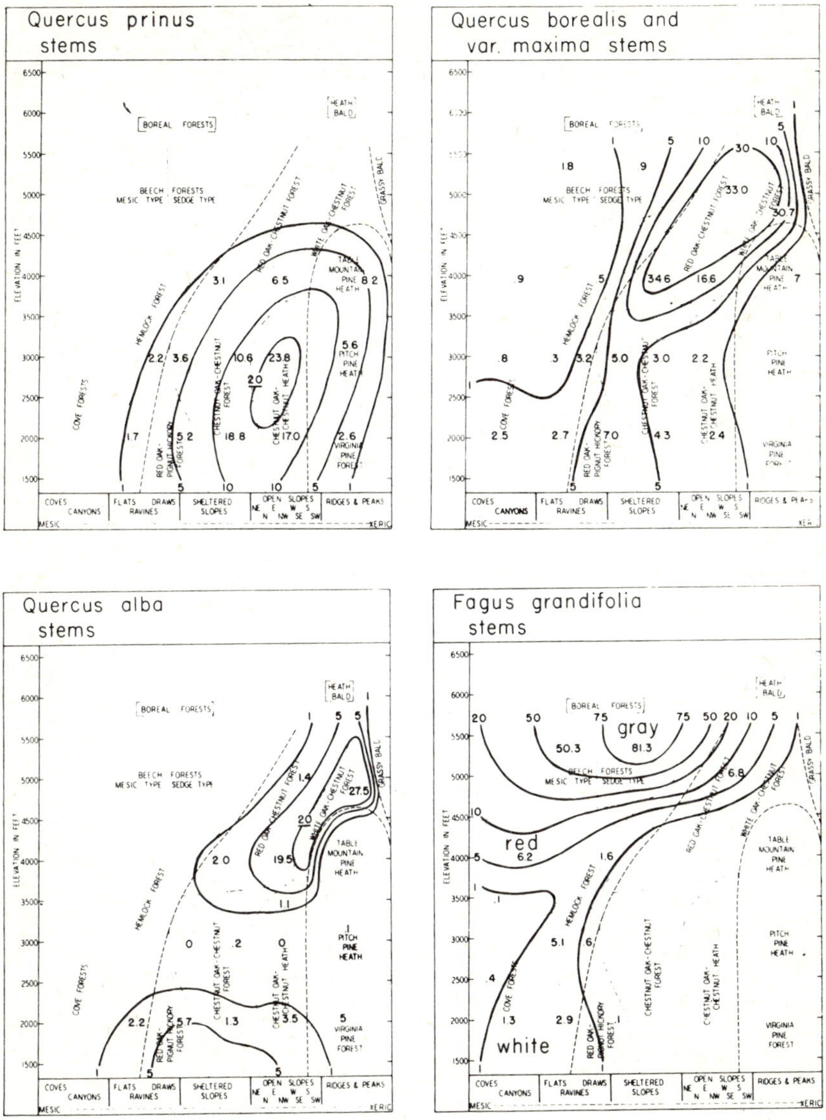

Fig. 7. Population nomograms for four tree species in the Great Smoky Mountains, Tennessee (Whittaker 1956). Per cents of tree stems in composite samples of about 1000 stems were plotted on a mosaic chart, and the values bounded by population lines (isodems). *Quercus prinus* shows a typical hill-shaped population, with a 'peak' in the chestnut oak heath, and densities decreasing in all directions away from this. The other three species show more complex distributions reflecting their genetic complexities.

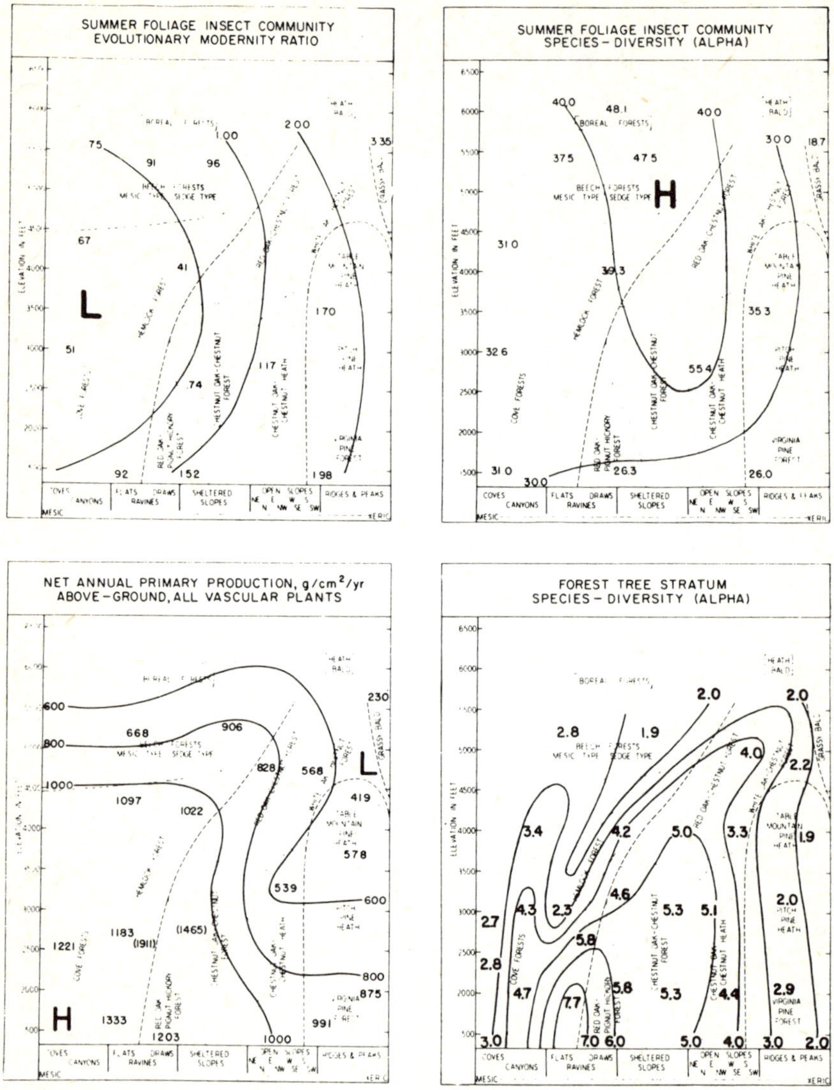

Fig. 8. Nomograms for community characteristics in the Great Smoky Mountains Tennessee. Upper left, the ratio of more to less modern insect groups in sweep samples, showing concentration of more primitive forms in the more humid environments; upper right, species diversities of sweep samples of insects (WHITTAKER 1952). Lower left, primary production of forest types (WHITTAKER 1966); lower right, species diversity of composite samples of the forest tree stratum (WHITTAKER 1956).

mation, a purpose shared with classification, though the ordering is based on gradients and continuities rather than classes. A transect table is a most effective means of organizing information about communities in relation to an environmental gradient and one another. A second purpose is revealing relationships to one another of gradients on three levels of ecological phenomena: environment, species populations, and community characteristics. The various types of figures are means of displaying some of these relationships in one direction of environmental variation, or two directions, among them: (i) species populations in relation to environment and one another, Figs. 1, 5, and 7; (ii) community characteristics in relation to environment, Fig. 11 in the following section and Fig. 8; (iii) community-types in relation to environment, Fig. 6; and (iv) species populations and community characteristics in relation to community-types, Figs. 7 and 8. Many of the figures in following articles represent analogous ways from different techniques, for illustrating these same relationships. Third, these techniques, in revealing relationships that are otherwise obscure in the complexity of vegetation, suggest directions for more intensive research into causal or functional meaning of the relationships. Finally, gradient analysis makes possible understanding of vegetation as a pattern of populations and communities in relation to environments, as developed further in the following section, 2.6.

2.5 **Summary of Techniques**

Direct gradient analysis is based on samples of vegetation, taken to represent the composition of plant communities and the characteristics of their environments at different points in the landscape. Samples should be taken from communities that are homogeneous, should be large enough to represent composition of communities effectively, and should be efficient in obtaining information. Three kinds of samples are in wide use: (i) the relevé of the school of BRAUN-BLANQUET (see article 20), (ii) quadrat samples based on a tenth hectare or other fixed area, used for many ordination studies, and (iii) the plotless or point-centered samples of the school of Wisconsin.

For direct gradient analysis samples are ordinated either on the basis of measurements of environment, or by means of other indices that use community composition to express relative position along environmental gradients. Two most useful means of ordination by composition are: (i) use of weighted averages based on 'ecological groups'–sets of species whose distributions are centered in the same parts of an environmental gradient, and (ii) comparison of samples with end-point samples representing the extremes of the environmental gradient. Relative effectiveness of different ordinations can be measured. By ordination samples may be arranged into a transect, and their data then compiled into a transect table representing change in species populations and community characteristics along the gradient.

By other techniques more complex vegetation relationships to more than one environmental gradient may be analyzed: (i) Crossed transects may be used to study the relation of species and communities to two environmental gradients affecting the vegetation of an area. (ii) Coenoclines or community-gradients, particularly those along topographic moisture gradients, may be compared from one elevation belt, parent material, or climate to another. (iii) Mosaic charts outline the distribution of community-types in relation to two environmental gradients as axes. (iv) Nomograms show the relation of species populations and community characteristics to the range of environment and communities represented in the mosaic chart. By these means it is possible to understand the vegetation of a landscape as a pattern of species populations and communities.

2.6 Results and Implications

2.6.1 Introduction

There may be reason to judge that research in gradient analysis has led to significant advances in the understanding of natural communities. Results from gradient analysis bear in several ways on theory of communities, for they offer insight into the manner in which species evolve in relation to one another in communities, the character of the organization of species into communities, and the way communities as assemblages of species respond to environmental gradients. The ways species are distributed, and communities change, along environmental gradients have in turn wide implications for the practical problems of both classification and ordination. Results from gradient analysis have been discussed in some detail in the reviews of Whittaker (1967, 1970b) and Mc Intosh (1967). The figures in the preceding section on techniques illustrate some of the essential findings, and some implications of these will be summarized here.

2.6.2 Forms of Species Distributions

The general form for distribution of a species population along an environmental complex-gradient is the bell-shaped, apparently Gaussian or binomial curve illustrated in Table I and Fig. 1 (Whittaker 1951, 1956, 1967, Brown & Curtis 1952). In response to two gradients, most populations form Gaussian surfaces, as illustrated in Fig. 7, upper left, with contours outlining a population "hill" any cut through which forms a Gaussian curve (Whittaker 1956). The form of distribution may be generalized to *n* dimensions as an "atmospheric distribution" (Bray & Curtis 1957) with a center of maximum density in a hyperspace of *n* complex-gradients as axes, away from which center density decreases in Gaussian curves in all directions. The center or mode of a species population along a complex gradient is not its physiological optimum but is a center of maximum population success in competition with other species populations. The population center may be in a less "favorable" part of the gradient than the physiological optimum defined by maximum growth-rate.

Some species depart from the simple Gaussian hill toward more complex forms (Fig. 7). The different centers of these bimodal or polymodal distributions represent different ecotypes, different adaptive peaks of genetic response to environment. As shown in Fig. 7, these peaks are in some cases connected and in some cases discontinuous with one another. In most species morphological differentiation can be detected when plants from different peaks are compared in detail with one another. When the peaks are connected by a "saddle" or col of intervening populations (Fig. 7, *Q. borealis*), the intermediate populations may show a continuous cline in morphological characteristics (Whittaker 1956). Width and form of species population distributions express some genetic characteristics of species, and results from gradient analysis are thus linked with those of genecology (Whittaker 1956).

The oblique relation of the population hills of Fig. 7 to the environmental complex-gradients may be noted. Proceeding in mountains from higher elevations toward lower elevations (usually with drier climates), the distribution of a species population is normally displaced toward the left (the more mesic) along the topographic moisture gradient. Similar displacements toward the mesic along topographic gradients affect species populations as one moves at low elevations from more humid toward drier climates (Boyko 1947, Walter & Walter 1953, Perring 1960). Such displacements appear also (Fig. 5) comparing areas of more and less favorable soil parent materials under the same climate–diorite versus serpentine, or diorite versus limestone (Whittaker 1960, Whittaker & Niering 1968a, 1968b). More complex displacements of their different ecotypes appear in species with bimodal or polymodal distributions (Whittaker 1960).

2.6.3 Species Individuality and Dispersion

The centers of species populations are scattered along a complex-gradient in an apparently random manner (Fig. 1). When species are arranged in relation to one another and more than one environmental gradient or axis, in indirect gradient analysis, their positions are scattered rather than clustered (Whittaker 1967, articles 6 and 8). Species do not form well-defined groups of associates with similar distributions, clearly separate from other such groups, but are distributed according to the principle of species individuality: Each species is distributed in its own manner, according to its own genetic, physiological, and population response to environmental factors that affect it (including effects of other species) (Ramensky 1924, cf. Gleason 1926, 1939, Lenoble 1926, 1928). Species can be assigned to ecological groups and character-species groups by similarity of distribution, but the limits of such groupings are essentially arbitrary.

Fig. 9. Population distributions on a habitat gradient. Coniferous tree species populations are plotted in relation to the elevation gradient on northfacing slopes in the Santa Catalina and Pinaleno Mountains, Arizona (data of Whittaker & Niering 1965). The tree species, competing as community dominants, form a replacement series along the gradient. Whereas the modes of species populations are apparently distributed randomly along the gradient when there are many competing species (Whittaker 1956), the distribution of modes of these competing dominants (arrows on bottom line) appears to be regular. Furthermore, in those cases in which the bars for habitat widths overlap, the species of an overlapping pair are differently distributed along the topographic moisture gradient. Species pairs of which the first occupies the more mesic and the second the more xeric topographic position at a given elevation are: Al and Pe, Ac and Ps,Pm and Ps, Pm and Pp, Pce and Jd. The bars below the abscissa give means and habitat widths expressed as one standard deviation, except for Pm, in which the apparent mode has been used as its center rather than its mean (also indicated). Tree species are indicated by genus and species initials: *Picea engelmanni, Abies lasiocarpa, Abies concolor, Pinus strobiformis, Pseudotsuga menziesii, Pinus ponderosa, Pinus chihuahuana, Pinus cembroides*, and *Juniperus deppeana.*

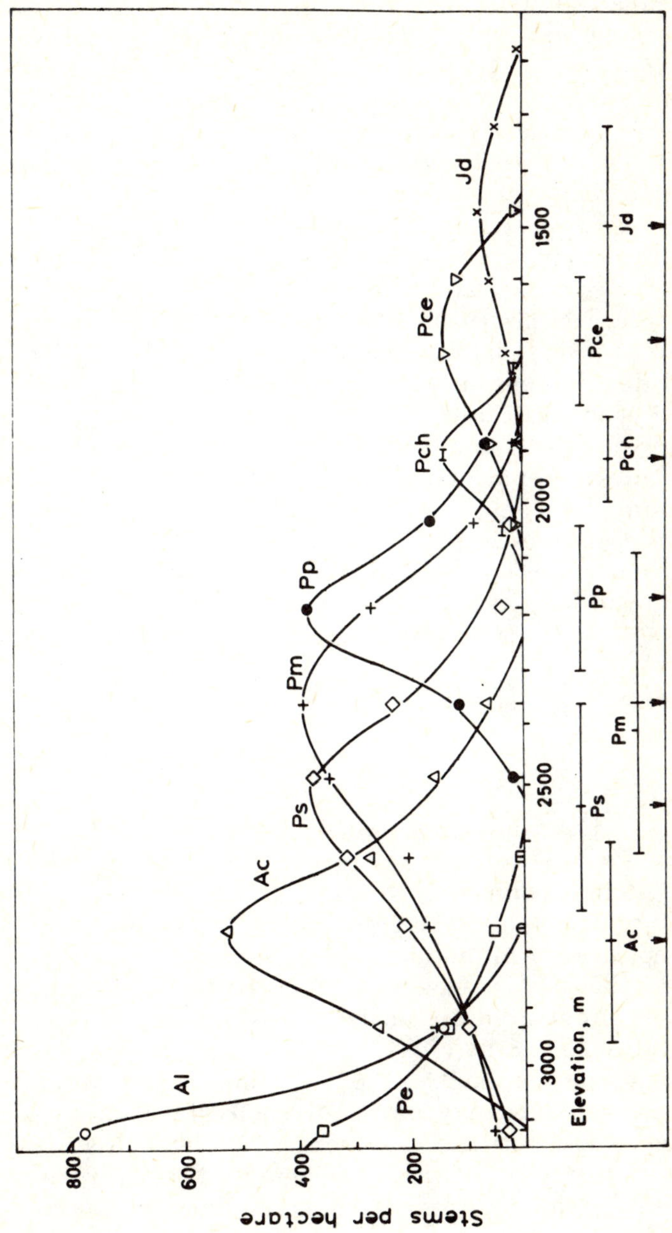
Stems per hectare
800
600
400
200
Elevation, m
3000
2500
2000
1500
Al
Pe
Ac
Ps
Pm
Pp
Pch
Pce
Jd

In some cases competing species appear to be not randomly, but regularly distributed along a gradient. Such regularity may best be observed when dominant species, few in number and close ecological equivalents, can be observed along a gradient. In the Santa Catalina Mountains, Arizona, both the broadleaf tree species (Whittaker 1969) and the conifers (Whittaker et al. 1973) form such replacement series. Fig. 9 illustrates the latter case, and the relatively even spacing of species modes may be observed on the bottom line of the figure. The overlap formulation developed by May & MacArthur (1972) for niche sequences—species dispersions, measured as standard deviations along a niche axis, are roughly equivalent to distances, measured as separation of modes of successive species along the axis—appears to apply also to these sequences along habitat gradients. Terborgh (1971) in a study of bird distributions on an elevation gradient, found that about one third of the distributional limits of species might be accounted for by direct competition between two species as ecological equivalents; gradual environmental changes (and possible diffuse competition) accounted for about half and ecotones for fewer than 20 % of the limits. Only sequences of species subject to direct competition would be expected to show regular spacing. Terborgh observed also that widths of distributions along the elevation gradient tended to increase toward higher elevations as species diversities decreased.

2.6.4 Community Continuity

Along a complex-gradient species populations, with their scattered centers and tapering, broadly overlapping distributions, form a population continuum or compositional gradient (Whittaker 1951, 1956, 1967, Curtis & McIntosh 1951, McIntosh 1967). Results from gradient analysis thus support a second principle stated by Ramensky (1924) and Gleason (1926) (see also article 4), that of the continuity of plant communities with one another (when that continuity is not interrupted by abrupt environmental change or disturbance). In relation to more than one complex-gradient, vegetation may be conceived as a *complex population continuum*, which to varying degrees in different landscapes is interrupted by discontinuities of environment and disturbance (Whittaker 1951, 1956, 1967). Physiognomic change in communities, as well as compositional change, is in general continuous along environmental gradients (Beard 1955, Whittaker & Niering 1965, 1968b). The response of communities to an environmental gradient may thus be approached either as a coenocline or community-gradient, or as a chain or sequence of (arbitrarily bounded) community-types that make up the gradient. A sequence of community-types along an environmental complex-gradient is an *ecological series* (Cajander 1903, Keller 1925-6, Sukachev 1928, 1932, articles 4, 15, 17). A sequence of formations (as physiognomically defined community-types along a complex-gradient) is a *formation-series* (Beard 1944, 1955, and article 13, Whittaker & Niering 1968b).

Certain situations occur in which, in the absence of environmental discontinuity or disturbance, the principle of community continuity must be modified (Fig.10). The distribution illustrated has been termed a "plateau distribution" (Whittaker 1956, 1967) and interpreted as a vertical truncation of the Gaussian distribution of a strongly dominant species. Such a distribution implies relative discontinuity of community composition (as regards the dominant species, at least) at the edge of the plateau distribution. Other observations on plateau distributions are: (i) Observed cases are

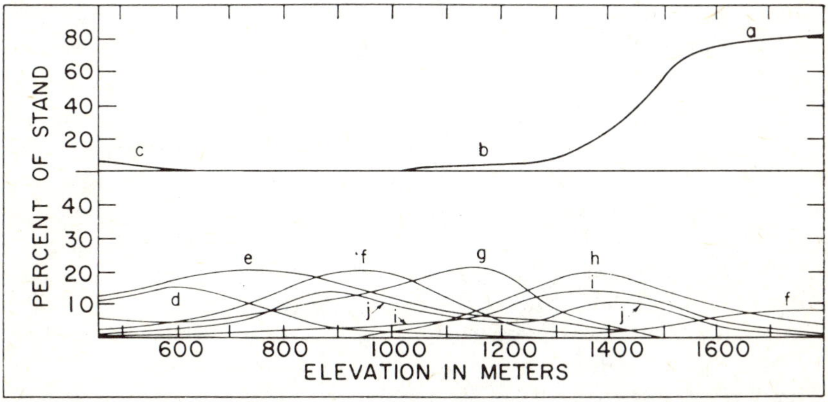

Fig.10. Tree species distributions in an elevation transect of mesic forests, Great Smoky Mountains, Tennessee (Whittaker 1956). Beech (*Fagus grandifolia*) shows a 'plateau' distribution, which may be contrasted with the more typical bell-shaped curves for other tree species in the lower panel. The three ecotypes of *Fagus* (cf. Fig. 7) are indicated – gray (a), red (b) and white (c). Other species – *Acer rubrum* (d), *Tsuga canadensis* (e), *Halesia monticola* (f, bimodal), *Tilia heterophylla* (g), *Acer spicatum* (h), *Aesculus octandra* (i) and *Betula allegheniensis* (j, bimodal). Data are percentages of tree stems over 1 cm dbh in composite samples at 100-meter elevation intervals. *Fagus* shows a relatively steep change in population density between 1350 and 1500m, in south-facing canyons, but the distributions of many other species extend continuously through these elevations.

very few in number, compared with the thousands of bell-shaped distributions observed in research in gradient analysis. (ii) These cases have been limited to more extreme environments, occupied by communities of strong single-species dominance. BEALS (1969a) observed a greater tendency toward such distributions along a steep elevation gradient in mountains than along a more gentle one. (iii) A community with a relatively discontinuous boundary along one complex-gradient may intergrade continuously with other communities along other gradients (WHITTAKER 1956). (iv) In some cases, such as the high-elevation beech forests (WHITTAKER 1956, 1967, cf. MARK 1963, SCOTT et al. 1964), the single dominant shows a relatively discontinuous, plateau distribution, but other species are continuously distributed through the edge of its distribution. The situation may then be one of apparent discontinuity, affecting a dominant population, superimposed on an underlying largely or fully continuous gradient of over-all floristic composition. Plateau distributions appear to represent local and partial qualifications of the principle of community continuity.

2.6.5 COMMUNITY TRENDS

Many community-level characteristics show continuous trends along complex-gradients. Fig. 11 shows relations of RAUNKIAER life-forms and areal types (geographic distributions) of species to the elevation gradient in the Santa Catalina Mountains, Arizona. A life-form composition that is characteristic for a cool-temperate forest, with hemicryptophytes predominant, changes toward low elevations to the chamaemphyte- (and therophyte-) dominated life-form composition of the Sonoran desert. In high-elevation forests species with northern and Rocky-Mountain distributions predominate; at middle elevations in woodlands Madrean species (with the main areas of their distributions in Mexican mountains) are important; at low elevations southwestern and Sonoran species (with their major areas in the deserts of Sonora, Arizona, and California) prevail. Coverages of plant strata show trends along coenoclines (Table I, WHITTAKER 1956, 1960, WHITTAKER & NIERING 1965). Some strata show more complex changes along a gradient than a simple rise or fall; and in some cases the coverages of the tree vs. the shrub or the shrub vs. the herb stratum are inversely related to one another (WHITTAKER 1956). Canopy height, community biomass, and primary productivity show trends along the elevation and topographic moisture gradients in mountains (WHITTAKER 1966).

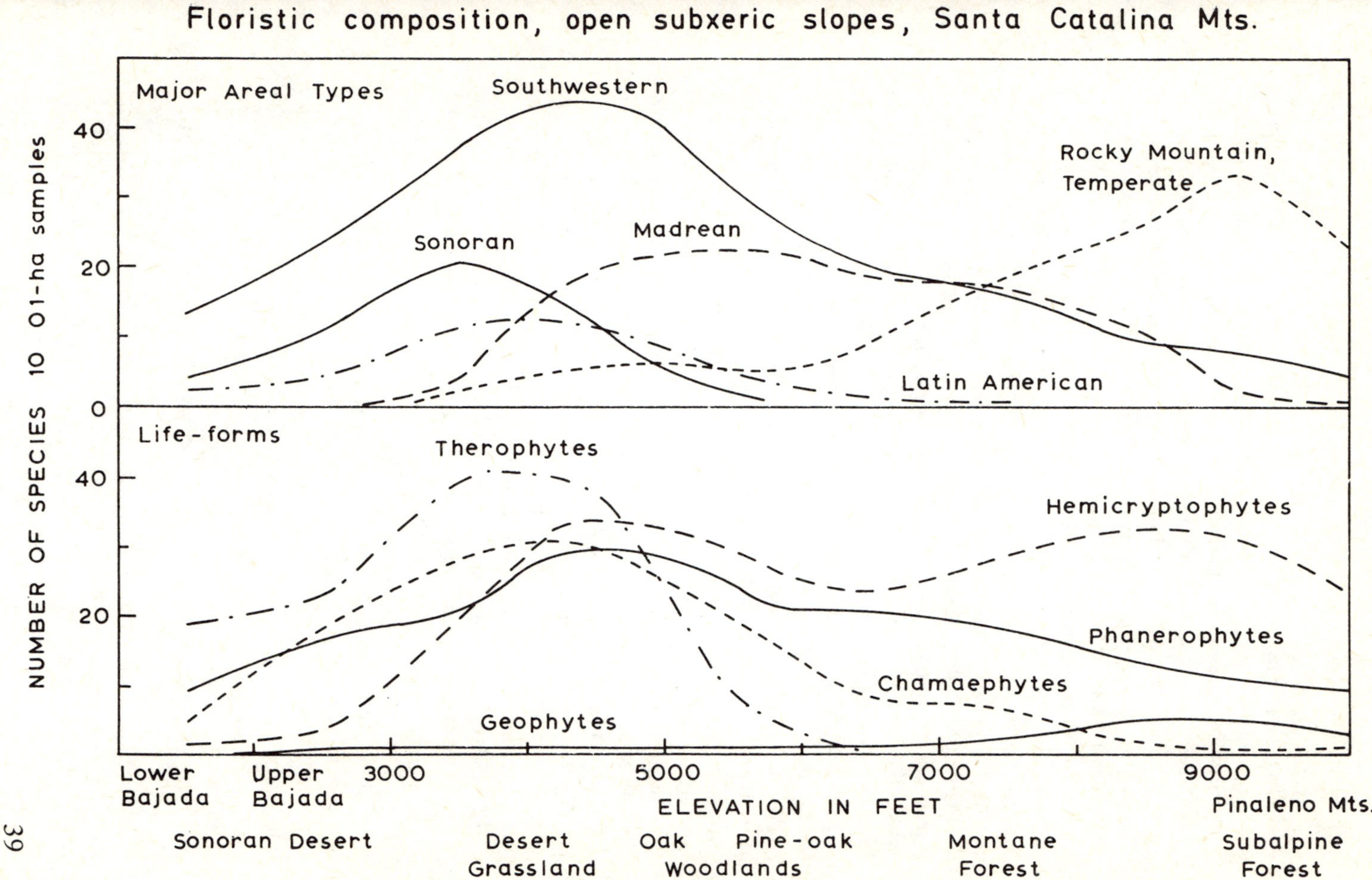

Fig.11. Trends in community composition in relation to elevation, Santa Catalina Mountains, southeastern Arizona (Whittaker & Niering 1964). Data are numbers of species by areal types and life-forms, in composite samples (of 10, 0.1 ha samples) from open intermediate and southerly slopes on the south side of the range.

2.6.6 Species Diversity

Diversity relations to gradients have special interest. *Alpha* or within-habitat diversity refers to richness in species of particular communities, or individual samples from them. The major geographic relationship of alpha diversity for many groups of organisms is its increase toward more favorable, and particularly more stable environments, especially those of tropical rainforests (Dobzhansky 1950, Ashton 1969, MacArthur 1969, Whittaker 1969). Along local environmental gradients, however, diversity change is often not a simple decrease toward less favorable environments. Maximum diversity of vascular plant species is in the middle, not at the mesic extreme, of some topographic moisture gradients (Whittaker 1956, 1969); in some cases maximum diversity is at intermediate pH values (Maarel & Leertouwer 1967). Diversity generally increases toward less acid and more fertile soils (Dahl 1957, Monk 1965, Frydman & Whittaker 1968, Whittaker 1969, Prusinkiewicz 1970), but in the Siskiyou Mountains diversity is higher on the chemically extreme serpentine soils than on the more normal diorite soils (Whittaker 1954b, 1960). Diversities of different life-forms and strata of plant communities are relatively independent of one another. In the Siskiyou Mountains maximum tree species diversities are at low, but maximum herb species diversities at middle, elevations (Whittaker 1960, cf. Daubenmire & Daubenmire 1968). Herb diversities are higher in broadleaved deciduous than in needleleaved evergreen forest in similar habitats, and they tend to be higher in open woodlands with small trees and an open canopy admitting light to the herb stratum than in closed forests (Whittaker 1965, Whittaker & Niering 1968b, Moral 1972). Over-all diversities of vascular plants are consequently generally higher in temperate woodlands than in temperate forests. Diversity shows no clear relation to productivity (Whittaker 1965, 1969, Whittaker & Niering 1975).

Measurements can be applied to determine beta diversity, or degree of floristic differentiation, of coenoclines (Whittaker 1960, Whittaker & Niering 1965). *Beta* or between-habitat diversity refers to the degree of change in species composition in the samples taken along a coenocline, or degree of contrast in species composition among samples of a set taken from a landscape. Beta, like alpha, diversity can be interpreted as a consequence of the accumulation through evolutionary time of species that have evolved toward reduction of competition between them by difference in the way they relate to environment and other species. Alpha diversity results from evolution into a community of species that differ in niche, that is in function, position, or relation to other species within that community. Beta diversity in a coenocline results from accumulation of species differing in habitat, that is in distributional relation to the environmental gradient, and from the narrowing of species amplitudes along the gradient by increasing competition from increasing numbers of species (WHITTAKER 1969, 1970a). Since both alpha and beta diversity express increase in species numbers through evolutionary time, they might be expected to vary in parallel along major environmental gradients. Both may be expected to decrease toward more rigorous, and especially more unstable, environments; but diversity is high in some deserts (WHITTAKER & NIERING 1965, 1975, WHITTAKER 1972).

Both alpha and beta diversity appear to increase (but with various complications affecting different communities and strata) from arctic and alpine to tropical lowland communities. Both alpha and beta diversity increase from maritime climates inland to continental climates in the Temperate Zone (WHITTAKER 1960, 1969, WHITTAKER & NIERING 1965). Responses of beta diversity (as coenocline differentiation along the topographic moisture gradient) to parent material contrasts are less consistent. On serpentine beta diversity is as high as, or higher than, on diorite in the Siskiyou Mountains (WHITTAKER 1960); but beta diversity is markedly lower on limestone than on diorite in the Santa Catalina Mountains (WHITTAKER & NIERING 1968b).

2.6.7 PATTERNS OF COMMUNITY CHARACTERISTICS

Measurements of community characteristics can be plotted on a mosaic chart, as in Fig. 8, to observe the manner in which these change through the vegetation pattern of a landscape (WHITTAKER 1952, 1956, WHITTAKER & NIERING 1965). Of these patterns of community characteristics one may observe: (i) They form (as suggested by their behavior along particular complex-gradients) surfaces of complex, but generally continuous slope. In some cases the surfaces are bimodal or polymodal. (ii) The community characteristics consequently form in many cases continuous gradients through the communities of the landscape, but a community characteristic may not show a simple and linear relation to any particular complex-gradient as normally recognized. (iii) Like species populations, growth-forms and life-forms, the community characteristics are "individualistic"; no two of these show the same pattern of variation in relation to communities and environments.

2.6.8 Criticisms of Gradient Analysis

Such patterns for species and community characteristics emphasize the coherence of the vegetation of a landscape, the interrelation and intergradation of its communities in the pattern as a whole. This conception of community intergradation, which is shared by some phytosociologists, has been rejected by others whose emphasis is on community-types as distinct assemblages of species separated by relatively sharp boundaries from other such assemblages. History of the argument has been reviewed elsewhere (Whittaker 1962, article 18); recent criticisms of gradient analysis are Daubenmire (1966), Langford & Buell (1969), and Gounot (1969), see also McIntosh (1967-8). Some of these criticisms are, first, in the area of method, with the arguments: (i) That the apparent continuity of communities is an artifact resulting from the circularity of method that is inherent in certain approaches to ordination (see section 2.2.3.1), but not in others; (ii) That the community samples used for gradient analysis are not sufficiently homogeneous, and that the apparent continuity of vegetation in a transect results from blurring of community boundaries by heterogeneous samples; (iii) That successional relationships are inadequately considered, and inclusion in ordinations of unstable stands obscures boundaries that would be evident in stable communities; and (iv) More broadly, that studies in gradient analysis have imposed on their research an assumption of vegetational continuity that has prevented recognition of discontinuity. Second, in the area of observation and theory it is asserted: (v) That the occurrence in the field of combinations of species that repeat themselves from place to place, and in some cases seem to be sharply bounded, is in conflict with the results of gradient analysis, and (vi) That the evolution of species toward adaptation to one another, and especially of subordinate species toward adaptation to the community dominants, must imply the formation of relatively discrete groups of co-adapted species that would appear in the field in this self-repeating manner.

The number of studies seems now adequate to establish that, in relation to points (i) to (iii), the most essential results of gradient analysis – the tapered and scattered curves of species populations, individuality of species distributions, and continuity of undisturbed communities along continuous environmental gradients – are not artifacts of any particular kind of sample or technique of ordination. Undisturbed communities are in general more clearly continuous than disturbed ones. It is, furthermore, possible to illustrate these essential results with tables compiled by phytosociologists employing classification when the tables arrange community-types along environmental gradients (Whittaker 1962). The last statement is true despite the fact that samples taken on the assumption of discontinuity should not be used to test for continuity. In contrast and in relation to point (iv), samples taken without bias and ordinated should, if discontinuities are present, reveal those discontinuities. Freedom from bias for or against discontinuity has been an objective of sample procedure in most research in gradient analysis, and some studies have had as an objective the unbiased testing of continuity vs. discontinuity of dominance-types or associations (Whittaker 1956, 1960, Beals 1969a, Beschel & Webber 1962) or of formations (Whittaker & Niering 1965). Such work supports the conclusion that the response of plant communities to environmental gradients is predominantly continuous even though, because of the effects of topographic, edaphic, and disturbance discontinuities in the field, vegetation in the field is often a complex mixture of continuity and discontinuity (Whittaker 1956, cf. Webb 1954, Dahl 1957, Scott 1974).

Repetition of similar species combinations in similar environments (v) is equally to be expected whether species are individualistic or form co-adapted groups, whether communities are continuous or discontinuous. The more significant question is that of the direction of species evolution. There is, at first glance, a paradox in the evolution of species both toward co-adaption within the community, and toward distributional individuality (Whittaker 1962). The paradox is resolved by recognition that species evolve toward the reduction of competition by both niche difference (difference in position or role in the community as a basis of co-adaption) and habitat difference (difference in distribution, or location of population centers, along environmental gradients). The fact species differ in niche implies that they will not in general exclude one another's populations along sharp boundaries, but may form the gradually tapered and broadly overlapping distributions observed in gradient analysis. Species do not in general evolve toward formation of discrete and clearly bounded groups as suggested by point (vi), but toward distributional difference and toward accomodations to one another that imply, for most species, continuously overlapping distributions (Whittaker 1962, 1967, 1970b). There is no conflict between such reasoning on theory of species evolution, and the results of gradient analysis.

2.7 Conclusion on Implications

From these results may be drawn a conception of the spatial structure of vegetation (WHITTAKER 1956, 1965, 1967, 1970b). Numerous plant species occur in the landscape and have evolved toward difference in their environmental relations. This evolution implies (i) difference in species' intracommunity positions, or niches, (ii) scattering of their population centers to different locations along environmental (habitat) gradients, and (iii) broadly overlapping distributions of these populations, with overlap in distribution made possible by niche difference. The species that make up the vegetation pattern freely and variously combine with one another in communities that predominantly intergrade with one another, forming a complex and potentially continuous but locally interrupted population pattern. It is a purpose of gradient analysis to understand this pattern in terms of gradient relations of species populations, environmental factors, and community characteristics, and as a design of interrelated parts.

The complexities of topography and other environmental factors in space make the pattern complicated and difficult to recognize in the field. A vegetation pattern is to some degree implicated—folded in upon itself— by the alternation and repetition of its parts from one valley or slope exposure to another. Gradient analysis is an approach to explication of the pattern, in the sense of unfolding its spatially complex relationships. Gradient analysis seeks to abstract from the landscape a coherent and comprehensible pattern of community composition in relation to environmental gradients. The axes of this abstract pattern are ecoclines–directions of variation in the landscape pattern, correlations of complex-gradients and coenoclines that repeat themselves in the landscape. The abstract pattern may be conceived also as a coordinate system, a field defined by two, or a hyperspace defined by *n* ecoclines as axes. Community samples, and centers of species populations, may be located in relation to one another in this field or hyperspace. The research approach of indirect gradient analysis, which derives a coordinate system from the compositional relations of samples and then seeks to relate the compositional axes to complex-gradients, will be discussed in articles that follow. The results of indirect gradient analysis in general converge with those of direct gradient analysis summarized here, and support the conception of vegetation suggested (MCINTOSH 1967, WHITTAKER 1967, 1970b).

2.8 **Summary of Implications**

Significant results from direct gradient analysis include: (i) Distributions of species populations along environmental gradients typically appear as broadly overlapping, bell-shaped curves. (ii) Along a given gradient the centers or modes of species populations are scattered, and no two species have the same distribution. (iii) Most natural communities (if they have not been disturbed) intergrade continuously along uninterrupted environmental gradients. (iv) Species diversity and other community characteristics change along environmental gradients through different kinds of communities as continuous trends, or more complex curves. (v) These characteristics form varied response surfaces on a mosaic chart representing a pattern of environment and vegetation, and in general no two community characteristics show the same pattern in relation to environment in such charts. (vi) Results from gradient analysis lead to a conception of the vegetation of a landscape as a complex continuum, an abstract pattern of interrelated gradients of environment, species populations, and community characteristics. The direction of species evolution is toward the scattered or "individualistic" and broadly overlapping distributions that are responsible for the complexity and continuity of vegetation patterns.

The complex continuum may be approached through a coordinate system using as axes the major directions of community variation in the landscape. The axes then define an abstract vegetational "space" in which samples, and species centers, may be located. Indirect ordination seeks to derive such a space from the characteristics of community samples as a basis for the arrangement of samples, and species, in relation to one another.

REFERENCES

ASHTON, P. S., – 1969 – Speciation among tropical forest trees: some deductions in the light of recent evidence. *Biol. J. linn. Soc. Lond.* 1: 155–196.

AYYAD, M. A. G. & R. L. DIX, – 1964 – An analysis of a vegetation-micro-environmental complex on prairie slopes in Saskatchewan. *Ecol. Monogr.* 34: 421–442.

BEALS, E. W., – 1960 – Forest bird communities in the Apostle Islands of Wisconsin. *Wilson Bull.* 72: 156–181.

BEALS, E. W., – 1969a – Vegetational change along altitudinal gradients. *Science, N.Y.* 165: 981–985.

BEALS, E. W., – 1969b – Vegetation of the Erer-Gota Plain, Ethiopia. *J. Ecol.* 57: 655–667.

BEARD, J. S., – 1944 – Climax vegetation in tropical America. *Ecology* 25: 127–158.

Beard, J. S., – 1955 – The classification of tropical American vegetation-types. *Ecology* 36: 89–100.

Beschel, R. E. & P. J. Webber, – 1962 – Gradient analysis in swamp forests. *Nature, Lond.* 194: 207–209.

Boyko, H., – 1947 – On the role of plants as quantitative climate indicators and the geo-ecological law of distribution. *J. Ecol.* 35: 138–157.

Braun-Blanquet, J., – 1932 – Plant Sociology, the Study of Plant Communities. McGraw-Hill, New York. 439 pp.

Braun-Blanquet, J., – 1951 – Pflanzensoziologie, Grundzüge der Vegetationskunde. 2nd ed. Springer, Wien. 631 pp. 3rd ed., 1964, 865 pp.

Bray, J. R., – 1956 – A study of mutual occurrence of plant species. *Ecology* 37: 21–28.

Bray, J. R., – 1960 – The composition of savanna vegetation in Wisconsin. *Ecology* 41: 721–732.

Bray, J. R., – 1961 – A test for estimating the relative informativeness of vegetation gradients. *J. Ecol.* 49: 631–642.

Bray, J. R. & J. T. Curtis, – 1957 – An ordination of the upland forest communities of southern Wisconsin. *Ecol. Monogr.* 27: 325–349

Brown, Dorothy., – 1954 – Methods of surveying and measuring vegetation. *Bull. Commonw. Bur. Past. Fld. Crops*, (Hurley, Berks, England) 42: 1–223.

Brown, R. T. & J. T. Curtis, – 1952 – The upland conifer-hardwood forests of northern Wisconsin. *Ecol. Monogr.* 22: 217–234.

Cain, S. A. & G. M. de Oliveira Castro, – 1959 – Manual of Vegetation Analysis. Harper, New York. 325 pp.

Cajander, A. K., – 1903 (1906) – Beiträge zur Kenntniss der Vegetation der Alluvionen des nördlichen Eurasiens. I. Die Alluvionen des unteren Lena-Thales. *Acta Soc. Sci. fenn.* 32(1): 1–182.

Clements, F. E., – 1936 – Nature and structure of the climax. *J. Ecol.* 24: 252–284.

Cottam, G. & J. T. Curtis, – 1949 – A method for making rapid surveys of woodlands by means of pairs of randomly selected trees. *Ecology* 30: 101–104.

Cottam, G. & J. T. Curtis, – 1956 – The use of distance measures in phytosociological sampling. *Ecology* 37: 451–460.

Curtis, J. T., – 1955 – A prairie continuum in Wisconsin. *Ecology* 36: 558–566.

Curtis, J. T., – 1959 – The Vegetation of Wisconsin: An Ordination of Plant Communities. Univ. Wisconsin, Madison. 657 pp.

Curtis, J. T. & R. P. McIntosh, – 1951 – An upland forest continuum in the prairie-forest border region of Wisconsin. *Ecology* 32: 476–496.

Dahl, E., – 1957 – Rondane: mountain vegetation in South Norway and its relation to the environment. *Skr. norske Vidensk-Akad.*, Mat.-naturv. Kl. 1956(3): 1–374.

Daubenmire, R., – 1966 – Vegetation: identification of typal communities. *Science, N.Y.* 151: 291–298.

Daubenmire, R. & Jean B. Daubenmire, – 1963 – Forest vegetation of eastern Washington and northern Idaho. *Tech. Bull. agric. Exp. Stn. Washington St.* 60: 1–104.

Dix, R. L., – 1959 – The influence of grazing on the thin-soil prairies of Wisconsin. *Ecology* 40: 36–49.

Dix, R. L. & J. E. Butler, – 1960 – A phytosociological study of a small prairie in Wisconsin. *Ecology* 41: 316–327.

Dix, R. L. & F. E. Smiens, – 1967 – The prairie, meadow, and marsh vegetation of Nelson County, North Dakota. *Can. J. Bot.* 45: 21–58.

Dobzhansky, T., – 1950 – Evolution in the tropics. *Am. Scient.* 38: 209–221.
Duvigneaud, P., – 1946 – La variabilité des associations végétales. *Bull. Soc. r. Bot. Belg.* 78: 107–134.
Dyksterhuis, E. J., – 1949 – Condition and management of range land based on quantitative ecology. *J. Range Mgmt.* 2: 104–115.
Dyksterhuis, E. J., – 1958 – Ecological principles of range evaluation. *Bot. Rev.* 24: 253–272.
Egler, R. E., – 1942 – Vegetation as an object of study. *Philosophy Sci.* 9: 245–260.
Ellenberg, H., – 1948 – Unkrautgesellschaften als Mass für den Säuregrad, die Verdichtung und andere Eigenschaften des Ackerbodens. *Ber. Landtech.* 4: 130–146.
Ellenberg, H., – 1950 – Landwirtschaftliche Pflanzensoziologie. I. Unkrautgemeinschaften als Zeiger für Klima und Boden. Ulmer, Stuttgart. 141 pp.
Ellenberg, H., – 1952 – Landwirtschaftliche Pflanzensoziologie. II. Wiesen und Weiden und ihre standörtliche Bewertung. Ulmer, Stuttgart. 143 pp.
Ellenberg, H., – 1956 – Aufgaben und Methoden der Vegetationskunde. In 'Einführung in die Phytologie,' ed. H. Walter, vol. IV, pt. 1. Ulmer, Stuttgart. 136 pp.
Frydman, I. & R. H. Whittaker, – 1968 – Forest associations of southeast Lublin Province, Poland (Germ. summ.) *Ecology* 49: 896–908.
Gittins, R., – 1965 – Multivariate approaches to a limestone grassland community. I. A stand ordination. *J. Ecol.* 53: 385–401.
Gleason, H. A., – 1926 – The individualistic concept of the plant association. *Bull. Torrey bot. Club* 53: 7–26.
Gleason, H. A., – 1939 – The individualistic concept of the plant association. *Am. Midl. Nat.* 21: 92–110.
Goff, F. G. & G. Cottam, – 1967 – Gradient analysis: the use of species and synthetic indices. *Ecology* 48: 793–806.
Goodall, D. W., – 1954 – Vegetational classification and vegetational continua. (Germ. summ.) *Angew. PflSoziol.* Wien, Festschr. Aichinger 1: 168–182.
Gounot, M., – 1969 – Méthodes d'étude quantitative de la végétation. Masson, Paris. 314 pp.
Greig-Smith, P., – 1964 – Quantitative Plant Ecology. 2nd ed. Butterworths, London. 256 pp.
Hill, M. O., – 1973 – Reciprocal averaging: an eigenvector method of ordination. *J. Ecol.* 61: 237–349.
Horikawa, Y. & K. Okutomi, – 1955 – The continuum of the vegetation on the slopes of Mt. Shiroyama, Iwakuni City, Prov. Suwo (Japan. with Engl. summ.) *Seibutsugakkaishi, Hiroshima* 6: 8–17.
Itow, S., – 1963 – Grassland vegetation in uplands of western Honshu, Japan. II. Succession and grazing indicators. *Jap. J. Bot.* 18: 133–167.
Keller, B. A., – 1925–6 – Die Vegetation auf den Salzböden der russischen Halbwüsten und Wüsten. *Z. Bot.* 18: 113–137.
Knight, D. H., – 1965 – A gradient analysis of Wisconsin prairie vegetation on the basis of plant structure and function. *Ecology* 46: 744–747.
Langford, A. N. & M. F. Buell, – 1969 – Integration, identity and stability in the plant association. *Adv. ecol. Res.* 6: 83–135.
Lenoble, F., – 1926 – À propos des associations végétales. *Bull. Soc. bot. Fr.* 73: 873–893.
Lenoble, F., – 1928 – Associations végétales et espèces. *Archs Bot.*, Caen, *Bull. mens.* 2(1): 1–14.
Lewin, D. C., – 1974 – The vegetation of the ravines of the southern Finger Lakes, New York, region. *Am. Midl. Nat.* 91: 315–342.

Loucks, O. L., – 1962 – Ordinating forest communities by means of environmental scalars and phytosociological indices. *Ecol. Monogr.* 32: 137–166.

Maarel, E. van der, – 1960 – Rapport inzake de vegetatie van het duingebied van de Stichting 'Het Zuid-Hollands Landschap' bij Oostvoorne. Report Z.H.L., Delft.

Maarel, E. van der, – 1969 – On the use of ordination models in phytosociology. (Germ. summ.) *Vegetatio* 19: 21–46.

Maarel, E. van der & J. Leertouwer, – 1967 – Variation in vegetation and species diversity along a local environmental gradient. *Acta bot. neerl.* 16: 211–221.

MacArthur, R. H., – 1969 – Patterns of communities in the tropics. *Biol. J. linn. Soc. Lond.* 1: 19–30.

Mark, A. F., – 1963 – Vegetation studies on Secretary Island, Fiordland. 3: The altitudinal gradient in forest composition, structure and regeneration. *N. Z. Jl. Bot.* 1: 188–202.

Mark, A. F. & F. R. Sanderson, – 1962 – The altitudinal gradient in forest composition, structure and regeneration in the Hollyford Valley, Fiordland. *Proc. N.Z. ecol. Soc.* 9: 17–26.

May, R. M. & R. H. MacArthur, – 1972 – Niche overlap as a function of environmental variability. *Proc. natn. Acad. Sci. USA* 69: 1109–1113.

Maycock, P. F. & J. T. Curtis, – 1960 – The phytosociology of boreal conifer–hardwood forests of the Great Lakes region. *Ecol. Monogr.* 30: 1–35.

McIntosh, R. P., – 1967 – The continuum concept of vegetation. *Bot. Rev.* 33: 130–187. Responses, *ibid.* 34: 253–332, 1968.

Monk, C. D., – 1965 – Southern mixed hardwood forest of northcentral Florida. *Ecol. Monogr.* 35: 335–354.

Moore, J. J., P. Fitzsimmons, E. Lambe & J. White, – 1970 – A comparison and evaluation of some phytosociological techniques. (Germ. summ.) *Vegetatio* 20: 1–20.

Moral, R. del, – 1972 – Diversity patterns in forest vegetation of the Wenatchee Mountains, Washington. *Bull. Torrey bot. Club* 99: 57–64.

Mowbray, T. B., – 1966 – Vegetational gradients in the Bearwallow Gorge of the Blue Ridge Escarpment. *J. Elisha Mitchell scient. Soc.* 82: 138–149.

Mowbray, T. B. & H. J. Oosting, – 1968 – Vegetation gradients in relation to environment and phenology in a southern Blue Ridge gorge. *Ecol. Monogr.* 38: 309–344.

Mueller-Dombois, D. & H. Ellenberg, – 1974 – Aims and Methods of Vegetation Ecology. Wiley, New York. 547 pp.

Okutomi, K., – 1958 – A forest continuum in Is. Ujina, Hiroshima. (Japan. with Engl. summ.) *Bull. Fukuoka Gakugei Univ.* 8: 75–83.

Peet, R. K., – 1976 – Forest vegetation of the east slope of the Colorado Front Range. Ph. D. Thesis, Cornell University.

Perring, F., – 1959 – Topographical gradients of chalk grassland. *J. Ecol.* 47: 447–481.

Perring, F., – 1960 – Climatic gradients of chalk grassland. *J. Ecol.* 48: 415–442.

Phillips, E. A., – 1959 – Methods of Vegetation Study. Holt, New York. 107 pp.

Prusinkiewicz, Z., – 1970 – Die Zahl der auf dem Minimi-Areal vorkommenden Gefässpflanzenarten als Mass für die Fruchtbarkeit der Waldböden. (Engl. summ.) In 'Gesellschaftsmorphologie (Stukturforschung)' ed. R. Tüxen. *Ber. Symp. int. Ver. Vegetationskunde*, Rinteln, 1966: 282–296.

Ramensky, L. G., – 1924 – Die Grundgesetzmässigkeiten im Aufbau der Vegetationsdecke. (In Russian) *Vêst. opỹt. Dêla sred.-chernoz. Obl.*, Voronezh (Abstr. *Bot. Zlb.* N.F. 7: 453–455, 1926).

Ramensky, L. G., – 1930 – Zur Methodik der vergleichenden Bearbeitung und Ordnung von Pflanzenlisten und anderen Objekten, die durch mehrere, verschiedenartig wirkende Faktoren bestimmt werden. *Beitr. Biol. Pfl.* 18: 269–304.

Rowe, J. S., – 1956 – Uses of undergrowth plant species in forestry. *Ecology* 37: 461–473.

Scott, G. A. M., A. F. Mark & F. R. Sanderson, – 1964 – Altitudinal variation in forest composition near Lake Hankinson, Fiordland. *N. Z. Jl. Bot.* 2: 310–323.

Scott, J. T., – 1974 – Correlation of vegetation with environment: a test of the continuum and community-type hypotheses. In 'Vegetation and Environment' ed. B. R. Strain & W. D. Billings. *Handb. Veget. Sci.* 6: 87–109.

Sukachev, V. N., – 1928 – Principles of classification of the spruce communities of European Russia. *J. Ecol.* 16: 1–18.

Sukatschew, W. N., – 1932 – Die Untersuchung der Waldtypen des osteuropäischen Flachlandes. *Handb. biol. ArbMeth.* 11, 6: 191–250.

Tanghe, M., – 1964 – Contribution à l'étude de la végétation forestière de la Haute Belgique. 2. Utilisation d'un transect topographique des environs de Muno, pour la délimitation des groupes écologiques forestiers de la Gaume et de l'Ardenne. *Bull. Soc. r. Bot. Belg.* 98: 5–22.

Terborgh, J., – 1971 – Distribution on environmental gradients: theory and a preliminary interpretation of distributional patterns in the avifauna of the Cordillera Vilcabamba, Peru. *Ecology* 52: 23–40.

Walter, H. & E. Walter, – 1953 – Einige allgemeine Ergebnisse unserer Forschungsreise nach Südwestafrika 1952/53: Das Gesetz der relativen Standortskonstanz; das Wesen der Pflanzengemeinschaften. *Ber. dt. bot. Ges.* 66: 228–236.

Waring, R. H. & J. Major, – 1964 – Some vegetation of the California coastal redwood region in relation to gradients of moisture, nutrients, light, and temperature. *Ecol. Monogr.* 34: 167–215.

Webb, D. A., – 1954 – Is the classification of plant communities either possible or desirable? *Bot. Tiddskr.* 51: 362–370.

Wells, Judith A. & A. F. Mark, – 1966 – The altitudinal sequence of climax vegetation on Mt Anglem, Stewart Island. I. The principal strata. *N.Z. Jl Bot.* 4: 267–282.

Wentworth, T. R., – 1976 – The vegetation of limestone and granite soils in the mountains of southeastern Arizona. Ph. D. Thesis, Cornell University.

Westhoff, V., – 1970 – Vegetation study as a branch of biological science. (Germ. summ.) In 'Vegetatiekunde als synthetische wetenschap,' ed. H. J. Venema, H. Doing, & I. S. Zonneveld. *Misc. Pap. Landbouwhogeschool Wageningen*, 5: 11–30.

Westman, W. E., – 1975 – Edaphic climax pattern at the pygmy forest region of California. *Ecol. Monogr.* 45: 109–135.

Whittaker, R. H., – 1951 – A criticism of the plant association and climatic climax concepts. *NW. Sci.* 25: 17–31.

Whittaker, R. H., – 1952 – A study of summer foliage insect communities in the Great Smoky Mountains. *Ecol. Monogr.* 22: 1–44.

Whittaker, R. H., – 1954a – Plant populations and the basis of plant indication. (Germ. summ.) *Angew. Pflsoziol.*, Wien, Festschr. Aichinger 1: 183–206.

Whittaker, R. H., – 1954b – The ecology of serpentine soils. IV. The vegetational response to serpentine soils. *Ecology* 35: 275–288.

Whittaker, R. H., – 1956 – Vegetation of the Great Smoky Mountains. *Ecol. Monogr.* 26: 1–80.

WHITTAKER, R. H., – 1960 – Vegetation of the Siskiyou Mountains, Oregon and California. *Ecol. Monogr.* 30: 279–338.

WHITTAKER, R. H., – 1962 – Classification of natural communities. *Bot. Rev.* 28: 1–239.

WHITTAKER, R. H., – 1965 – Dominance and diversity in land plant communities. *Science, N.Y.* 147: 250–260.

WHITTAKER, R. H., – 1966 – Forest dimensions and production in the Great Smoky Mountains. *Ecology* 47: 103–121.

WHITTAKER, R. H., 1967 – Gradient analysis of vegetation. *Biol. Rev.* 42: 207–264.

WHITTAKER, R. H., – 1969 – Evolution of diversity in plant communities. *Brookhaven Symp. Biol.* 22: 178–196.

WHITTAKER, R. H., – 1970a – Communities and Ecosystems. Macmillan, New York. 162 pp. 2nd ed., 1975, 385 pp.

WHITTAKER, R. H., – 1970b – The population structure of vegetation (Germ. summ.) In 'Gesellschaftsmorphologie (Strukturforschung)' ed. R. Tüxen. *Ber. Symp. int. Ver. Vegetationskunde*, Rinteln 1966: 39–62.

WHITTAKER, R. H., – 1972 – Evolution and measurement of species diversity. *Taxon* 21: 213–251.

WHITTAKER, R. H., S. A. LEVIN & R. B. ROOT, – 1973 – Niche, habitat, and ecotope. *Amer. Nat.* 107: 321–338.

WHITTAKER, R. H. & W. A. NIERING, – 1964 – Vegetation of the Santa Catalina Mountains, Arizona. I. Ecological classification and distribution of species. *J. Ariz. Acad. Sci.* 3: 9–34.

WHITTAKER, R. H. & W. A. NIERING, – 1965 – Vegetation of the Santa Catalina Mountains, Arizona. (II.) A gradient analysis of the south slope. *Ecology* 46: 429–452.

WHITTAKER, R. H. & W. A. NIERING, – 1968a – Vegetation of the Santa Catalina Mountains, Arizona. III. Species distribution and floristic relations on the north slope. *J. Ariz. Acad. Sci.* 5: 3–21.

WHITTAKER, R. H. & W. A. NIERING, – 1968b – Vegetation of the Santa Catalina Mountains, Arizona. IV. Limestone and acid soils. *J. Ecol.* 56: 523–544.

WHITTAKER, R. H. & W. A. NIERING, – 1975 – Vegetation of the Santa Catalina Mountains, Arizona. V. Biomass, production, and diversity along the elevation gradient. *Ecology* 56: 771–790.

YANG, T. W. & C. H. LOWE, JR., – 1956 – Correlation of major vegetation climaxes with soil characteristics in the Sonoran Desert. *Science, N.Y.* 123: 542.

3 RETROGRESSION AND COENOCLINE DISTANCE

Robert H. Whittaker and George M. Woodwell

Contents page

3 RETROGRESSION AND COENOCLINE DISTANCE

3.1 Introduction

In succession community and environment are coupled together as parts of a functional system in which the interplay among species populations and the interactions between community and environment determine the course of development. Development continues until a stable community or climax, adapted to steady-state function and self-maintenance of its species in its biotope as affected by itself, results. It seems reasonable to consider the community changes during succession a community-gradient or coenocline (even though some replacements of species populations may be relatively abrupt). The progressive changes in environment during succession may be considered a seral complex-gradient (though successions differ widely in the degree of environmental modification they entail). A succession is thus an ecocline in time, comprising a parallel and functionally related coenocline and complex-gradient.

The perspectives of gradient analysis are appropriate to research on succession. There has been little application of them to such problems, except for the appearance of successional or disturbance factors among the axes of indirect ordinations. In one area of research in ecology, however, this direction can be pursued. This is the study of retrogression: effects of chronic stress in reducing community structure and changing species composition in ways suggestive of succession in reverse. Technological civilization is producing widespread retrogressive changes in communities—toward simplified structure, loss of specialist species and reduction of species diversity, loosening of nutrient cycles, and decrease in stability (Woodwell 1970). We shall consider the character of retrogressions, measurements of degree of retrogression, and the nature of ecological distance along retrogressive and other coenoclines.

3.2 Observations on Retrogression

3.2.1 Grazing

From overgrazing of grasslands a familiar sequence of changes may result: (i) First, the more sensitive grass and herb species are decreased in cover or eliminated, with some compensatory increase in less sensitive species. (ii) Second, the more sensitive species are largely eliminated, while less sensitive ones assume dominance and weedy or woody invaders appear in a community now of reduced total coverage. (iii) Third, continued grazing reduces these less sensitive dominants, until the invaders assume dominance of a community of gravely reduced coverage and productivity, with much of its soil surface exposed to erosion. (iv) Fourth, combined effects of grazing, trampling, and erosion reduce the cover of even the invaders that remain until (v) Finally, on vulnerable slopes, there are few plants left on an eroded rocky slope or mudfield. One notes, as expressions of 'retrogression' in this process, the reduction of community coverage, mass and productivity, and ultimately of soil quality. Of the extensive American literature on the evaluation of range condition and trend (ELLISON 1949, 1954, STODDART & SMITH 1955), most to the point of our purpose is the approach of DYKSTERHUIS (1949). Three groups of plants are recognized: (i) decreasers: the the more sensitive climax species that are reduced in coverage by even moderate grazing pressure, (ii) increasers: species present in the climax grassland that increase in coverage with moderate grazing, and (iii) invaders: species that are not present (or are restricted to local disturbed areas) in the climax grassland and that enter the community in response to grazing. It is reasonable to regard the sequence of changes in response to grazing as a community-gradient or coenocline, and to treat these three groups as ecological groups in the sense of DUVIGNEAUD (1946), ELLENBERG, (1950, 1952) and others. Weighted averages using these three groups can be used as indices of position along the gradient of community disturbance. DIX (1959) and ITOW (1963) have used weighted averages for indication of grazing effects. ITOW's study shows also the decrease in species diversity with increasing pressure, although light or moderate grazing in some cases increases diversity (HARPER 1969). DYKSTERHUIS (1949), used a form of percentage similarity as a measurement of retrogression under grazing. Given composition of the climax as a standard of comparison, he summed for a given sample the percentage of its cover made up of decreasers, plus the percentage of increasers up to but not beyond their coverage in the climax, as an index of degree of departure from climax composition.

3.2.2 POLLUTION

A second group of observations involve responses of terrestrial communities to air pollution. It has long been known that epiphytic

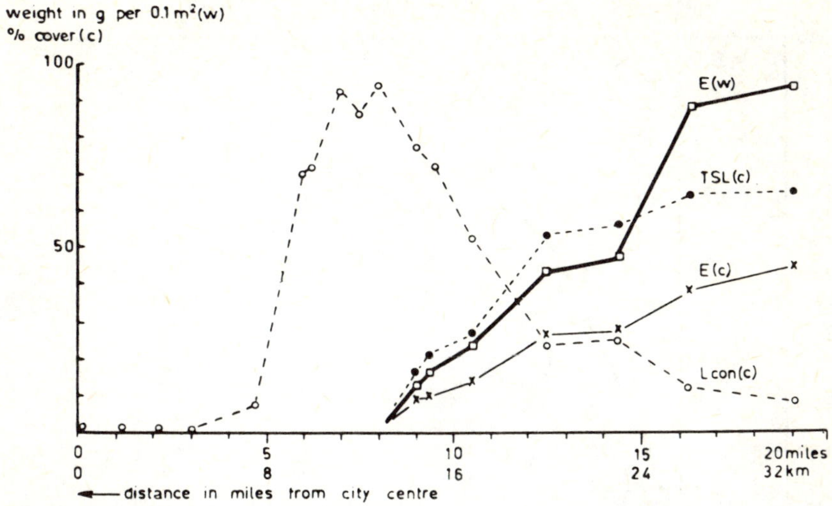

Fig. 1. Reduction in numbers of lichen and bryophyte species along a gradient of increasing sulfur dioxide pollution in England (GILBERT 1969). The lichen species *Lecanora conizaeoides* increases as other species are reduced at intermediate pollution intensities.

lichens and mosses are affected by the air pollution of cities; these effects, familiar in Western Europe (HILITZER 1925, BARKMAN 1958, 1969, SKYE 1958, 1968, GILBERT 1965, 1968), are known also from Montreal (LEBLANC 1969, 1971) and other American cities. Indicator species – decreasers and increasers – can be recognized among lichens and mosses. Particularly to the point are such observations as those of GILBERT (1969) in Fig. 1, showing both decrease in numbers of species and the behavior of an increaser, *Lecanora conizaeoides*, in response to an air pollution gradient. This, and other data for species populations along successional gradients, give bell-shaped curves suggestive of those for stable populations along environmental gradients (preceding article). The forms and dynamics of the successional curves, the extent to which they are skewed on a time-axis, and the possible occurrence of plateau distributions among them, have apparently not been studied. GORDON and GORHAM (1963) describe effects of an iron sintering plant releasing sulfur dioxide in Ontario. Along a gradient of increasing damage to forest communities, the dominant trees (*Picea glauca* and *P. mariana*) were eliminated about 25 km from the plant; and between 25 and 7 km the plant species diversity of the remaining vegetation was progressively reduced, until from 2 to 7 km from the plant only two resistant species (*Polygonum cilinode* and *Sambucus pubens*) remained in most samples. Fig. 2 represents

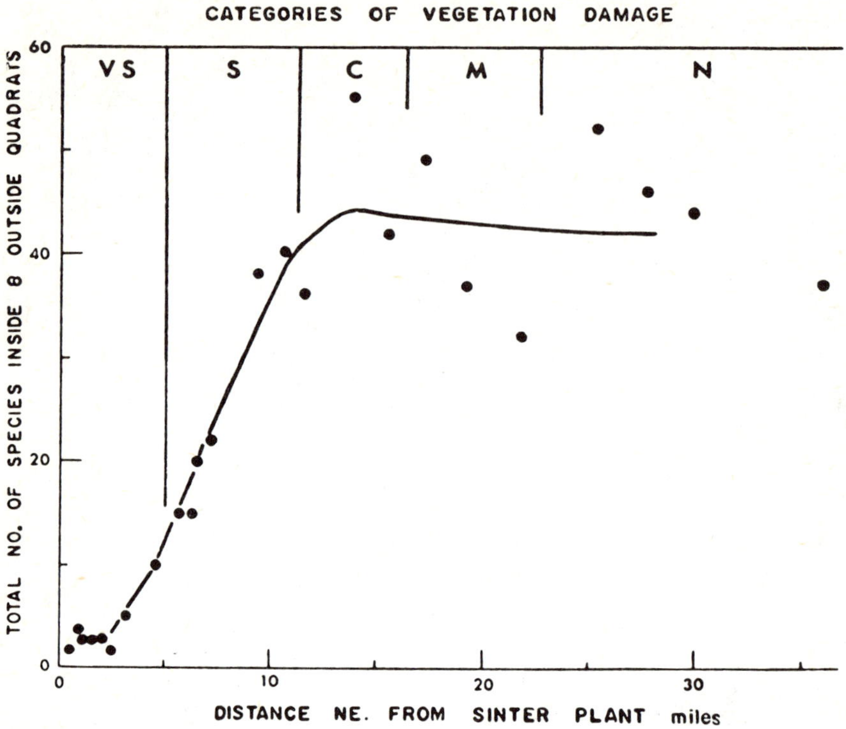

Fig. 2. Reduction in species diversity of vascular plants along a gradient of increased sulfur dioxide pollution in Ontario (GORDON & GORHAM 1963).

the effects on species-diversity with increasing proximity to the sulfur dioxide source. RAO and LEBLANC (1967) studied effects of the same pollution gradient on epiphytic lichens and mosses and observed increasing sensitivity with plant height, from crustose to foliose to fruticose forms among lichens. HADJÚK (1970) and HADJÚK & RUŽIČKA (1959) describe reductions in species diversity by air pollution. Retrogression of terrestrial communities under pollution is thus expressed in reduction of stature and coverage, living mass and productivity, and species diversity (see also NASH 1975).

A third group of observations concern the effects of pollution on aquatic communities. The most extensive literature deals with effects of various kinds of pollution on animal groups, particularly benthic forms (PATRICK 1951, HYNES 1963, INGRAM et al. 1966, WILHM & DORRIS 1968). GORHAM and GORDON (1963) have shown the decrease in species diversity of aquatic vegetation with sulphate pollution. Effects on phytoplankton communities have been studied by PATRICK (1963, PATRICK et al. 1954) and others (STAUB et al. 1970). PATRICK has shown that plankton diatom communities may

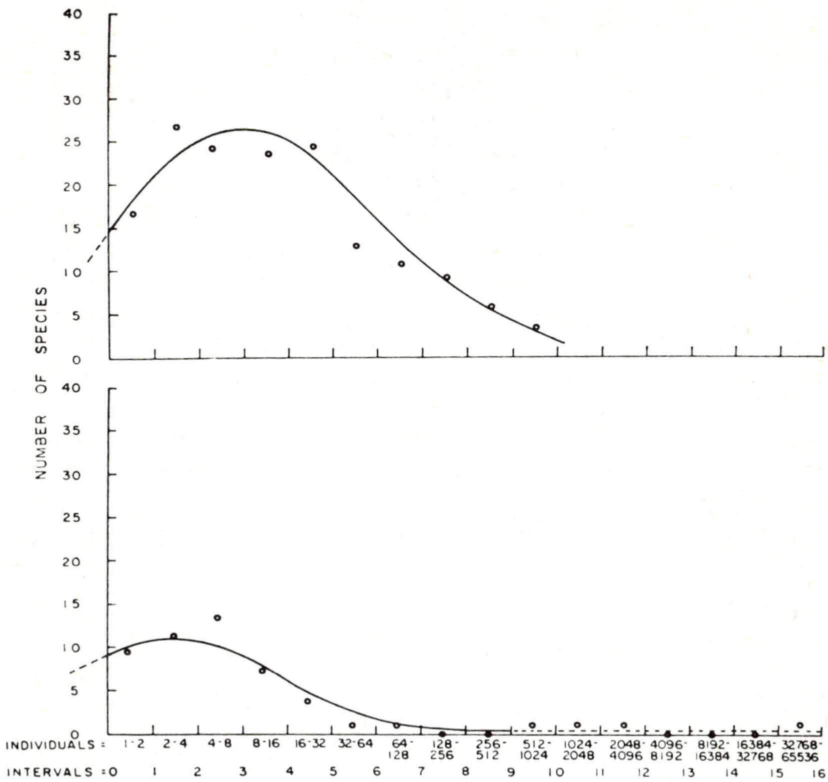

Fig. 3. Lognormal distributions of diatom species on glass slides suspended in streams, an unpolluted stream (Ridley Creek, Pennsylvania) above, and a polluted stream (Back River, Maryland) below (PATRICK 1963).

be described by the lognormal distribution of PRESTON (1948). When species in a large community sample are classed into octaves or doubling ranges of importance values, they form a frequency distribution with many species of intermediate abundance, few very common ones, and few very rare ones. The curves of Fig. 3 are truncated or broken by the limits of sample adequacy on the left, but one may estimate the total numbers of species in communities by extending the curves to the left. The upper curve is for a diatom community established on glass slides immersed in an unpolluted stream; with it may be compared the curve for a polluted stream below. The polluted stream has a lower number of species per octave, a wider dispersion of the species through octaves, and a much lower total richness (PATRICK 1963 cf. CAIRNS et al. 1974).

3.2.3 Irradiation

An oak-pine forest at Brookhaven National Laboratory was subjected to chronic gamma irradiation (Woodwell 1967, 1970, Woodwell & Whittaker 1968). With increasing intensity of irradiation the forest was progressively reduced. First the pine (*Pinus rigida*) was eliminated, then the canopy oaks (*Quercus coccinea*, *Q. alba*) to leave a community dominated by vacciniaceous shrubs (*Gaylussacia baccata*, *Vaccinium vacillans*). With more intensive irradiation the shrubs were eliminated and a sedge (*Carex pensylvanica*) normally present in the forest at low coverage acted as an increaser, forming a sedge-dominated zone. At still more intense irradiation the sedge was eliminated, and of the plant community close to the gamma source only lichens on the bark of dead trees remained. Fig. 4 represents the effect of irradiation on vascular plant species diversity of the forest. Fig. 5 shows by coefficients of community the increasing degree of change in forest composition with increasing irradiation.

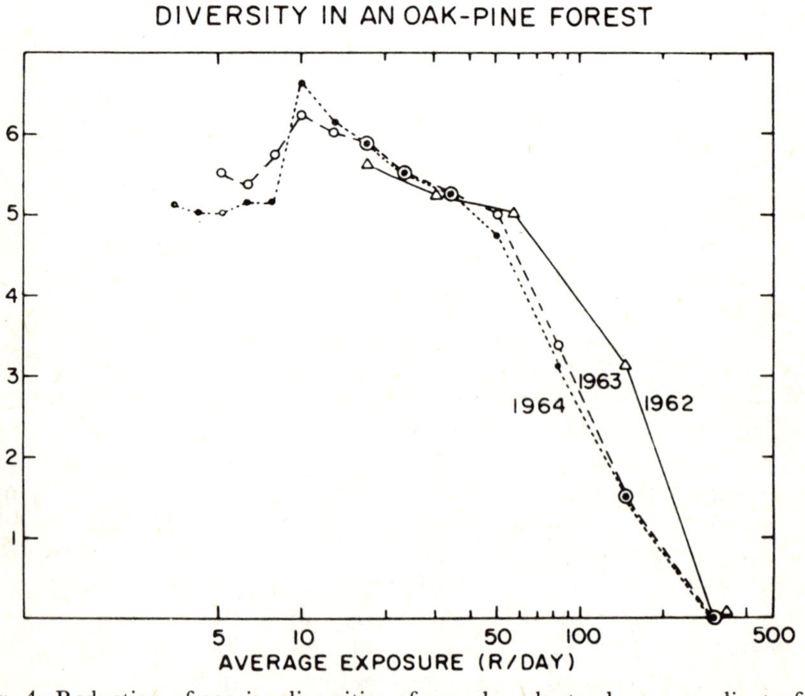

Fig. 4. Reduction of species diversities of vascular plants along a gradient of increasing gamma irradiation in the Brookhaven forest after one, two, and three years' exposure (Woodwell & Rebuck 1967).

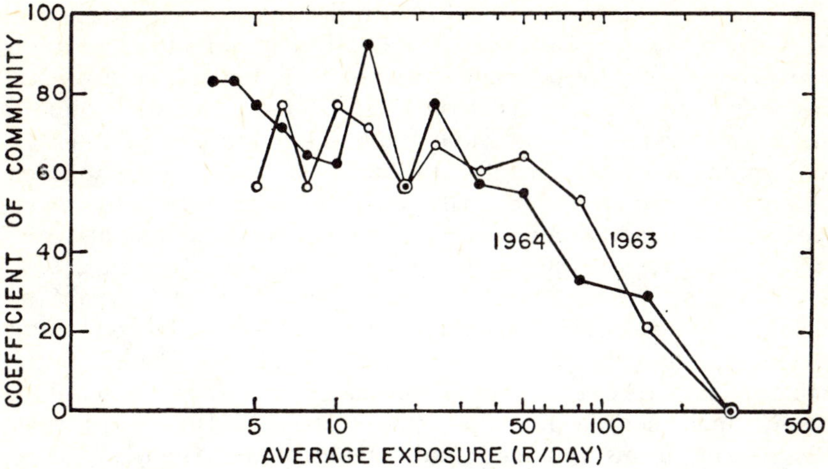

Fig. 5. Decrease in sample similarity (coefficient of community for vascular plants) along a gradient of increasing gamma irradiation in the Brookhaven forest (WOODWELL & REBUCK 1967).

The Brookhaven studies included observations on three different plant communities in response to irradiation: the forest vascular plants, the forest lichens, and the herbs of old-field successions (WOODWELL & OOSTING 1965, WOODWELL & GANNUTZ 1967, WOODWELL & REBUCK 1967, WOODWELL & WHITTAKER 1968). In the responses of these communities some parallelisms and contrasts of interest were observed. (i) The coenocline of forest response to irradiation showed marked similarity to the retrogressive effects of repeated fires, to the successional sequence for forest recovery from fire, and to the climatic gradient from Brookhaven north to the Arctic (WOODWELL 1967, 1970). Implications of plant stature (and chromosome volumes) for sensitivity to stress may underlie these parallels. (ii) In all three of the communities, there is correlation of plant height with sensitivity, and consequent replacement of taller forms – trees, erect herbs, fruticose and foliose lichens – with lower or prostrate forms with increasing intensity of irradiation. (iii) Species diversity and coefficient of community approach a linear relation to the logarithm of radiation intensity. This relation implies that the community response is proportional to the relative increase in radiation ($\Delta R/R$) rather than to the absolute value of the increase (ΔR) in Roentgen units. The community relation to a logarithmic, rather than a linear environmental gradient has been observed for salinity in inland water bodies (WHITTAKER & FAIRBANKS 1958) and may apply to chemical pollutants (GORHAM & GORDON 1963, Fig. 2, HADJÚK 1970, Fig. 1).

(iv) The three kinds of communities differ markedly in sensitivity. As a criterion of sensitivity we used the exposure to irradiation that produced a fifty per cent reduction in either species diversity, or coefficient of community, from that of the unirradiated community (Woodwell & Whittaker 1968). The approach parallels on the community-level the 50 % lethal dose (LD_{50}) for expressing relative toxicity of different substances for organisms of a given species. The irradiation intensities for fifty-per cent reduction were: for species diversity 150 R/day for the forest vascular plants, 1000 R/day for the first-year old field herbs, and 2700 R/day for the lichens; for coefficient of community 120 R/day for the forest vascular plants, 200 R/day for the herbs, and 1500 R/day for the lichens. These three groups as wholes thus show increasing sensitivity with increasing height above the substrate. Lichens, with their low-growing forms, relatively simple tissue organization, and small chromosomes are most resistant to irradiation among these groups. They are, in contrast, because of their direct subjection to the chemistry of precipitation, most sensitive to air pollution.

3.3 Measurement of Retrogression

3.3.1 Successional Trends

Successions normally involve trends of increase in a number of community dimensions (Whittaker 1953, Margalef 1963, Odum 1969). During succession on land we expect: (i) increasing productivity, supporting increasing biomass, (ii) a consequent increase in the ratio of biomass to productivity, and increase in the ratio of total community respiration to gross productivity toward 1.0, (iii) increasing stock of nutrients held in the organisms and organic matter, and increasing ratio of the community's nutrient stock to that in environment, (iv) increasing soil organic content and horizon development, increasing microclimatic modification by the increasingly massive plant community, (v) increasing community height and stratal differentiation, increasing species diversity, (vi) increasing longevity of dominant populations and increasing stability of the community. Most of these reverse themselves during some successions, and it may often be the case that productivity and species diversity are higher in a subclimax stage than in the climax. Probably biomass and the biomass accumulation ratio (of biomass to net annual primary production) are least subject to reversal, most reliably maximal in the climax. Recent discussions of successional trends are by Drury & Nisbet (1973), Horn (1974), Whittaker (1975), and Pickett (1976).

In principle a decrease in any of these trends, particularly decrease in biomass, might serve as an index of retrogression. In practice most of these may be set aside for difficulty of measurement or unreliability as indicators of retrogression. In terrestrial communities biomass may be an insensitive measurement of retrogression because killing of more sensitive species may occur before the mass of tissue has been significantly reduced. In aquatic communities both biomass and production may be ineffective measurements of disturbance or retrogression, for both may increase in response to some, but decrease in response to other kinds of pollution. We desire measurements that can express any pollution disturbance including unnatural eutrophication that increases productivity, whether or not we interpret eutrophication as retrogression.

3.3.2 Taxon Sensitivities

Such generalized approaches to the measurement of retrogression must probably rely on species composition. Many pollution indices depend on calibration by species sensitivities, on specification of particular species or broader taxa that may be used to express effects for a particular stress and kind of community. Two purposes of retrogression measurement are in this connection to be distinguished. The first is detection of early retrogressive effects; such detection must generally depend on knowledge of sensitive species. The second is measurement of degrees of retrogression, which may be well advanced. The measurement should be independent of calibrations by the sensitivity of particular taxa, and should permit comparisons of degrees of retrogression for different kinds of communities and stresses. Three approaches that may serve in this way follow.

3.3.3 Weighted Averages

The standard ordination technique of weighted average calculation (Ellenberg 1950, Whittaker 1954, 1967) can be used as an effective index. The index values are dependent on definition of the ecological groups for a particular retrogression gradient. Weighted average indices consequently may not fit the stipulation of independence of the species of particular communities, except as the definitions of decreasers, increasers, and invaders provide a kind of standard that is consistent from one retrogressive coenocline to another.

3.3.4 Species Diversity

Reduction of richness in species is a general and effectively measurable expression of retrogression. Numbers of species in standard quadrats or sample counts may provide more sensitive measurement than the Shannon-Wiener information index and other indirect expressions. Reduction of species diversity to one half (or some other fraction) of that of the undisturbed community or control sample provides a standard by which one can compare relative sensitivities – of different communities to the same stress, or of the same community to different stresses. It should be recognized as a limitation of diversity as an index, that there are cases in which light or moderate disturbance increases species diversity before heavier disturbance decreases it.

3.3.5 Sample Similarities

Degree of departure from composition of undisturbed communities can be measured by various similarity indices. The work at Brookhaven (Woodwell 1967) suggests that coefficient of community, based on species presence and absence, may be preferable to percentage similarity and Euclidean distance, which are based on species importance values and are much affected by irregularity of unstable populations. (These measurements are defined and discussed in articles 5 and 10). Reduction of coefficient of community to 50 % of that for comparisons of controls provides a standard by which sensitivities may be compared. Fifty per cent of control does not imply a coefficient of community of 50 %. When replicate samples from undisturbed communities are compared, their coefficients of community are not 100 %, but are lower values, in this case about 60 %. The standard of comparison, CC_{50}, should thus be a reduction of coefficient of community of samples along the retrogression coenocline to one half the value for undisturbed or control samples compared with one another. A standard of comparison other than 50 % of control similarity might, of course, be used for other studies.

3.4 **Coenocline Distance**

3.4.1 Measurement

The measurement of retrogression by sample similarity relates to a more general problem of ordination: the nature of 'distance' in coenoclines and community patterns. WHITTAKER (1952) stated the concept of ecological distance as degree of community separation, expressed in change of community composition in response to environmental change. We shall term the idea "coenocline distance" here to distinguish this, as a measurement applied to communities, from separation along an environmental complex-gradient as expressed by an environmental measurement.

Two approaches to coenocline distance suggest themselves. It may, first, be expressed as a multiple or fraction of mean species amplitude in the coenocline. The mean spread of species populations through the ten steps of Table I, in article 2, is 5.3 steps (including the incomplete distributions of species whose full distributions extend beyond the ends of the transect). Using this amplitude as a unit, the coenocline distance from one extreme of this transect to the other would be 10/5.3 steps or 1.9. The amplitudes of species populations in transects are, however, affected by sample size; as sample size increases the apparent amplitudes increase. It may be better, when feasible, to use as a unit mean standard deviation for these bell-shaped curves (Fig. 1, article 2). Distances in the simulated coenoclines to be discussed below (article 10) are thus defined .It is difficult, however, to compute standard deviations for the species in transects from field data, for the distributions of many species are cut off by the ends of the transects.

The alternative is use of sample similarity measurements, as applied above to retrogression. The simplest use of such measurements— the similarity value for samples from the extremes of a coenocline— is inadequate as an expression of coenocline distance. It is not the case that a similarity measurement of 25 % represents twice the coenocline distance of one of 50 %, because: (i) It is the complement of the similarity measurement that expresses distance hence (if 100 % is used to calculate the complements) the similarities just given would become percentage differences of 75 % and 50 %. (ii) But 100 % is not the appropriate basis of calculation. As indicated above, similarity values for replicate samples are less than 100 %; such similarity or "internal association" values are often in the range of 60 to 90 % (WHITTAKER 1952, 1960, BRAY & CURTIS 1957). In general replicate similarities are lower, the smaller the sample and the higher the species diversity (WHITTAKER 1952). If 85 % is the replicate similarity, sample similarities of 25 % and 50 % become percentage differences of 60 and 35 %. (iii) Even if the appropriate similarity is used, percentage difference of samples bears a curvilinear relation to environmental separation (Fig. 6). Use of sample similarities for coenocline distance has a further limitation (iv) the fact that similarity mea-

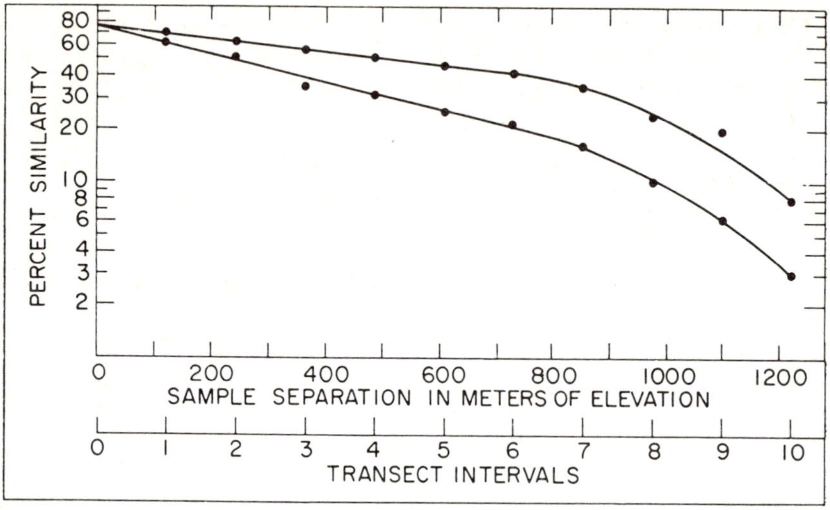

Fig. 6. Decrease in mean sample similarity measurements with increasing sample separation along the elevation gradient in the Great Smoky Mountains (WHITTAKER 1960, Fig. 18, 1967, Fig. 9). The upper curve is for coefficient of community, the lower for percentage similarity, applied to samples of forest trees, on the same logarithmic scale on the ordinate. The abscissa indicates the separation of the samples being compared in terms of elevation in meters and transect intervals of 122 m. The curves are smoothed by averaging a number of sample pairs for a given number of transect intervals by which the samples are separated.

surements become increasingly insensitive expressions of coenocline distance as the values become small until, when the values are zero, no measurement of coenocline distance is possible.

Fig. 6 illustrates the form of sample similarity decrease with separation along an environmental gradient. The first parts of many such curves approach straight lines when the logarithm of sample similarity is plotted against environmental separation. All such curves, however, slope off with increasing steepness for coenocline distances that exceed mean species amplitudes. The forms of the curves were used by WHITTAKER (1960, WHITTAKER & NIERING 1965) to obtain "half-changes" as units of coenocline distance. A straight line is fitted to the initial portion of the curve. The y intercept of this line, for zero separation along the environmental gradient, is used as the replicate similarity value. The line is extrapolated also to the full length of the coenocline for which a distance value is desired; the upper curve of Fig. 6 thus extrapolated for 10 transect intervals gives a similarity value of 30 % (in contrast with the actual similarity measurement for extreme samples of 8 %). Half-changes (β), as a measurement of beta diversity, may then be computed as, $z = a \times 2^{-\beta}$, $\beta = (\log a - \log z) / \log 2$, in which *a* is the replicate similarity (75 %) and *z* the estimate for extreme similarity (30 %). On this basis half-changes for the elevation transect illustrated are 1.32 for the upper curve (and 3.06 for the lower curve), inplying rates of community change in half-change units per 1000 m of 1.08 when measured through coefficient of community, 2.5 when measured through percentage similarity. The latter rate is steeper because it reflects change in quantitative representation of species, whereas coefficient of community expresses only species presence and absence. The technique has been used to compare beta diversities for topographic moisture gradients (WHITTAKER 1960, WHITTAKER & NIERING 1965, 1968) and, as indicated, degrees of retrogression.

3.4.2 THE FORM OF SAMPLE SIMILARITY DECREASE

More recently the form of sample similarity decrease in relation to environmental separation has been investigated (GAUCH 1973). Computations are based on the assumption that the coenocline includes numerous species with Gaussian distributions, their modes in scattered positions along the gradient. The curve formed by similarity values is then a complement of the non-standardized error function,

$$PS_{(s)} = 100 \left[1 - ERF \frac{s}{2\sqrt{2}}\right] \text{ where } ERF_{(x)} = \frac{2}{\sqrt{\pi}} \int_0^x e^{-u^2}\, du,$$

and *s* is separation in units of average standard deviation of species along the coenocline. The curves are of the same form for the different sample similarity measurements investigated (coefficient of community, percentage similarity, and the complement of Euclidean distance). An ideal curve, for coenocline data unaffected by sample error, is indicated in Fig. 7, *a*; simulated coenocline data have been used to check the fit of the relationship given.

A curve for simulated coenocline data affected by sample error or population irregularity is given as Fig. 7, *b*. It is a fair approximation of the forms observed for field data. One notes that the 50 % point is not the same for the ideal curve and that affected by sample error. The half-change technique uses a quasi-linear initial slope that is a consequence of the effect of sample error on field data. In principle the error function should be used to correct sample similarity measurements to give coenocline distances (article 10); but the form of field data is not that of the ideal error function curve.

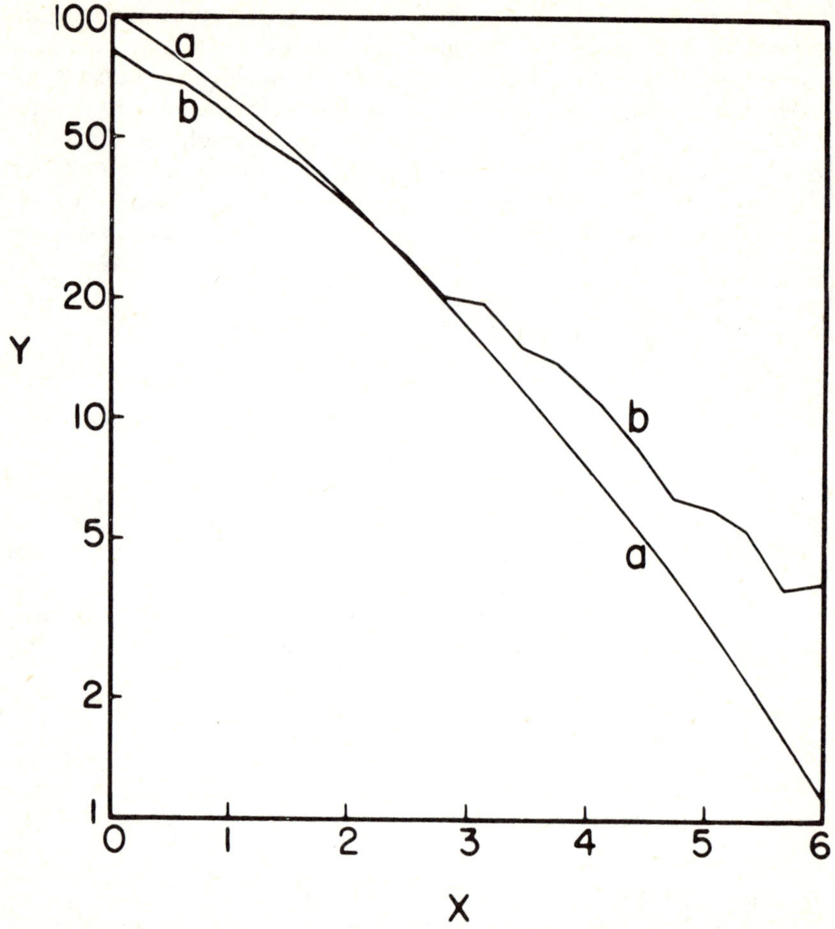

Fig. 7. Decrease in mean sample similarity, *Y*, with increasing sample separation in a simulated coenocline along a hypothetical gradient, *X* (GAUCH 1972). The smooth curve fits an error function complement to points based on averaging samples without sample error; the irregular curve shows the effect of sample error.

3.5 Conclusion

Limitations inherent in the character of community data affect all these measurements of retrogression and coenocline distance. The use of such measurements for ordination is to be further discussed below. For retrogression there seems to be no single, fully satisfactory measurement. We find instead a group of measurements that are reasonably suitable, affected by different merits, limitations, and appropriateness to different research problems. If such measurements are not ideal, they may still be sufficiently effective to express the degrees of the disturbance and retrogression that man is increasingly imposing upon many of the natural and seminatural communities of the world.

3.6 **Summary**

Chronic disturbances or stresses applied to natural communities by man can produce changes that are in some respects the reverse of succession, changes termed "retrogression". Among these are the reduction of grasslands by overgrazing, of terrestrial and aquatic communities by air and water pollution, and of forests by ionizing irradiation.

Reduction in any community characteristic that increases through succession might be used as a measure of retrogression, but many such measurements are ineffective. Reduction of particular sensitive species or other taxa may indicate beginning retrogression. For more general measurements three approaches seem most useful: (i) weighted averages, which may use "decreasers, increasers, and invaders" as ecological groups, (ii) reduction of species-diversity from that of a control or undisturbed sample, and (iii) coefficients of community for disturbed compared with undisturbed samples. The second and third of these make possible definition of a 50 % change in the community with retrogression, by which may be compared the responses of a given community to different stresses, or of different communities to a given stress.

Measurement of retrogression relates to the problem of coenocline distance— degree of difference in community composition in response to degree of separation along an environmental (or successional) gradient. Coenocline distance can be approached as extent of change in community composition, as expressed in measurements of sample similarity. These similarity measurements bear a complex and curvilinear relation to sample separation; the relation can be fitted by the complement of a non-standardized error function. Mean amplitudes of species populations and half-changes in community composition provide units by which coenocline distance can be expressed.

Acknowledgment

Research carried out at Brookhaven National Laboratory un-the auspices of the U.S. Atomic Energy Commission.

BARKMAN, J. J., – 1958 – Phytosociology and Ecology of Cryptogamic Epiphytes. Van Gorcum, Assen. 628 pp.

BARKMAN, J. J., – 1969 – The influence of air pollution on bryophytes and lichens. (French & Germ. summs.) In 'Air Pollution,' *Proc. First European Congr. on the Influence of Air Pollution on Plants and Animals*, Wageningen 1968, pp. 197–209. Centre for Agricultural Publishing and Documentation, Wageningen.

BRAY, J. R. & J. T. CURTIS, – 1957 – An ordination of the upland forest communities of southern Wisconsin. *Ecol. Monogr.* 27: 325–349.

CAIRNS, J. JR., D. W. ALBAUGH, F. BUSEY & M. D. CHANAY, – 1974 – The sequential comparison index – a simplified method for non-biologists to estimate relative differences in biological diversity in stream pollution studies. *J. Water Pollution Control Fed.* 40: 1607–1613.

DIX, R. L., – 1959 – The influence of grazing on the thin-soil prairies of Wisconsin. *Ecology* 40: 36–49.

DRURY, W. H. & I. C. T. NISBET, – 1973 – Succession. *J. Arnold Arbor*, Harvard Univ. 54: 331–368.

DUVIGNEAUD, P., – 1946 – La variabilité des associations végétales. *Bull. Soc. r. Bot. Belg.* 78: 107–134.

DYKSTERHUIS, E. J., – 1949 – Condition and management of range land based on quantitative ecology. *J. Range Mgmt* 2: 104–115.

ELLENBERG, H., – 1950 – Landwirtschaftliche Pflanzensoziologie. I. Unkrautgemeinschaften als Zeiger für Klima und Boden. Ulmer, Stuttgart. 141 pp.

ELLENBERG, H., – 1952 – Landwirtschaftliche Pflanzensoziologie. II. Wiesen und Weiden und ihre standörtliche Bewertung. Ulmer, Stuttgart. 143 pp.

ELLISON, L., – 1949 – The ecological basis for judging condition and trend on mountain range land. *J. For.* 47: 786–795.

ELLISON, L., – 1954 – Subalpine vegetation of the Wasatch Plateau, Utah. *Ecol. Monogr.* 24: 89–184.

GAUCH, H., – 1973 – The relationship between sample similarity and ecological distance. *Ecology.* 54: 618–622.

GILBERT, O. L., – 1965 – Lichens as indicators of air pollution in the Tyne Valley. In 'Ecology and the Industrial Society,' ed. G. T. Goodman, R. W. Edwards, & J. M. Lambert. *Brit. Ecol. Soc. Symp.* 5: 35–47.

GILBERT, O. L., – 1968 – Bryophytes as indicators of air pollution in the Tyne Valley. *New Phytol.* 67: 15–30.

GILBERT, O. L., – 1969 – The effect of SO_2 on lichens and bryophytes around Newcastle upon Tyne. (French & Germ. summs.) In 'Air Pollution' *Proc. First European Congr. on the Influence of Air Pollution on Plants and Animals*, Wageningen 1968, pp. 223–235. Centre for Agricultural Publishing and Documentation, Wageningen.

GORDON, A. G. & E. GORHAM, – 1963 – Ecological aspects of air pollution from an iron-sintering plant at Wawa, Ontario. *Can. J. Bot.* 41: 1063–1078.

GORHAM, E. & A. G. GORDON, 1963 – Some effects of smelter pollution upon aquatic vegetation near Sudbury, Ontario. *Can. J. Bot.* 41: 371–378.

HADJÚK, J., – 1970 Einwirkung von Industrie-Exhalationen auf die Struktur der Phytocoenosen. (Engl. summ.) In 'Gesellschaftsmorphologie (Strukturforschung),' ed. R. Tüxen. *Ber. Symp. int. Ver. Vegetskunde*, Rinteln 1966: 340–349.

HAJDÚK, J. & M. RUŽIČKA, – 1969 – Das Studium der Schäden an Wildpflanzen und Pflanzengesellschaften verursacht durch Luftverunreinigung. (French & Engl. summs.) In 'Air Pollution,' *Proc. First European Congr. on the Influence of Air Pollution on Plants and Animals*, Wageningen 1968, pp. 183–192. Centre for Agricultural Publishing and Documentation, Wageningen.

Harper, J. L., – 1969 – The role of predation in vegetational diversity. *Brookhaven Symp. Biol.* 22: 48–62.

Hilitzer, A., – 1925 – Étude sur la végétation épiphyte de la Bohême. *Spisy vydáv. přír. Fak. Karl. Univ.* (*Publ. Fac. Sci. Univ. Charles*), Praha 41; 1–202.

Horn, H. S., – 1974 – The ecology of secondary succession. *Ann. Rev. Ecol. Syst.* 5: 25–37.

Hynes, H. B. N., – 1963 – The Biology of Polluted Waters. Liverpool Univ. Press, Liverpool. 202 pp.

Ingram, W. M., K. M. Mackenthun, & A. F. Bartsch, – 1966 – Biological field investigative data for water pollution surveys. U.S. Dept. Interior, Washington. 139 pp.

Itow, S., – 1963 – Grassland vegetation in uplands of western Honshu, Japan. II. Succession and grazing indicators. *Jap. J. Bot.* 18: 133–167.

LeBlanc, F., – 1969 – Epiphytes and air pollution. (French & Germ. summs.) In 'Air Pollution.' *Proc. First European Congr. on the Influence of Air Pollution on Plants and Animals*, Wageningen 1968, pp. 211–221. Centre for Agricultural Publishing and Documentation, Wageningen.

LeBlanc, F., – 1971 – Possibilities and methods for mapping air pollution on the basis of lichen sensitivity. In 'Methoden zur Erkennung und Beurteilung forstschädlicher Luftverunreinigungen' XV. IUFRO-Kongress, 1971. *Mitt. Forstl. BundVersAnst. Wien* 92: 103–126.

Margalef, R., – 1963 – On certain unifying principles in ecology. *Am. Nat.* 97: 357–374.

Nash, T. H. III, – 1975 – Influence of effluents from a zinc factory on lichens. *Ecol. Monogr.* 45: 183–198.

Odum, E. P., – 1969 – The strategy of ecosystem development. *Science, N.Y.* 164: 262–270.

Patrick, Ruth, – 1951 – A proposed biological measure of stream conditions. *Verhandl. int. Ver. theor. angew. Limnol.* 11: 299–307.

Patrick, Ruth, – 1963 – The structure of diatom communities under varying ecological conditions. *Ann. N. Y. Acad. Sci.* 108: 359–365.

Patrick, Ruth, M. H. Hohn & J. H. Wallace, – 1954 – A new method of determining the pattern of the diatom flora. *Notul. Nat.*, Acad. Nat. Sci. Philad. 259: 1–12.

Pickett, S. T. A., – 1976 – Succession: an evolutionary interpretation. *Amer. Nat.* 110: 107–119.

Preston, F. W., – 1948 – The commonness, and rarity, of species. *Ecology* 29: 254–283.

Rao, D. N. & F. LeBlanc, – 1967 – Influence of an iron-sintering plant on corticolous epiphytes in Wawa, Ontario. *Bryologist* 70: 141–157.

Skye, E., – 1958 – Luftföroreningars inverkan på busk- och bladlavfloran kring skifferoljeverket i Närkes Kvarntorp. (Engl. summ.) *Svensk bot. Tidskr.* 52: 133–190.

Skye, E., – 1968 – Lichens and air pollution: a study of cryptogamic epiphytes and environment in the Stockholm region. *Acta phytogeogr. suec.* 52: 1–123.

Staub, R., J. W. Appling, A. M. Hofstetter, & I. J. Haas, – 1970 – The effects of industrial wastes of Memphis and Shelby County on primary planktonic producers. *BioScience* 20: 905–912.

Stoddart, L. A. & A. D. Smith, – 1955 – Range Management. 2nd ed. McGraw-Hill, New York. 433 pp.

Whittaker, R. H., – 1952 – A study of summer foliage insect communities in the Great Smoky Mountains. *Ecol. Monogr.* 22: 1–44.

WHITTAKER, R. H., – 1953 – A consideration of climax theory: the climax as a population and pattern. *Ecol. Monogr.* 23: 41–78.
WHITTAKER, R. H., – 1954 – Plant populations and the basis of plant indication. (Germ. summ.) *Angew. PflSoziol.*, Wien, Festschr. Aichinger 1: 183–206.
WHITTAKER, R. H., – 1960 – Vegetation of the Siskiyou Mountains, Oregon and California. *Ecol. Monogr.* 30: 279–338.
WHITTAKER, R. H., – 1967 – Gradient analysis of vegetation. *Biol. Rev.* 42: 207–264.
WHITTAKER, R. H., – 1975 – Functional aspects of succession in deciduous forests (Germ. summ.) In 'Sukzessionsforschung' ed. W. Schmidt. *Ber. Symp. int. Ver. Vegetationskunde*, Rinteln 1973: 377–405.
WHITTAKER, R. H. & C. W. FAIRBANKS, – 1958 – A study of plankton copepod communities in the Columbia Basin, southeastern Washington. *Ecology* 39: 46–65.
WHITTAKER, R. H. & W. A. NIERING, – 1965 – Vegetation of the Santa Catalina Mountains, Arizona. (II.) A gradient analysis of the south slope. *Ecology* **46**: 429–452.
WHITTAKER, R. H. & W. A. NIERING, – 1968 – Vegetation of the Santa Catalina Mountains, Arizona. IV. Limestone and acid soils. *J. Ecol.* 56: 523–544.
WILHM, J. L. & T. C. DORRIS, – 1968 – Biological parameters for water quality criteria. *BioScience* 18: 477–481.
WOODWELL, G. M., – 1967 – Radiation and the patterns of nature. *Science N.Y.* 156: 461–470.
WOODWELL, G. M., – 1970 – Effects of pollution on the structure and physiology of ecosystems. *Science, N.Y.* 168: 429–433.
WOODWELL, G. M. & J. K. OOSTING, – 1965 – Effects of chronic gamma irradiation on the development of old field plant communities. *Radiat. Bot.* 5: 205–222.
WOODWELL, G. M. & T. P. GANNUTZ, – 1967 – Effect of chronic gamma irradiation on lichen communities of a forest. *Am. J. Bot.* 54: 1210–1215.
WOODWELL, G. M. & A. L. REBUCK, – 1967 – Effects of chronic gamma radiation on the structure and diversity of an oak-pine forest. *Ecol. Monogr.* 37: 53–69.
WOODWELL, G. M. & R. H. WHITTAKER, – 1968 – Effects of chronic gamma irradiation on plant communities. *Q. Rev. Biol.* 43: 42–55.

4 RUSSIAN (RAMENSKY) APPROACHES TO COMMUNITY SYSTEMATIZATION

L. N. SOBOLEV AND V. D. UTEKHIN

Contents page

4 RUSSIAN (RAMENSKY) APPROACHES TO COMMUNITY SYSTEMATIZATION

4.1 Introduction

In this paper we summarize views of L. G. RAMENSKY, one of the greatest Russian geobotanists, on the systematic analysis of vegetation. RAMENSKY's views were underlain by the concept of continuity of the vegetation cover. Since such ideas were expressed by him as far back as the beginning of the twentieth century (RAMENSKY 1910), he has clear priority over American authors (WHITTAKER 1951, 1956, CURTIS & MCINTOSH 1951) for the concept of the vegetation continuum (ALEKSANDROVA 1969). RAMENSKY considered arrangement of plant communities into gradient series to be the best method for systematizing data on continuous vegetation. So far as the second way of systematizing objects – namely classification – is concerned, RAMENSKY did not rule it out; but he criticized contemporary methods of classification of plant communities and asserted that it must not be separated from classification of other components of the biocoenose-ecosystem.

It is noteworthy that RAMENSKY considered vegetation a part of the natural complex and opposed separation of vegetation from the whole for fully independent study. He possessed the knowledge of both a geobotanist and a soil-scientist, and applied all his ideas equally to vegetation and to soils. It is not by chance that his main work (1938) is entitled "*Introduction to the Complex Soil-Geobotanic Investigation of Lands*." These ideas are very close to what WHITTAKER (1962) characterized as the "tradition of multifactoral or landscape units." RAMENSKY connected his teaching with practical evaluation of productivity of lands for agricultural purposes, and in this respect he was close to ELLENBERG's (1952) "*Landwirtschaftliche Pflanzensoziologie*." He aspired to create a "production typology of lands" for the foundation of which he developed the "natural types of land-units," the latter representing an idea very close to the ecosystem of TANSLEY or biogeocoenose of SUKACHEV, although the latter concept was strongly criticized by RAMENSKY in many respects. RAMENSKY objected particularly to SUKACHEV's idea of the biogeocoenose as a sum of three independent parts: phytocoenose, zoocoenose and biotope; he pointed out that there existed interrelations between every element of these parts.

Despite the range and importance of RAMENSKY's work, much of that work has not been made available in Western languages (see also, however, RAMENSKY 1926, 1930, 1932). We shall here treat his concepts and techniques in application to geobotanic material but it should be borne in mind that they apply to the broader problems of natural complexes.

4.2 **Ideas on the Plant Community**

4.2.1 MINIMAL AREA

RAMENSKY was one of the first investigators to pay attention to the statistical character of species abundance. Even in a homogeneous habitat the abundances of species can vary substantially due to chances and conditions that influence the result of competition (RAMENSKY 1924). As size of the sample area is enlarged, effect of such variation on the sample decreases, and in the end it becomes negligible. The minimum area for which variation of abundance of a species in samples does not exceed some value was called by RAMENSKY a minimal area of abundance for that species. Generalizing this notion he spoke of minimal areas for quantitative proportions of species, for crop production measurements, etc. (RAMENSKY 1924, 1937a). This idea of minimal area may be parental to those in use among Western authors (DU RIETZ 1921, GREIG-SMITH 1964), article 18).

4.2.2 DEPENDENCE OF VEGETATION UPON BIOTOPE

When investigating vegetation with samples of at least minimal area, one would expect vegetation to show strict dependence upon conditions of the biotope or habitat. A number of circumstances should, however, be taken into account here. First, vegetation response to the biotope is complicated by the influence of two factors – historical causes and availability of disseminules. The presence of traces of a former state of the vegetation, or the absence of some normal components of a community because disseminules fail to reach it, can weaken tightness of the bond of the vegetation composition with conditions of the environment. From normal historical causes (involving past community development) may be distinguished violation or destruction of vegetation by irregular influences of hazardous forces of nature or man. Regular influences, for example annual spring flood or haymaking, may be regarded as factors of the biotope.

Second, different characteristics of external environment have different implications for vegetation. RAMENSKY warned against confusing 'ecology with topology' and insisted upon separating 'conditions of the biotope' (habitat or ecotope) from 'conditions of the locality' (entopion). Briefly speaking, the distinction between these lies in the difference between directly acting factors and the factors that act indirectly. Heat, light, water, salt and other such qualities of the location that act directly on plants are conditions of the biotope. Geomorphological conditions, geological, hydrological, macroclimatic and other conditions, for example, relief, mechanical composition of the soil, parent material, the depth of ground waters below plant roots, and so on, are conditions of the locality. The importance of distinguishing between factors of the biotope and factors of the locality is conditioned by the fact that there is no simple dependence between the first factors and the second ones: various combinations of factors of the locality can stipulate the same biotope conditions, and dissimilar biotope conditions can occur in similar localities.

Thus, vegetation samples should display close relation with conditions of the environment, but only when, (i) the environment is characterized by effective factors of the biotope, (ii) the vegetation sample is adequate as regards minimal area, and (iii) the character of the phytocoenose is not unduly distorted by the influence of historical causes, or as a result of inaccessibility to disseminules or violation by disturbance. However, even when these conditions are observed the connection between vegetation and conditions of the environment is not always obvious. This can be explained by the complexity of the connection itself, when the vegetation shows multidimensional changeability in relation to a large number of environmental factors (RAMENSKY 1924, 1926).

4.2.3 ECOLOGICAL INDIVIDUALITY OF SPECIES

Biotopes, in the main, change continuously; moreover, they change differently in respect to different factors. This creates a continuous diversity of biotopes, full use of which by plants is possible only because of the diversity of environmental requirements of species. This last diversity should be such that any combination of conditions of the biotope will be favorable for the development of some species, from the many present in the landscape. Species differ in their requirements and consequent distributions in relation to biotopes; species show ecological peculiarity or ecological individuality. Ramensky illustrated the ecological individuality of species with numerous examples, particularly for meadow vegetation (RAMENSKY 1924, 1938, RAMENSKY et al. 1956).

4.2.4 CONTINUITY OF THE VEGETATION COVER

The idea of continuity of the vegetation cover is a logical outcome from the ecological individuality of species. Vegetation must continuously change with continuous change of the environment. Hence it follows that vegetation can be the best integrator and indicator of conditions of the habitat (RAMENSKY 1924). Indicator properties of vegetation were a major interest of RAMENSKY 's (1938, RAMENSKY et al. 1956) research.

We must ask how absolute was his view of the continuity of the vegetation cover. Two kinds or levels of continuity are to be distinguished: the spatial (between communities actually in contact along a habitat gradient in the field) and typological (between samples, that may have been taken from different areas not in contact, and are compared in the laboratory). RAMENSKY considered continuity to be a natural feature of the vegetation cover. He did not, however, deny the existence of gaps in continuity where sharp changes in the environment occur. At the same time he pointed out that even in such cases interpenetration of species from neighboring communities produces a distinctive transitional zone ("marginal deviations"). Marginal communities can be well developed and of extent surpassing the minimal area. The idea of marginal deviations seems to be very close to the ideas of "ecotone" and "edge effect" among American authors (ODUM 1953, WHITTAKER 1956).

So far as the typological continuum is concerned, RAMENSKY speculated that some breaks might occur in this continuum in ways suggestive of, as the author puts it, "inversion points of physico-chemical systems (water-ice-vapour, two hydrates of Glauber salt in one solution)" (RAMENSKY 1924). Later on Ramensky suggested a simple method of detecting whether breaks in continuity of gradient series of communities actually occur (see 4.4.4.).

4.3 **Ordination as a Method**

4.3.1 SAMPLE CHARACTERISTICS

The vegetation is to be analyzed through a set of samples (lists, relevés) that represent its variation. The selection of samples from various types of localities in an area should be as uniform as possible. At the same time RAMENSKY considers sample sets which are not too heterogeneous to be the most suitable for analysis. Separate investigations of rather small territories with relatively minor variations of the environment are most effective.

The size of sample quadrats (plots) must correspond as far as possible to the minimal area for a sufficient number of species. The species number and minimal area are established empirically and can be different in different landscapes. For meadows RAMENSKY recommended quadrats of 100 m^2, for desert communities he quotes the figure of 1000 m^2, and for forest 2500 m^2 (RAMENSKY 1937a). The shape of the test area can be different, and instead of one large plot a series of smaller ones can be used. Intentional selection of 'typical' or 'average' conditions is accepted. The number of sites, equal in area to the minimal area, for each type of locality should be no less than 8 to 10.

RAMENSKY demanded that determination of plants in the samples described be as full and exact as possible. To make possible determination of plants at life stages other than the flowering, he compiled a guide-determiner by vegetative parts (FLOROVA & RAMENSKY 1932, 1938).

4.3.2 COVERAGE MEASUREMENT

Out of several measures of the role of plants in a community (importance values – density, frequency, biomass, coverage, etc.) RAMENSKY preferred to single out coverage or "projective abundance," i. e. proportion of the earth's surface covered by plant parts, as vertically projected onto the ground (RAMENSKY 1929b, 1932), as a characteristic with a number of advantages. First, coverage has biological significance in relation to area of light-use by plants (RAMENSKY 1937b); second, it is easier to determine than biomass or density and can be better determined than these; third, in case of necessity it is possible to estimate biomass and density from coverage since the three indices are interrelated. The RAMENSKY school applies visual evaluation of coverage on a wide scale. A number of devices were developed to make visual determinations more objective: scales in the form of a net through which the vegetation cover is viewed, forks placed amid the standing grass, and artificial standards of coverage degrees based on depiction of plants' leaves, and so on (RAMENSKY 1929a,b, 1937b, 1938). Applications of such devices can provide quite reliable coverage determinations. For further treatment of samples, however, actual coverage percentages are replaced by coverage classes. The scale of coverage degrees (RAMENSKY et al. 1956) is the following: *s* is less than 0,1 %; *p* is equal to 0,1—0,2 %; *n* is 0,3—2,5 %; *c* is 2,5—8 %; and *m* is more than 8 %. Ranges of coverage values increase approximately in a geometrical progression, as the "octaves of numbers" of PRESTON (1948) might suggest is appropriate.

4.3.3 ORDINATION

The thesis that the vegetation cover is predominantly continuous suggests the need for ways of systematizing plant communities consistent with this thesis. This method, the basic principles of which were formulated by RAMENSKY in the 1920's, was named by the author arrangement into rows or row coordination (RAMENSKY 1924, 1929a 1930). Later in the West it was called *ordination* (Latin 'ordo' – row, order; 'ordination' hence denotes 'arrangement into rows'). Of late, the term ordination has gained wide use in Russia owing to its brevity; and it will be used here. The simplest methods of ordination (ecological series) were used in geobotany long before RAMENSKY, but the history of systematic ordination begins with this investigator.

The essence of ordination as a method lies in locating plant communities along abstract gradients of environmental factors or in a theoretical space. This location can in principle be found with any precision desired. It is in this that ordination has advantage over classification; the precision with which plant communities may be located in a classification is limited by the fact that each unit of the classification comprises a range of conditions. Classification and ordination are by no means mutually exclusive; and RAMENSKY did not seek to refute classification, while decisively preferring ordination. It is worthwhile to note that, even though vegetational continuity implies the appropriateness of ordination, RAMENSKY (1924) did not consider continuity a necessary condition for ordination to be effective.

4.3.4 THE IDEA OF STANDARD LANDMARKS

Since ordination is location of samples (or species) in relation to coordinates, a question arises: what shall the axes for such a system be? It is natural that indices of biotope conditions should be involved in the coordinates, but these indices are not always to be identified with elementary physical characteristics (temperature, moisture content of soil, light intensity in luxes, pH and so on). Some of these characteristics (e. g. year-around moisture regime) are very difficult to measure in a large number of samples to be ordinated. There are both theoretical and practical difficulties that prevent all-sided investigation of the environments of many phytocoenoses as a basis of ordination.

Proceeding from this difficulty, Ramensky (1938) suggested an approach to ordination based on *standard landmarks*. The essential idea is that there are to be found for a given area of vegetation a certain number of standard, normal, or typical community samples, to which correspond certain intensities of environmental factors. These standards may be either actual samples, or provisional approximations to them for positions along environmental gradients. Given these standards, then the samples of a set may be arranged according to the degree of their proximity to them, and departure from them along axes of environmental and community variation. Such arrangement is done separately for every one of the major factors of the environment. As a result it becomes possible to show the position of each community sample in the coordinate axes as described by a first, second, third factor, etc., and, in the end, in a multidimensional coordinate system. This approach both has some things in common with, and in other respects is quite different from, that of the American school of Curtis (Bray & Curtis 1957, Curtis 1959, see also Greig-Smith 1964, Whittaker 1967).

4.3.5 Summarized Ecological Series

Our description of the method may be clearer if we begin, not with the multidimensional relationships, but with a particular environmental gradient and ordination axis. We shall come back (4.3.9) to the origin of the standard landmarks. A sequence of standard landmarks, that is of samples lying directly along a given axis of the ordination system, is called a *summarized monofactor ecological series* (Ramensky et al. 1956). Making such series is, according to Ramensky, the principal basis of ordination as a method. To explain the meaning of a summarized ecological series it is useful to compare it with the ideas of "concrete" and "assembled" ecological series (Ramensky 1938, cf. the "field transect" and "composite transect" of Whittaker 1956, 1967, article 2). The term *ecological series* was introduced into Russian scientific literature by Keller (1909), who so designated a sequence of communities following one another along a gradient of the environment (for instance, a series from desert through littoral to aquatic communities on a particular shore of a desert lake). Such a series can be called a concrete ecological series. It is possible to make an abstract series by grouping into each step of the series several similar but non-neighboring communities; the result is an assembled ecological series. A generalized ecological series can, finally, describe a community gradation by mean characteristics of communities at a number of points along the gradient. Concrete ecological series very often imply changes not of one but of several interrelated factors. RAMENSKY reasoned that it would be most effective to build the ordination system so that the axes of the coordinates corresponded as much as possible to changes of each factor of the biotope separately, with relative constancy of other factors.

Ordination is possible not only by gradients of factors of biotopes but also by the position of communities in successional series (RAMENSKY et al. 1956). In fact, successions are regarded as a variety of ecological series ('temporal ecological series'), for in the time-gradient of succession changes in vegetation occur in parallel with changes in environment (cf. preceding article).

4.3.6 THE IDEA OF THE FUNCTIONAL MEAN

Perhaps the most important theoretical idea of RAMENSKY is that of the *functional mean* (RAMENSKY 1929a, 1930). The concept of the functional mean is rather simple in principle: if one index depends

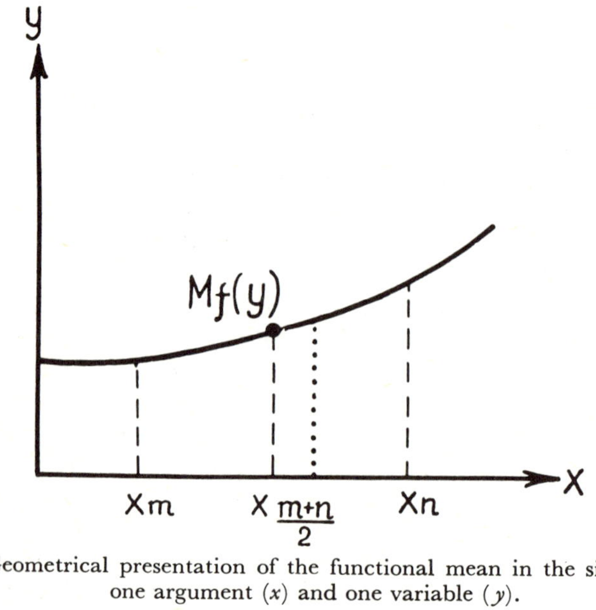

Fig. 1. Geometrical presentation of the functional mean in the simplest case: one argument (x) and one variable (y).

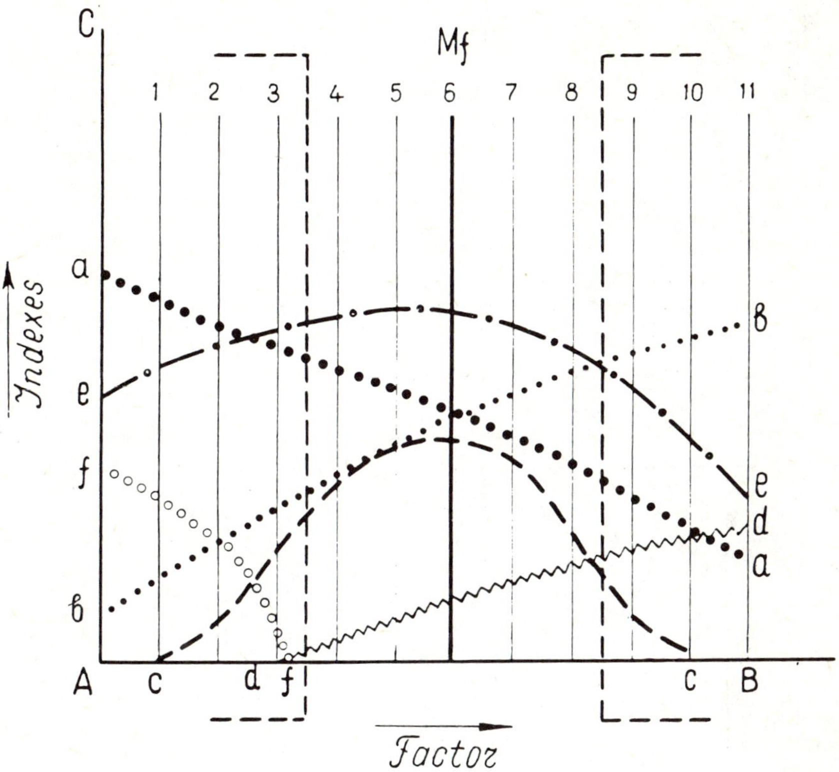

Fig. 2. The position of the functional mean M_f in the series of eleven plant communities varying in response to change in an environmental factor. Species *a* and *f* decrease in abundance; species *b* and *d* increase, whereas species *e* and *c* find their optima near the middle of the range of communities and the functional mean that represents that median position.

on a second index, then the value of the first index which corresponds to the mean value of the second index is its functional mean (Fig. 1). In other words, the functional mean is the value of a function or dependent variable, corresponding to the mean value of the argument or independent variable, for a set or subset of samples for which the latter is averaged.

The functional mean for a series of vegetation variants is depicted by a simple sketch (Fig. 2). On the vertical axis are indicated the abundance indices for species in a series of community variants located along the horizontal axis of the factor AB. The median position among these variants is occupied by sample 6, which represents the functional mean.

For the further account we introduce a mathematical symbol

Table I. Determination of the elective mean with the help of the rejection method

Species	First evaluation and rejection																							
	Number of lists																				Occurrence given cove			
	1	(2)	3	(4)	(5)	6	(7)	8	(9)	(10)	11	12	(13)	14	15	16	(17)	18	(19)	20	m	c	n	p
Andromeda polifolia	s	0	n	0	0	s	s	0	n	0	s	0	p	s	s	s	s	p	0	s	0	0	10	10
Arctostaphylos uva ursi	0	n	s	c	n	0	0	0	s	n	0	0	0	0	0	0	s	0	0	0	0	5	15	0
Carex limosa	c	0	p	0	0	c	m	c	n	0	n	0	n	m	c	m	p	c	0	m	20	25	15	10
Cetraria islandica	n	m	m	m	m	n	p	p	p	m	p	m	p	p	p	n	c	p	m	n	30	5	25	40
Cladonia rangiferina	0	n	p	m	m	n	p	0	0	m	p	c	p	n	m	s	c	p	m	p	25	10	15	30
Eriophorum angustifolium	n	0	0	0	0	p	c	c	s	0	n	0	s	n	p	n	s	c	0	c	0	20	20	10
Festuca supina	0	n	0	n	c	0	0	0	0	m	0	0	0	0	0	0	0	0	c	0	5	10	10	0
Ledum palustre	n	s	n	0	p	n	p	n	c	s	n	n	p	p	p	s	p	c	n	n	0	10	40	30
Pleurozium schreberi	c	n	c	c	c	n	0	n	c	n	n	c	m	n	n	n	m	c	p	0	10	35	40	5
Polygonum viviparum	0	p	0	s	0	0	0	0	0	s	0	0	0	0	0	0	s	0	s	0	0	0	0	5
Polytrichum commune	n	n	n	n	p	0	0	m	m	c	n	m	m	n	n	n	c	0	c	m	25	15	40	5
Rubus chamaemorus	c	s	c	p	c	c	n	c	c	s	n	n	s	c	n	n	s	n	0	m	5	30	30	5
Sphagnum angustifolium	m	0	m	0	0	m	m	m	m	0	m	0	c	m	m	m	p	m	0	n	55	5	5	5
Sphagnum fuscum	c	0	c	0	0	m	m	m	m	0	m	0	0	m	c	m	0	0	0	n	35	15	5	0
Vaccinium uliginosum	n	0	c	0	0	c	n	p	c	s	c	c	s	c	c	c	m	c	0	n	5	45	15	5
Vaccinum vitis idaea	s	p	p	n	n	0	s	0	s	n	0	c	p	s	0	s	c	0	c	s	0	15	15	15
The number of positive indications	2	0	2	0	1	2	2	2	2	0	2	2	2	2	2	2	1	2	0	2				
The number of negative indications	4	6	4	7	5	3	5	3	6	7	2	2	7	3	3	4	11	4	7	7				
The difference	-2	-6	-2	-7	-4	-1	-3	-1	-4	-7	0	0	-5	-1	-1	-2	-10	-2	-7	-5				

Second evaluation and rejection															
Number of lists										Occurrence with the given coverage					
(1)	(3)	6	(8)	11	(12)	14	15	16	(18)	m	c	n	p	s	0
s	n	s	0	s	0	s	s	s	p	0	0	10	10	60	20
0	s	0	0	0	0	0	0	0	0	0	0	0	0	10	90
c	p	c	c	n	0	m	c	m	c	20	50	10	10	0	10
n	n	n	p	p	m	p	p	n	p	10	0	40	50	0	0
0	p	n	0	n	c	n	n	s	p	0	10	30	30	10	20
n	o	p	c	n	0	n	p	n	c	0	20	40	20	0	20
0	0	0	0	0	0	0	0	0	0	0	0	0	0	0	100
n	n	n	n	n	n	p	p	s	c	0	10	60	20	10	0
c	c	n	n	n	c	n	n	n	c	0	40	60	0	0	0
0	0	0	0	0	0	0	0	0	0	0	0	0	0	0	100
n	n	0	m	n	m	n	n	n	0	20	0	60	0	0	20
c	c	c	c	n	n	c	n	n	n	0	50	50	0	0	0
m	m	m	m	m	0	m	m	m	m	90	0	0	0	0	10
c	c	m	m	m	0	m	c	m	0	50	30	0	0	0	20
n	c	c	p	c	c	c	c	c	c	0	80	10	10	0	0
s	p	0	0	0	c	s	0	s	0	0	10	0	10	30	50
4	4	5	4	5	4	5	5	5	5						
3	6	2	5	1	9	2	3	3	5						
+1	-2	+3	-1	+4	-5	+3	+2	+2	0						

Third evaluation and rejection					The ranged series for deduction of the median					The median
Number of lists										
6	11	14	15	16						
s	s	s	s	s	s	s	s	s	s	s
0	0	0	0	0	0	0	0	0	0	
c	n	m	c	m	m	m	c	c	n	c
n	p	p	p	n	n	n	p	p	p	p
n	p	n	n	s	n	n	n	p	s	n
p	n	n	p	n	n	n	n	p	p	n
0	0	0	0	0	0	0	0	0	0	
n	n	p	p	s	n	n	p	p	s	p
n	n	n	n	n	n	n	n	n	n	n
0	0	0	0	0	0	0	0	0	0	
0	n	n	n	n	n	n	n	n	0	n
c	n	c	n	n	c	c	n	n	n	n
m	m	m	m	m	m	m	m	m	m	m
m	m	m	c	m	m	m	m	m	c	m
c	c	c	c	c	c	c	c	c	c	c
0	0	s	0	s	s	s	0	0	0	

that is not RAMENSKY's. What is said above can be written:

$$y = f(x)$$

If x is to be found in the interval between "m" and "n" then

$$y = f\left(\frac{m+n}{2}\right) = M_f,$$

where M_f is the functional mean.

The concept of the functional mean can be extended to the dependence of one index on several variables, i. e., when

$$y = f(x_1, x_2, \ldots x_n),$$

and to the case when several indices are dependent upon several variables, namely when

$$(y_1, y_2, \ldots y_n) = f(x_1, x_2, \ldots x_n).$$

The last case with the number of arguments over three is impossible to imagine graphically. In Fig. 3 there is shown the position of functional mean for the dependence

$$(y_1, y_2, y_3, y_4) = f(x_1, x_2),$$

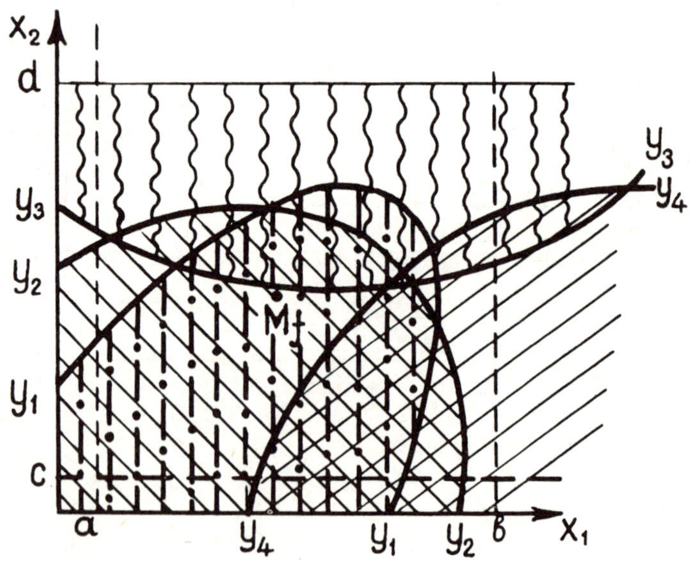

Fig. 3. The position of the functional mean for the function $\{y_1, y_2, y_3, y_4\} = f\{x_1, x_2\}$ in the interval of x_1 from a to b, x_2 from c to d. Areas are hatched for each dependent variable where it is not equal to zero.

where x_1 assumes any value from a to b, x_2 from c to d, and y_1, y_2, y_3, y_4 can assume only one out of two possible values (equal to zero, non-equal to zero).

The concept of the functional mean was used to prepare a monofactor summarized ecological series in the following way. As was noted above, a monofactor series is a series of communities corresponding to the gradation of a certain factor of the habitat, with other factors being constant. This condition will be met when a series of functional mean values are derived for successive sections of the gradient of a given environmental factor, while other factors remain invariable along the whole gradient. Change in communities along a monofactor ecological series is represented in a series of functional mean values for sections or steps of the gradient. RAMENSKY suggested an "elective" method for finding a functional mean when the values of independent variables are not known.

4.3.7 THE IDEA OF THE ELECTIVE MEAN

The election method is based upon several considerations (RAMENSKY 1929a, 1930). The functional mean should, for one thing, occur in the geometrical centre of the (abstract, compositional) space for a set or subset of samples (Fig. 3). Thus, samples of vegetation which correspond to the functional mean must have the largest sum of similarity indices with all other samples, of those under consideration. Consequently, the sample with the highest similarity sum with other samples must approximate the functional mean and is termed the "elective mean." Without a computer calculation of community indices for all possible pairs of samples is laborious. RAMENSKY offered also a simpler, semi-empirical method by successive removal of "peripheral" samples. Besides election RAMENSKY suggested that for a homogeneous set of samples the medians of abundance values for species in the set may be taken as representing the functional mean.

RAMENSKY (1929a, 1930) did not provide a detailed mathematical substantiation of the election methods, and two questions were left open for investigation. The first concerns the dependence of the position of the elective mean on the form of the function $y=f(x)$. Must the sample corresponding to the functional mean in all cases have the highest sum of affinity indices with the other samples, that have other values of the function? The second question is connected with the dependence of the elective mean on the character of a sample set for which it is being calculated. RAMENSKY (1929a, 1938) proceeded from the assumption that all values of the function for which an elective mean is determined are equally probable. But some variants of environmental conditions occur more often in the field than others, and with random or systematic selection of samples some combinations of species will occur in sample sets more often than others. RAMENSKY was conscious of the possibility of biasing the election results in this way, and he therefore recommended selecting samples so that the range of environmental conditions is filled uniformly.

4.3.8 ELECTION BY REJECTING DEVIANTS

As was mentioned, the essence of the method is in successive removal from a set of samples of those which are in some way deviant from the rest. The technique under description includes also some simplifications (inter- and extrapolation) at the initial stage of forming a generalized ecological series.

We shall now regard modes of work on a simple specific example. Let us imagine that we have a group of only 20 samples collected in relatively similar biotope conditions. Among them we must find the most "characteristic" or "central" samples, i. e. samples corresponding to the average conditions of the habitat for the group. The following operations are performed with these samples:

i) Projective coverage of every species in a group should be transformed into classes of coverage as mentioned above.

ii) All samples of the set should be compiled into a table (TableI).

iii) All the abundance indices of the species should be divided into three groups. The first group comprises indices which occur often, in no less than in 75 % of the samples as does, for example the index 0 for the species *Festuca supina*. (The percentage boundary to be used is arbitrary and may depend on homogeneity of the material; a value other than 75 % may be preferred.) Occurrence in a sample of one of these indices is regarded as a positive indication implying that species in that abundance is characteristic for the sample set. A second group of indices occur rarely, in no more than in 25 % of lists, say, for example the indices *n* and *p* for the species *Andromeda polifolia.* Occurrence in a sample of one of these indices is considered a negative indication, implying that species in that abundance is not characteristic for the sample set. The remaining indices form a third group considered neutral as indicators.

iv) Positive data are underlined in one way (solid line), negative data in another way (dotted line), or the use of colored pencils is helpful.

v) After this it is possible to reject samples with larger numbers of negative indications and smaller numbers of positive ones. For this purpose, the number of negative indications is subtracted from the number of positive ones, and the difference is entered as a positive or negative value below each sample in the table.

vi) Samples are now rejected, beginning with those most deviant as indicated by the largest negative values at the bottom of the table. The samples deleted may be no more than 30 % of the total if the material is fairly homogeneous, but up to 50 %, if the set is less homogeneous, as in our example. The numbers of the samples rejected are in parentheses.

vii) The remaining samples are compiled into a new table. It will be seen that the abundance indices of the species are more consistent in the samples, and that this new sample set is more homogeneous or homotonous.

viii) The same operations are performed with the new table, after which we have left 5 samples which are quite consistent in composition. We can now derive a typical index of coverage for each species in these samples using the arithmetical mean, even better, the median as such an index.

Groups of lists considerably larger than the example can be subjected to similar treatment.

4.3.9 Ordination on the Basis of Election

We now return to the problem of ordering a larger set of samples, representing a broader vegetation pattern and a number of directions of community variation in response to environment. For each environmental factor we shall partition the samples into steps, by ranges of magnitude of that biotope factor (irrespective of the value of other factors). The sequence of steps represents the gradation of the factor, and if we derive the elective mean for each step we then obtain a monofactor summarized ecological series. These elective means serve as standard landmarks by which the whole set of samples may now be ordinated by relative positions along this series as an ordination axis. The problem lies in how to partition the samples into steps when effective measurements of the environmental variable are not available. However, when investigating vegetation it seems always possible to single out several or, at least two, rough gradations of a major biotope factor from the description of the locality (conditions of the relief, proximity of ground waters, etc.) or other ecological data (Ramensky 1929a, 1938). Given two (or more) groups of samples representing difference in community response to the environmental gradation, refinement of the community gradation (ecological series) can be accomplished by election.

In a large number of samples collected in a region, at least two groups *A* and *B* are singled out as of contrasting composition in response to a given environmental factor. Elective means with their standards are derived for these groups. We can now characterize an artificial intermediate sample *C* between them. For instance, if the coverage standard for a given species is *m* in group *A* and *n* in group *B*, then the index should be *c* for the intermediate *C* between *A* and *B*. We may choose further to characterize more extreme samples by extrapolation. The same species should have index values that extrapolate down to *p* for a sample *E*, more extreme than *B*; in the opposite direction the index, that is *m* or maximal in sample *A*, should be *m* also in a more extreme sample *D*.

As a result we obtain 5 standard landmark samples (*D*, *A*, *C*, *B*, *E*). Three samples, obtained artificially, are only provisional; but the sequence characterizes an ecological series in relation to the environmental factor with which we are concerned. All the samples available are compared with these five and classified into five groups according to which one they most nearly correspond to. The elective mean is again determined for each group. (If groups are homogeneous the abundance medians or means for species may be determined without going through the rejection process.) We have now an ecological series of five steps of which none are artificial. Between the 5 elective means as actual, not provisional, standards we can establish new provisional intermediates and new actual intermediate elective means. The procedure can be repeated to divide the ecological series as finely as the number of samples allows.

4.3.10 Multidimensional Ordination

After such ordination by one factor we may ordinate by another factor, and still a third and so on with the use of the same technique. With this there will be obtained series of values of elective means for a number of biotope factors, by which samples may be located in the multidimensional ordination system. Not only intensity of each factor taken separately, but also the effects of their combinations are often of importance for vegetation. In this connection Ramensky suggested a further method of multidimensional ordination: building up of monofactor series for successive factors separately, for various intensities of the preceding factors (Ramensky 1938, cf. the transect grid of Whittaker 1956, article 2). Fig. 4 represents the approach to two factors, soil humidity and fertility (Ramensky 1938). On the basis of such an approach for two factors

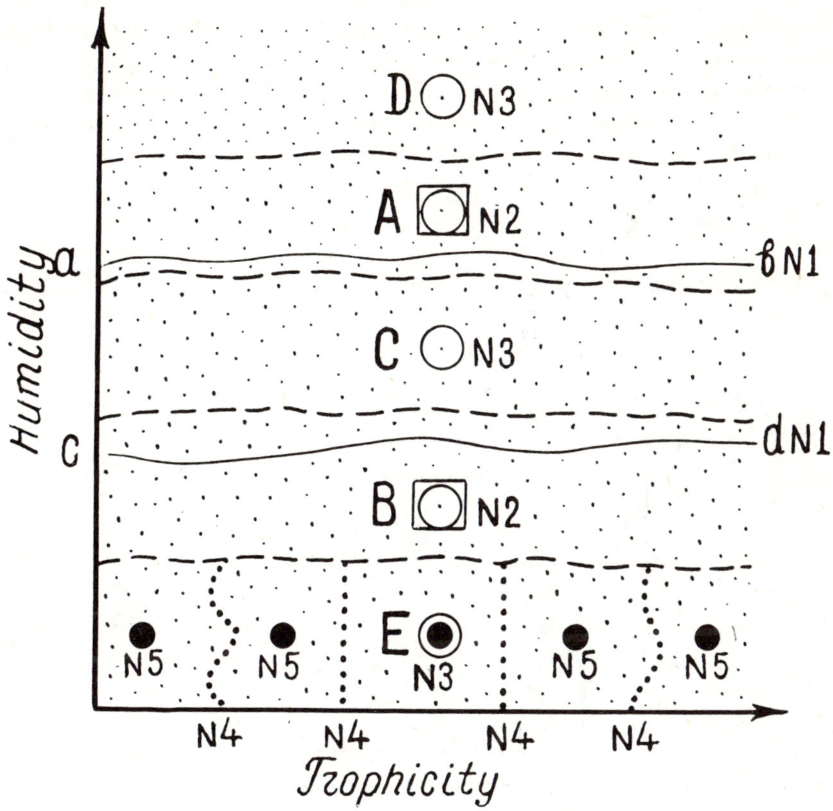

Fig. 4. Scheme for the successive deduction of ecological series for two factors. The positions of separate samples of plant communities in the 'space' of two factors (of humidity and trophicity of soils) are marked with points. Numbers show the succession of actions. Number I (N1) shows singling out of two groups by contrast of soil moisture (their boundaries, which are very approximate, are are suggested by lines ab and cd). Number 2 is the determination of the functional mean by the election method in each of these groups (the mean value is marked with a square). Number 3 is the division of all lists into 5 groups with the help of interpolation (*C*) and extrapolation (*D* and *E*). With further determination of the elective means or medians (marked with circles) a monofactor ecological series for the humidity gradient is obtained. Following this a similar division of the samples by the second factor (trophicity, fertility) is possible. It can be done irrespective of the division by the first factor or, as shown for group *E* (Number 4), for each of the gradations of soil moisture separately. Determination of the functional means of this division is shown, finally, by black circles (Number 5) which form a monofactor ecological series for the fertility gradient.

one can prepare charts on which samples can be located and species distributions outlined by isopleths (isodems, lines bounding equal importance values for a species population). Relations of three species populations to a coordinate system are illustrated in Fig. 5.

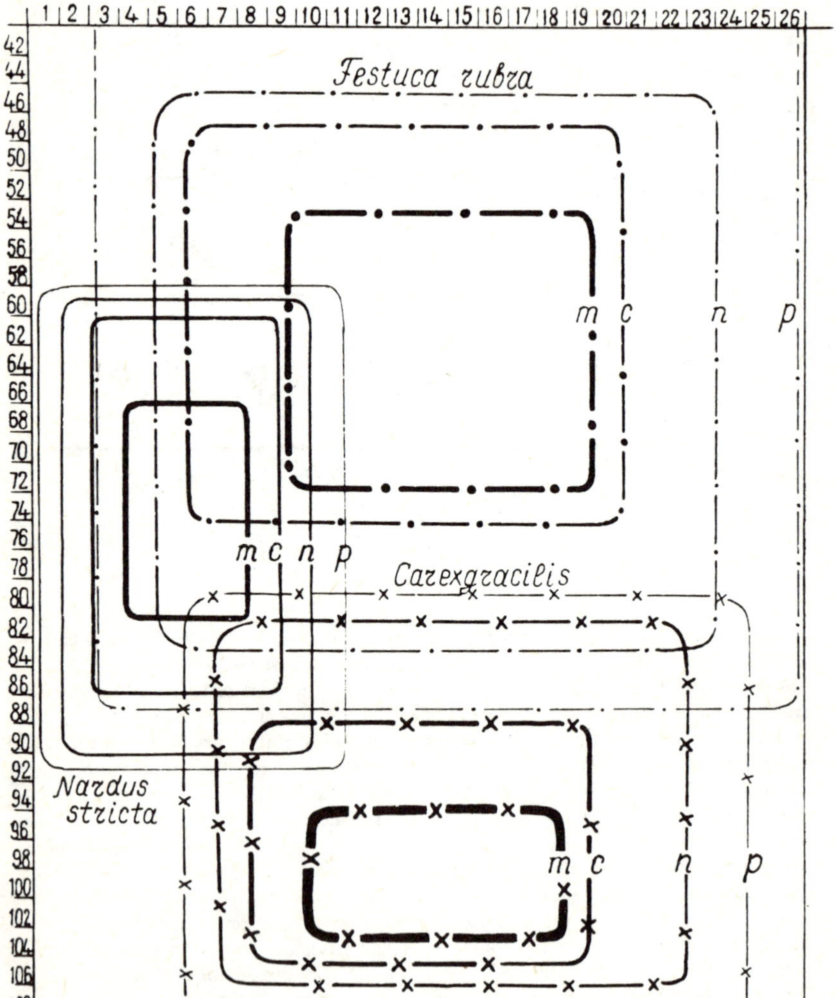

Fig. 5. Isopleths (isodems) of three species depicting their distribution in ecological series of humidity and richness of the soil or trophicity (*m* - mass distribution, *c* - abundant, *n* - moderate, *p* - small).

4.3.11 Location of Samples in an Ordination

In a monofactor ordination series each species, and each level of abundance for a species, has a definite band distribution. If there is a sample in which species *a, b, c* and *d* are represented (for the sake of simplicity we shall think here that only presence or absence of a species is taken into account), and these species have the distribution amplitudes equal to AB, CD, EF and GH, then the position of this sample should be delimited by the range of overlap of the distributional amplitudes of all the species. What is said above can be written down with the help of symbols of mathematical logic:

$$R_{\text{phe}} = R\ (a, b, c, d) = R(a) \cap R(b) \cap R(c) \cap R(d),$$

where R is a segment of an ecological series in which the phytocoenose under investigation can be located. This approach permits one to locate in the ordination system samples taken after this system has been developed. For the system of multifactor ordination the position of a sample will be determined by the overlap in two or more dimensions of the ecological areas of species (see Fig. 5).

Table 11 Determination of the stage of an ordination series for the cenosis by the scale of richness of the soil with the help of the notch method.

Species	coverage	Notches on stages of richness of the soil 1	2	3	4	5	6	7	8	9	10	. . . .
Vaccinium vitis idaea	m											
Vaccinium uliginosum	m											
Ledum palustre	n											
Carex limosa	n											
Rubus chamaemorus	n											
Sphagnum fuseum	c											
Deduction of the stage for the cenosis		2.5										

Two techniques for locating an individual sample along an ecological series may be briefly described.

Table II gives a list of 6 species and, on the right side the stages of an ecological series (in this case richness of soil). If for the first species (*Vaccinium vitis idaea*) its coverage (*m*) is characteristic of stages 2-6 we mark this range with a line or "notch." We further mark with such lines ranges for all species of the sample. Evidently, the sample fits into the ecological series where there is the largest number of overlapping lines-hence at 2.5.

Alternatively, coverage values for the species are converted into index categories *m*, *c*, *n* (species with abundance *p*, *s* can be best left unused). Ranges of an ecological scale for which these indices are characteristic are written as two vertical columns (Table III). These numbers are now rearranged in descending and ascending orders, and the first pairs are averaged to derive the sample position as shown in Table III.

Table III Determination of the stage of an ordination series by the limitation method.

Species	coverage	Limitating stages		Limitating stages in the order of subsiding and increasing		Deduction of the mean from the closest stages
Vaccinium vitis idaea		2	6	2	3	(2+3) : 2=2.5
Vaccinium uliginosum		1	3	2	3	(2+3) : 2=2.5
Ledum palustre		1	6	1	5	
Carex limosa		2	9	1	6	
Rubus chamaemorus		1	5	1	6	
Sphagnum fuseum		1	3	1	9	

Here the account of principal foundations of the method of RAMENSKY comes to an end. In the next sections we give accounts of some practical research methods worked out by RAMENSKY and his scholars.

Table IV Fragment of a standard ecological scale for humidity and richness of soil. The values show the ranges of scales for environmental factors, in which the species may occur in the coverage index at the head of column.

Species	Factors	Gradations of projective coverage and stages of factors corresponding to them				
		m (more than 8%)	c (2.5-8%)	n (0.3-2.5%)	p (0.1-0.2%)	s equal to 1 (single)
Vaccinium uliginosum	Humidity	80-89	74-89	72-90	67-96	65-10
	Trophicity	1-3	1-4	1-5	1-6	?
Vaccinium vitis idaea	Humidity	61-83	59-87	58-89	55-90	? -92
	Trophicity	2-6	1-7	1-9	1-10	1-10

4.4 **Other Applications**

4.4.1 Ecological Scales

The monofactor generalized ecological series obtained is a series of standards showing the influence of different gradations of an environmental factor upon the composition of communities. Such data compiled for different species (Table IV) have been termed *standard ecological scales* (Ramensky 1938, Ramensky et al. 1956). For making series of standard scales for different biotope factors Ramensky used very extensive material collected from all zones of the plains of the Soviet Union, except northern taiga, tundra, and forest tundra, in all about 20,000 samples. Results of this work by a group of collaborators between 1925 and 1950 were published in 1956, after death of the main author (Ramensky et al. 1956). This book also contains an important theoretical introduction. The book gives scales of humidity, richness of the soil (trophicity and salinity, the latter regarded as a surplus trophicity), changeability of humidity, alluviality (intensiveness of silt deposition), and pasture retrogression. V. A. Sorokina (1953, 1955) worked out the scale of altitude (regarding altitude as a complex-gradient of environment, not as a purely topographical characteristic) for pastures of the Northern Caucasus. Of late standard ecological scales for various factors of the biotope, including altitude, have been published by TSATSENKIN (1967) for mountains of Middle Asia, the Urals, and the Altai Mountains, as well as for the Caucasus (1968). Such scales have value for purposes other than the ordination of particular samples. They permit one to locate in the system of ecological coordinates any unit of geobotanical classification that is of interest, and they permit separate evalution of ecological relations of strata or other structural parts of communities.

4.4.2 APPLICATIONS TO SOILS

RAMENSKY justly considered his technique of ordination applicable not only to plant communities; he regarded it as a general method of systematization of "objects defined by several factors which act differently" (RAMENSKY 1929a, 1930). Of other categories of objects RAMENSKY was especially concerned with soils, for which a series of experiments in ordination and making of ecological scales were undertaken by him and his scholars (RAMENSKY 1938). RAMENSKY also used mean values of soil characteristics for the groups of samples on which were based the standards in ecological series. Thus correspondence of vegetation gradation and soil factors was established (RAMENSKY 1937a, 1938, SOBOLEV 1969).

4.4.3 INVESTIGATION OF VEGETATION DYNAMICS

The method which enables us to determine the position of any geobotanic object in a system of ecological coordinates allows investigation of vegetation changes through time. Evaluation by ecological scales of different time-stages of the same site provides us with an indication of direction and degree of change in the community of that site. Analyses of meadow vegetation in the Voronezh Reservation on this basis were carried out by NIKOLAYEVSKAYA (1949). The author annually evaluated samples from the same sites by ecological scales; changes in the (humidity factor) positions of sites in the ordination system were shown and clearly expressed climatic fluctuation. Later UTEKHIN (1969) studied vegetation changes during longer-term development, using manuscript materials of NIKOLAYEVSKAYA. Samples taken in 1936-7 and 1965-6, were compared, both treated by ecological scales. Significant changes in positions of communities in relation to the humi-

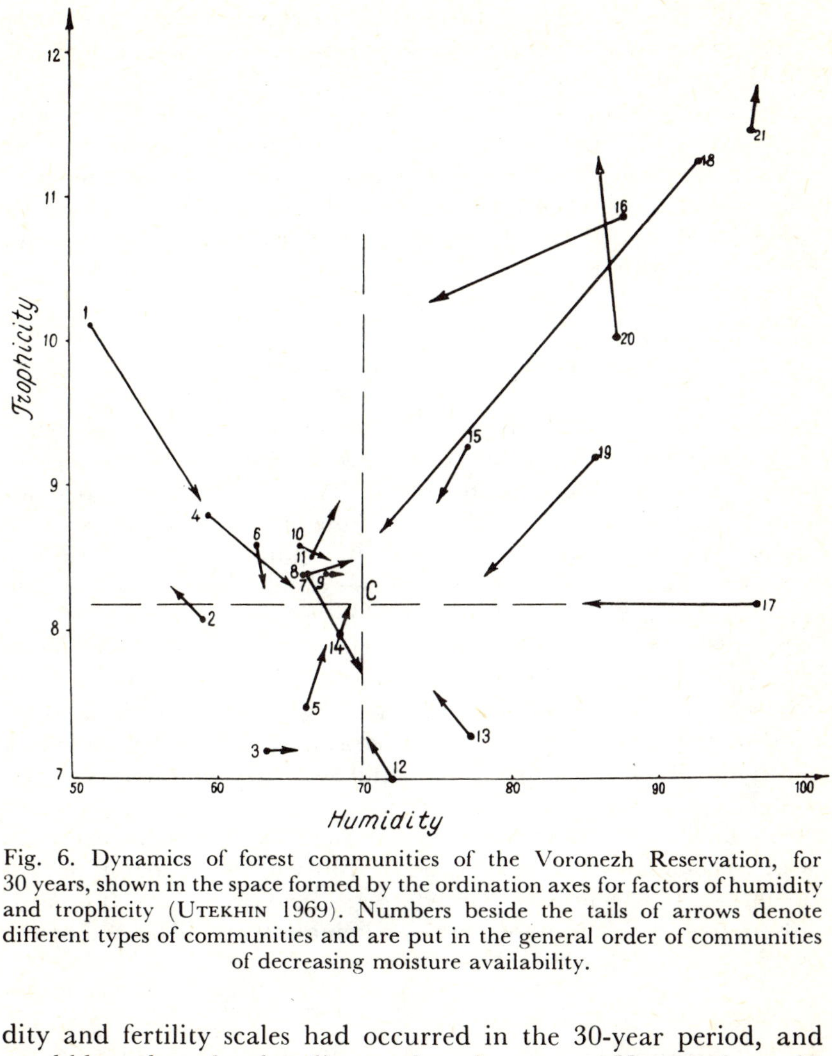

Fig. 6. Dynamics of forest communities of the Voronezh Reservation, for 30 years, shown in the space formed by the ordination axes for factors of humidity and trophicity (UTEKHIN 1969). Numbers beside the tails of arrows denote different types of communities and are put in the general order of communities of decreasing moisture availability.

dity and fertility scales had occurred in the 30-year period, and would have been hard to discover by other means. UTEKHIN's results are shown in Fig, 6 and indicate, in particular, rapprochement of different communities in the ordination system with time when man does not interfere.

4.4.4 AN EXPERIMENT IN APPLICATION TO CLASSIFICATION

While not simply opposed to classification of plant communities RAMENSKY preferred that it be based on ordination. Continuity of the vegetation cover is not monotonous. In some cases vegetation may change more rapidly in certain sections of ecological series than in others. If such sections of more rapid change occur, they may naturally be used as boundaries between types of plant communities. RAMENSKY (1938, cf. WHITTAKER 1960) suggested a simple method for locating such boundaries: summing the differences in projective coverage (by the RAMENSKY scale) of species in the standard samples for successive pairs of stages of an ecological series. This method of establishing types of communities was applied by UTEKHIN (1967) to vegetation of the Central-Chernozem Reservation near Kursk.

4.4.5 METHODS OF RAMENSKY IN PHYTOGEOGRAPHY

Application of RAMENSKY's methods in phytogeography is being worked out by L. N. SOBOLEV (KUBANSKAYA & SOBOLEV 1969). SOBELEV's approach involves ordering the full ranges of community-types peculiar to different regions in a universial system of ecological coordinates, one embracing all habitats that are possible in regions compared. The system of coordinates in this case is made rough and generalized, and it is divided into discrete sections forming a grid in the cells of which plant communities are placed. Such a method, though strongly schematic, has advantage in allowing interrelation of geographically diverse vegetation cover in a single coordinate system, facilitating comparison of vegetation units, and clarifying the relations of zonal and localized community-types.

The extent of this paper does not permit us to develop further the application of RAMENSKY's concepts and methods in diverse areas of phytocoenology and related fields. But we wish recognition that RAMENSKY's ideas not only antedate gradient analysis in the West, but are part of a continuing tradition being actively advanced by the investigations of his disciples.

4.5 Summary

The approach to vegetation of L. G. RAMENSKY was based upon the ecological individuality of species, the continuity of communities, and the systematization of communities by their arrangement into ecological series. The essential concepts and method are those of gradient analysis, as it is called in the West, but the Russian development of this approach preceded the American by WHITTAKER, CURTIS, et al.

Vegetation samples are arranged (ordinated) along an environmental gradient; communities thus arranged in sequence form an ecological series. Positions along an environmental gradient may be described by the compositions of actual or ideal samples, used as standard landmarks. A sequence of such landmarks along a given ordination axis (representing an environmental gradient) describes major features of an ecological series. When a set of samples have been ordinated by comparing them with standard landmarks, each section of the environmental gradient will contain a number of samples. For each section of the gradient a mean or most representative sample is to be chosen. The concept underlying the choice is that of the functional mean: One seeks the value of a dependent (vegetational) variable corresponding to the mean value of an independent (environmental) variable. For a section of the gradient the sample(s) most nearly representing the functional mean is chosen by a process of election, by which deviant samples are progressively removed until only the most central or characteristic samples remain. Such a sample is an elective mean, and its composition (serving as a standard landmark) describes a position along the environmental gradient. A sequence of elective means along a given environmental gradient form a monofactor summarized ecological series.

For analysis of a large number of samples from a region, these may first be divided into two or more groups by vegetational responses to a given environmental factor. If the set is divided into two groups, elective means may be determined for each of these; and the means may then be used to specify characteristics of five standard landmarks (of which one is intermediate and two more extreme than the two means). These five standards (or these and additional intermediates) may be used to ordinate the sample set in relation to the first factor, and for each section of the gradient elective means may be derived to replace the first sequence of standards and form a monofactor summarized ecological series. The sample set may now be treated in the same way for a second and additional environmental factors to produce a multidimensional ordination, in which individual samples may be located by their species compositions in relation to each environmental factor as an ordination axis.

Other applications include: (i) Standard ecological scales that summarize species' responses to intensities of environmental factors, (ii) Correlation of soils characteristics with ecological series, (iii) Treatment of successional sequences as ecological series, (iv) Expression of relative rates of change in community composition along the environmental gradient in different sections of an ecological series, and (v) Broad systematization of community-types in relation to gradients over wide geographic areas.

ALEKSANDROVA, VERA D., – 1969 – Classification of Vegetation: Principles of Classification and Classification Systems of Various Phytocoenological Schools (In Russian). Publ. House Nauka, Leningrad. 275 pp.

BRAY, J. R. & J. T. CURTIS, – 1957 – An ordination of the upland forest communities of southern Wisconsin. *Ecol. Monogr.* 27: 325–349.

CURTIS, J. T., – 1959 – The Vegetation of Wisconsin: An Ordination of Plant Communities. Univ. Wisconsin, Madison. 657 pp.

CURTIS, J. T. & R. P. MCINTOSH, – 1951 – An upland forest continuum in the prairie–forest border region of Wisconsin. *Ecology* 32: 476–496.

DU RIETZ, G. E., – 1921 – Zur methodologischen Grundlage der modernen Pflanzensoziologie. Holzhausen, Wien. 267 pp.

ELLENBERG, H., – 1952 – Landwirtschaftliche Pflanzensoziologie. II. Wiesen und Weiden und ihre standörtliche Bewertung. Ulmer, Stuttgart. 143 pp.

FLOROVA, V. M. & L. G. RAMENSKI, – 1932 – Identification Guide to Plants in a Non-Flowering State for the Middle Part of the USSR (In Russian). Moscow-Leningrad. 256 pp. 2nd ed. 1938.

GREIG-SMITH, P., – 1964 – Quantitative Plant Ecology. 2nd ed. Butterworths, London. 256 pp.

KELLER, B. A., – 1909 – Grassland, semidesert, and desert formations, pp. 7–17 in: Preliminary Programs of Botanical-Geographical Investigations (In Russian). Pechatnyi Trud, Petrograd.

KUBANSKAYA, Z. V. & L. N. SOBOLEV, – 1966 – The system of plants and soils of the Betpak-Dala desert (In Russian). *Trudy̆ Inst. Bot. Akad. Nauk Kazakh. SSSR*, Alma-Ata 23: 54–80.

NIKOLAYEVSKAYA, M. V., – 1949 – Dynamics of the vegetation cover of the Voronezh Reservation (In Russian). *Nauchno-Metod. Zapiski Glavn. Uprav. po zapovednikam* 12: 126–135.

ODUM, E. P., – 1953 – Fundamentals of Ecology. Saunders, Philadelphia. 384 pp. 2nd ed., 1959, 546 pp. 3rd ed., 1971, 574 pp.

PRESTON, F. W., – 1948 – The commonness, and rarity, of species. *Ecology* 29: 254–283.

RAMENSKI, L. G., – 1910 – On the comparative method for ecological study of plant communities (In Russian). *Dnevnik S'êzda russk. Estestvoisp. Vrach.* 2(9): 389–390.

RAMENSKI, L. G., – 1924 – The main regularities of the vegetation cover (In Russian). *Vêstnik opy̆tnogo dêla Sredne-Chernoz. Obl.*, Voronezh pp. 37–73.

RAMENSKY, L. G., – 1926 – Die Grundgesetzmässigkeiten im Aufbau der Vegetationsdecke. *Bot. Zbl.*, N.F. 7: 453–455.

RAMENSKI, L. G., – 1929a – On the method of comparative treatment and systematization of lists of plants and other objects determined by several factors with unlike actions (In Russian). *Trudy̆ Sovesch. geobot.-lugov., sozvan. (Gos.) lugovoi Inst.* 15-20, 1928, pp. 11–36. (Transl. in *Beitr. Biol. Pfl.* 18: 269–304, 1930).

RAMENSKI, L. G. – 1929b – Projective counts and description of vegetation (In Russian) Moscow. (Transl. in *Handb. biol. ArbMeth.* 11, 6: 137–190, 1932).

RAMENSKY, L. G., – 1930 – Zur Methodik der vergleichenden Bearbeitung und Ordnung von Pflanzenlisten und anderen Objekten, die durch mehrere, verschiedenartig wirkende Faktoren bestimmt werden. *Beitr. Biol. Pfl.* 18: 269–304.

RAMENSKY, L. G., – 1932 – Die Projektionsaufnahme und Beschreibung der Pflanzendecke. *Handb. Biol. ArbMeth.* 11, 6: 137–190.

RAMENSKI, L. G., – 1937a – Inventarization of natural hay-lands and pastures and methodological foundations of nature-productivity typology of lands (In Russian). *Trudȳ Vses. Akad. Sel'. Khoz. Nauk*, Moscow 31(2): 11–36.

RAMENSKI, L. G., – 1937b – Counts and Description of Vegetation (on the Basis of the Projective Method) (In Russian). Selkhozgiz, Moscow. 100 pp.

RAMENSKI, L. G., – 1938 – Introduction to the Complex Soil-Geobotanical Investigation of Lands (In Russian). Selkhozgiz, Moscow. 620 pp.

RAMENSKI, L. G., – 1953 – On ecological study and systematization of plant communities (In Russian). *Byull. mosk. Obshch. Ispȳt. Prir.*, otd. biol. 58(1): 35–54.

RAMENSKI, L. G., I. A. TSATSENKIN, O. N. CHIZHIKOV & N. A. ANTIPIN, – 1956 – Ecological Evaluation of Pasture and Fodder Lands by the Vegetation Cover (In Russian). Selkhozgiz, Moscow. 472 pp.

SOBOLEV, L. N., – 1969 – Dependence of productivity of plants, communities, and their geographical systems on factors of the environment (three levels of investigation of productivity) (In Russian). In, 'Obshchije teoreticheskije Problemy biologicheskoj Produktivnosti', pp. 70–74. Nauka, Leningrad.

SOROKINA, V. A., – 1953 – The experience in application of the methods of L. G. Ramensky (In Russian). *Bot. Zh. SSSR* 38 (5) 718–728.

SOROKINA, V. A., – 1955 – Some results of applying the methods of L. G. Ramensky *Herbage Abstr.* 25(4): 210–218.

TSATSENKIN, I. A., – 1967 – Ecological Scales for Plant Species of Pastures and Haylands of Mountainous and Plains Regions of the Middle Asia, Altai, and Urals (In Russian). 'Donish,' Dushanbe. 226 pp.

TSATSENKIN, I. A., ed., – 1968 – Ecological Evaluation of Pasture and Range and Fodder Lands of the Caucasus by the Vegetation Cover (In Russian). Moscow.

UTEKHIN, V. D., – 1967 – The vegetation of the Central-Chernozem Reservation and its productivity (In Russian). *Mater. Moskov. Fil. Geogr. Obshch. SSSR*, Biogeography, phenology, vyp 1: 18–21. Moscow.

UTEKHIN, V. D., – 1969 – Vegetation changes in the Voronezh Reservation during 1936–1966 (In Russian). *Mater. Moskov. Fil. Geogr. Obshch. SSSR*, Biogeography, vyp 3: 15–17.

WHITTAKER, R. H., – 1951 – A criticism of the plant association and climatic climax concepts. *NW. Sci.* 25: 17–31.

WHITTAKER, R. H., – 1956 – Vegetation of the Great Smoky Mountains. *Ecol. Monogr.* 26: 1–80.

WHITTAKER, R. H., – 1960 – Vegetation of the Siskiyou Mountains, Oregon and California. *Ecol. Monogr.* 30: 279–338.

WHITTAKER, R. H., – 1962 – Classification of natural communities. *Bot. Rev.* 28: 1–239.

WHITTAKER, R. H., – 1967 – Gradient analysis of vegetation. *Biol. Rev.* 42: 207–264.

5 SAMPLE SIMILARITY AND SPECIES CORRELATION

David W. Goodall

Contents

5 SAMPLE SIMILARITY AND SPECIES CORRELATION

5.1 Introduction

Correlations between species most often arise from biotope differences affecting their distributions. They are of interest for the light they throw on species biology, as a tool in the classification or ordination of plant communities, and as a means of testing the homogeneity of vegetation (GOODALL 1953, DAGNELIE 1962a, b). Similarity between vegetation samples may also be used as a tool in ordination or classification. Similarity between samples and correlation between species are two sides of the same coin. It is by their occurrence in a group of similar communities that a group of correlated species may be recognized, and it is through the fact that stands include a number of correlated species that they are recognized as similar. General discussions of these questions will be found in a number of books, such as those of GREIG-SMITH (1964), PIELOU (1969b), GOUNOT (1969) and VASILEVICH (1969).

As suggested in the previous paragraph, measures of correlation and similarity find their main use in application to the ordination and classification of vegetation samples. Other articles are devoted to these topics. Here we shall deal with measures of similarity and correlation in their own right, irrespective of the particular uses to which they may be put.

5.2 Distributional Similarity of Species

5.2.1 RELATIONS BETWEEN PAIRS OF SPECIES

5.2.1.1 *Interpretation of Interspecific Relationships*

Two species may be positively associated because one is dependent on the other (as parasite and host, tree and epiphyte) or because both are similarly affected by some third factor in the biotic or inanimate environment. They may be negatively associated through mere spatial exclusion, through competition or allelopathic effects, or through differing environmental requirements.

Whatever the nature and causes of the relationship between them, it is essentially dependent on the area of study. It is virtually meaningless to say simply that two species are correlated without indicating the area covered by this generalization. Taking the earth's surface as a whole, all European plants, however diverse in their habitat requirements, will be positively correlated in the sense (for instance) that the distance from a plant of one of these species to the nearest plant of another will be, on average, much shorter than an arc joining two random points. On the other hand, if one is studying a particular rock surface by a grid of small samples, two lichen species may be negatively correlated through spatial exclusion, even though they are both confined to limestone and consequently normally found side by side. Thus the degree of correlation between two species, and even its sign, may depend on the range of biotopes occurring within the area studied.

Moreover, if association is being studied by a sampling method, the size of sample will also affect the results (BOURDEAU 1961, BYER 1970). If samples in a bog are small compared with the size of the hollows, two *Sphagnum* species may be negatively associated, whereas quadrats comparable in size with the pattern of hummocks and hollows might show them as positively associated.

In homogeneous vegetation, one may expect interspecific correlation to change with increases in sample size up to a certain point, beyond which it would remain constant. This sample size, reflecting the scale of local heterogeneity in the vegetation, might be a suitable choice as a minimal area for sampling (GOUNOT 1961). It does not follow, though, as GOUNOT suggests, that correlation studies on a smaller scale are not useful. They can in fact be valuable in illuminating the pattern of the vegetation. This is illustrated by KERSHAW's (1961) application of analysis of covariance to the study of joint distribution of pairs of species in nested samples of different sizes. In consequence of these effects of sample size, a statement about the correlation of two species can have precision only if it incorporates a statement of the universe sampled and of the sampling method – such as, "When Tregaron Bog was sampled by random metre-square quadrats, a negative correlation was shown between *Calluna vulgaris* and *Erica tetralix*."

Correlations among species are most commonly studied by recording their presence or absence in a series of samples, or by estimating some quantitative measure of their abundance in each sample. A third approach, less commonly used, depends on measurement of distance between individuals of different species. The relative importance of presence on the one hand and quantitative measures (density, cover, basal area, etc.) on the other has been somewhat debated. KASSAS & EL-ABYAD (1962/3) point to the inadequacy of presence records to distinguish types of desert vegetation, and many workers have expressed the view that quantitative measures add appreciably to the useful information about vegetation samples contained in species lists. DAGNELIE (1962b) has stated that correlation results based on presence and on quantitative measures are usually concordant. On the other hand, KOLBE (1956), for instance, regards correlation studies on the quantities of two species as appropriate only if they have first been shown to be correlated in respect of presence.

5.2.1.2 *Presence and Absence Data*

We first consider measures of species association (SA) based on records of presence or absence only. For two species, these may be expressed in the form

		Species A		
		Present	Absent	Total
Species B	Present	a	b	$a+b$
	Absent	c	d	$c+d$
	Total	$a+c$	$b+d$	M

where the total number of samples M, is partitioned into those with both species (a), those without both (d), and those with one or the other only (b and c).

Numerous ways of expressing the degree of association between the two species have been proposed and used. The data are in fact the usual 2×2 contingency table, and the measures of association based on such tables for use in completely different contexts are also available here. Most of these measures range from 1 (in the case of perfect positive association) to 0 or -1 (in the case of perfect negative association).

Measures of association and tests of their statistical significance are often confused. It may be important to determine whether two species are distributed independently of one another, or whether on the other hand they are positively or negatively associated. In this case, a significance test is required. More often, the hypothesis of independence is unrealistic, and a test of it has little interest; one is concerned rather with expressing the *degree* of association than with questioning whether it exists. The two questions are discussed separately in the following sub-sections.

5.2.1.2.1 Significance tests

A significance test for interspecific association is based on the null hypothesis that the probability of encountering each species is uniform throughout the set of samples, irrespective of whether the other species is present or not. It goes without saying that significance tests are relevant only if the samples have been taken randomly from the universe they represent.

The null hypothesis has normally been tested by calculating χ^2 with YATES's (1934) correction for continuity:

$$\chi^2 = \frac{M(|ad-bc|-M/2)^2}{(a+b)(c+d)(a+c)(b+d)} \tag{1}$$

or from the rather more accurate alternative expression (GILBERT & WELLS 1966)

$$\chi^2 = \ln \frac{M^3}{2\pi\,(a+b)(c+d)(a+c)(b+d)} + 2\ln \frac{M!\,a!\,b!\,c!\,d!}{(a+b)!\,(a+c)!\,(b+d)!\,(c+d)!} \tag{2}$$

and comparing with the theoretical distribution of χ^2 with one degree of freedom. This is an approximation to the hypergeometric distribution, which may also be calculated exactly for the same purpose. The probability of the particular set of figures observed, given the marginal totals, is

$$\frac{(a+b)!\,(c+d)!\,(a+c)!\,(b+d)!}{a!\,b!\,c!\,d!\,M!} \tag{3}$$

and the probabilities for the observed set, and other less probable sets consistent with the marginal totals, are summed to provide a significance test. The use of this exact distribution has been recommended when the minimum product of the marginal totals is less than $5M$.

It has been pointed out by PIELOU (1969a,b) that the assumptions on which these tests are based are, to say the least, questionable. They assume that the marginal totals are fixed — in other words they give the probability that, in a set of M samples of which $(a+c)$ contained Species A and $(a+b)$ contained Species B, as many (or as few) as a would contain both if their occurrence were independent. This is not really the question the investigator usually wants answered. He wants to know whether, if Species A and Species B were independently distributed with their actual population frequencies and patterns, they would give by chance a set of samples with this apparent degree of association. In other words, the marginal totals are themselves random variables — sample estimates of population frequencies — and not fixed as assumed in the usual tests. The probability of the contingency table as calculated from (1), (2), or (3) above is greater than the true probability taking variation in marginal frequencies into account, and the test as normally performed has consequently a conservative bias. PIELOU describes a modified form of the χ^2 test taking variation in marginal frequencies into account. In most cases, the adjustment in the probability estimate will be relatively small.

PIELOU (1969a,b) has pointed out that, if the mean distance between samples is less than the scale of pattern in the vegetation, significance tests of association may be vitiated by lack of independence among the samples. If, for instance, two species occur in patches large compared with the mean distance between samples, the fact that one sample contains both species enhances the probability that neighbouring samples will also contain both species, and the association between them may appear significant even where the patches are in fact random and independently distributed.

5.2.1.2.2 Measures of interspecific association

A significance test such as is described in the preceding section indicates only *whether* the two species are associated, and provides no measure of the *degree* to which they are associated.

If M is large, a relatively small degree of association may be demonstrated at a high significance level. If M is small, even a large degree of association may not be demonstrable. On the other hand, most measures of association bear little or no relation to the significance of the relationship in the sample observations, and it is perhaps debatable whether this should be regarded as a drawback. It is perhaps best to regard the significance test and the measure of association as being separate, and in the latter to seek rather the property of being independent of the intensity of sampling – unbiased, with sampling variance progressively reduced as the size of sample increase. The various indices[1)] used to express species correlation have been reviewed by GOODALL (1952), DAGNELIE (1960a, 1965), GREIG-SMITH (1964), and WHITTAKER (1967), and statistical aspects are considered by GOODMAN & KRUSKAL (1954, 1959).

We will first consider the non-centred indices, as DAGNELIE (1960a) calls them — those taking only non-negative values. Among

[1)] Most of these indices have from time to time occurred in two forms – as a proportion, and as a percentage. The expressions given will all be in the proportion form.

them are the association coefficient of AGRELL (1945) and IVERSON (1954), also called 'percentage co-occurrence' by WHITTAKER & FAIRBANKS (1958) and 'spezifische Gemeinschaftskoeffizient' by ELLENBERG (1956):

$$a/(a+b+c) \tag{4}$$

and the coincidence index introduced by DICE (1945) — though he subsequently (1952) discarded it — and also used by BRAY (1956):

$$2a/(2a+b+c) \tag{5}$$

both of which take into account the samples in which either species is present, and ignore those in which both are absent. MORISITA (1959) has shown that both these indices are markedly affected by sample size.

Another worker who considered that samples from which both species were absent should be ignored was FAGER (1957, 1963), who proposed the index:

$$\frac{a}{\sqrt{(a+b)(a+c)}} - \frac{1}{2\sqrt{\min\{(a+b), (a+c)\}}} \tag{6}$$

as a means of expressing the extent to which each of the species was normally a part of the other's environment. GILBERT & WELLS (1966) also argued that samples from which both species were absent were irrelevant, and fitted a model (GILBERT, 1963) which we may express:

$$\log a/M = \log q_i + \log q_j + p_i p_j \tag{7}$$

a and M having the meanings with which we are familiar, while q_i and q_j are the population frequencies of the two species, and p_i, p_j are parameters expressing their degree of association.

FORBES (1907), MOZLEY (1936) and BARNARD (1964) used

$$Ma/\{(a+b)\ (a+c)\} \tag{8}$$

which is the ratio of the observed to the expected number of joint occurrences. TARWID (1960) put forward a similar index in which the divisor was the integral part of the expected number of joint occurrences.

Closely related is the expression

$$C = (a+b)\ (a+c)/a \tag{9}$$

used by MARGALEF (1958) as an inverse measure of association. Since only positive association is of interest, the fact that the index could take values of infinity was not regarded as important. The centred form

$$\log (M/C) \tag{10}$$

was also proposed.

HACKER's (1921) "association value"

$$a/(a+b), \text{ and } a/(a+c) \tag{11}$$

also used by DICE (1945, 1952), GARDNER (1951) and CASSAGNAU (1961), takes different values according to which species is being considered as the independent variable, and expresses the probability that one will occur where the other is present, rather than their association as usually understood.

The indices mentioned so far use joint presence (a) but not joint absence (d) as evidence of association between the species. Other indices treat joint presence and joint absence equally. The matching coefficient of SOKAL & MICHENER (1958)

$$(a+d)/M \tag{12}$$

widely used in numerical taxonomy, does not yet seem to have been used as a measure of interspecific association. BEALS (1965b) used

$$5-\sqrt{\chi^2} \tag{13}$$

where χ^2 is derived from expression (1), as an inverse measure of association ranging from a low value where the species are very closely associated (the actual minimum depends on M and on the marginal totals) to a maximum close to 5 if their relationship is negative.

Other indices taking joint presence and absence into account are centred — taking negative or positive values according to the sign of the association between the species. One of the first was that proposed by MICHAEL (1920):

$$\frac{4\{Ma - (a+b)(a+c)\}}{(a+d)^2 + (b+c+2d)^2} \tag{14}$$

but it has been little used since. Much more common is the point correlation coefficient

$$\phi = \frac{ad - bc}{\sqrt{(a+b)(a+c)(b+d)(c+d)}} \tag{15}$$

which has been used in ecology by DAGNELIE (1962a) and FREY (1963), among others.

Since where b and c are unequal ϕ can never attain unity, COLE (1949, 1957) proposed that the point correlation coefficient should be modified by using the ratio

$$\phi/\phi_{\max} \tag{16}$$

where $\phi_{\max}$ is the largest value of ϕ, with the same sign, consistent with the marginal totals. Similar considerations led FORBES (1925) to propose the expression

$$\frac{aM - (a+b)(a+c)}{M \min\{(a+b), (a+c)\} - (a+b)(a+c)} \tag{17}$$

and KOLBE (1956) to suggest the ratio of the deviation of a from expectation to the maximum possible deviation in the same direction consistent with the marginal totals.

HURLBERT (1969) tested COLE's index (16) in the form

$$\sqrt{\text{obs } \chi^2/\max \chi^2} \tag{18}$$

with the sign of $(ad-bc)$, where obs χ^2 is calculated in the usual way for 2×2 contingency tables and max χ^2 is the largest value the marginal totals permit. He found this expression to be markedly influenced by the species frequencies, and suggested it be improved by taking into account the fact that, in general, the lower limit for χ^2 is not zero but a somewhat higher figure (min χ^2), owing to the data being integral. His expression accordingly became

$$\sqrt{\frac{\text{obs } \chi^2 - \min \chi^2}{\max \chi^2 - \min \chi^2}} \tag{19}$$

again with the sign of $(ad-bc)$, and this he found a much more sastisfactory expression of association than COLE's index.

Information theory has contributed ideas in this field, as to other aspects of statistical ecology, and GODRON (1966) has suggested the information statistic

$$\log_2 \frac{M}{a!\,b!\,c!\,d!} \tag{20}$$

as a measure of association between the two species.

To a large extent, the choice of index is a matter of taste and of one's particular purpose, for each index has its own properties, and they may give very different results (see, for instance, MORISITA 1959, MIYATA 1966, CANCELA DA FONSECA 1966). A point for debate is, however, whether joint absence should be taken into account — whether the degree of association should be assessed solely on those quadrats where at least one of the species is present, (4)—(11), or whether the whole set of samples should be taken into account, including perhaps some that are completely bare of vegetation, (12)—(20). As GILBERT & WELLS (1966) pointed out, association measures including d may be greatly affected by its magnitude, which seems to introduce an arbitrary element into their determination. But the arbitrary element intrudes only in the delimitation of the universe to which the conclusions apply; once this is decided, measures including d will be no more arbitrary than those excluding it.

There are advantages in selecting an index whose possible range is independent of whether the frequencies of the two species are equal or not. COLE's index (16) and that of KOLBE, for instance (which take joint absence into account) range from −1 through 0 (where the two species are unrelated) to +1. Of the indices proposed which ignore joint absence, none has this property except perhaps that of GILBERT (7). One could suggest, for instance,

$$\frac{a}{a + \min(b, c)} \qquad (21)$$

PIELOU (1969b) distinguishes between "complete" and "absolute" association, the latter meaning that the two species occur only in one another's presence, while in the former case they may differ in frequency, so that Species A occurs only in the presence of Species B but not *vice versa*. Some indices, like that of COLE (16), can attain their maximum value if association is complete, while most of those used in ecology require it to be absolute. PIELOU emphasises the importance of making this distinction. While not quarreling with this point of view, one might perhaps suggest that it would be useful to supplement the symmetrical indices which have this property with asymmetrical indices more suitable for handling relations between common and infrequent species. An example is the "association value" (11) of HACKER (1921). It may be of much greater interest to know whether, and how much, the chance of encountering the infrequent species differs according to whether the common one is present or not, than to have a value expressing their mutual association. The distinction is parallel with that between correlation and regression, the latter expressing the quantity of Species A in terms of Species B (or *vice versa*), the former the closeness of relationship between the two.

Little information is to be found in the literature on the sampling properties of these indices. Accordingly, an empirical study of the distribution of some of them has been carried out, by computing them for repeated samples from a population with known and constant cell frequencies a, b, c, and d. For 20 such sets of cell frequencies, 100 samples of ten quadrats were taken, 100 of 20 quadrats, 100 of 50, and 100 of 100 each. The results are given in Table I. It will be seen that the IVERSON index (4) and the point correlation

Table I. Empirical estimates of mean and standard deviation of different association indices based on sets of 100 samples from known populations.

Population Proportions		a	.01	.02	.02	.05	.10	.10	.10	.20	.20	.30
		b	.29	.03	.03	.20	.02	.05	.10	.05	.10	.10
		c	.30	.45	.30	.35	.18	.25	.10	.25	.30	.10
		d	.40	.50	.60	.40	.70	.60	.70	.50	.40	.50
Simple Matching Coefficient (12)	N=10	Mean	.372	.223	.394	.371	.551	.610	.674	.684	.572	.804
		SD	.118	.269	.302	.200	.367	.310	.326	.222	.201	.119
	N=20	Mean	.414	.339	.575	.431	.711	.693	.749	.706	.585	.812
		SD	.102	.255	.213	.122	.252	.161	.183	.097	.113	.083
	N=50	Mean	.404	.472	.614	.442	.797	.693	.799	.707	.593	.789
		SD	.062	.148	.067	.065	.062	.066	.060	.069	.082	.065
	N=100	Mean	.405	.516	.626	.452	.797	.703	.798	.705	.698	.806
		SD	.048	.050	.044	.049	.037	.047	.043	.045	.053	.038
	Population Value		.410	.520	.620	.450	.800	.700	.800	.700	.600	.800
Iverson Coefficient (4)	N=10	Mean	.025	.050	.039	.065	.292	.295	.385	.430	.346	.622
		SD	.066	.103	.109	.108	.273	.252	.323	.253	.210	.312
	N=20	Mean	.016	.042	.057	.076	.300	.276	.327	.400	.321	.627
		SD	.035	.063	.084	.077	.202	.190	.202	.160	.134	.164
	N=50	Mean	.014	.041	.058	.080	.329	.246	.323	.413	.335	.581
		SD	.023	.034	.048	.045	.118	.093	.126	.109	.090	.106
	N=100	Mean	.013	.043	.050	.087	.329	.254	.344	.410	.328	.611
		SD	.013	.026	.036	.038	.092	.068	.084	.073	.064	.067
	Population Value		.013	.040	.050	.083	.333	.250	.333	.400	.333	.600
Dice Coefficient (5)	N=10	Mean	.041	.080	.058	.106	.387	.401	.478	.557	.478	.744
		SD	.108	.159	.154	.167	.319	.293	.347	.261	.236	.183
	N=20	Mean	.028	.073	.097	.132	.427	.399	.458	.553	.470	.757
		SD	.063	.108	.140	.127	.235	.227	.236	.168	.158	.137
	N=50	Mean	.026	.076	.106	.144	.484	.386	.475	.576	.495	.729
		SD	.043	.061	.084	.077	.134	.120	.143	.110	.102	.088
	N=100	Mean	.026	.081	.094	.157	.488	.400	.507	.578	.491	.757
		SD	.026	.048	.065	.064	.104	.085	.095	.075	.072	.052
	Population Value		.033	.077	.095	.154	.500	.400	.500	.571	.500	.750
Point Correlation Coefficient (15)	N=10	Mean	-.354	-.013	-.096	-.269	.344	.320	.411	.436	.202	.601
		SD	.237	.215	.212	.288	.323	.332	.376	.301	.351	.259
	N=20	Mean	-.382	-.034	-.051	-.258	.402	.295	.355	.415	.183	.611
		SD	.127	.178	.197	.216	.242	.261	.274	.204	.231	.183
	N=50	Mean	-.400	-.041	-.068	-.250	.450	.268	.361	.421	.211	.559
		SD	.085	.123	.129	.105	.138	.142	.173	.129	.158	.135
	N=100	Mean	-.400	-.031	-.082	-.228	.447	.282	.384	.413	.216	.597
		SD	.061	.097	.095	.087	.103	.105	.113	.091	.104	.079
	Population Value		-.392	-.032	-.086	-.236	.455	.279	.375	.406	.218	.583
Cole Coefficient (16)	N=10	Mean	-.806	-.058	-.454	-.579	.524	.465	.477	.676	.324	.755
		SD	.409	.623	.620	.544	.589	.665	.658	.508	.612	.379
	N=20	Mean	-.907	-.140	-.414	-.552	.713	.443	.434	.673	.287	.725
		SD	.213	.688	.705	.430	.441	.587	.531	.352	.402	.214
	N=50	Mean	-.913	-.157	-.341	-.532	.788	.478	.426	.659	.330	.625
		SD	.146	.589	.503	.233	.209	.256	.237	.190	.254	.161
	N=100	Mean	-.913	-.134	-.403	-.489	.768	.493	.430	.637	.328	.641
		SD	.087	.459	.390	.192	.155	.187	.125	.135	.162	.095
	Population Value		-.893	-.149	-.375	-.500	.768	.487	.375	.636	.333	.583

coefficient (15) are in general unbiased, even with samples of only ten quadrats; the simple matching coefficient (12) and the DICE index (5) estimated from small samples tend to underestimate the population value; in the COLE (16) index, significant discrepancies between sample and population values occur even up to sample sizes of 50, but the sign and extent of the discrepancy vary and appear to depend on the degree of association of the species, and the inequality of their total frequencies.

5.2.1.3 *Quantitative Data*

Where information about the samples includes some quantitative measure for the various species, and is not limited to their presence and absence, a variety of other measures of association become available. Such measures are referred to elsewhere in this volume as species correlation (SC).

In the first place, one may sacrifice part of the information by converting the quantitative data into a 2×2 table, the range of values for each species being divided at a convenient point, and the allocation of the sample to a cell of the table being determined by whether the quantity of each of the two species exceeds this arbitrary point or not. As a suitable point of division Kolbe (1956) used the geometric mean ignoring zero values, the Dutch grassland workers (Vries 1953, 1954, Vries et al. 1954) used the median, while Tarwid (1960) selected a density value exceeded by 30-40 % of the samples.

Where an arbitrary division of this sort is used to produce a 2×2 table for testing, it is conceivable that the degree or sign of association could be affected by the choice of a point at which this arbitrary division is made. Consider, for instance, the following hypothetical data for numbers of quadrats in different cover categories for two species:

Table II

% cover, Species A

		0	1—10	11—20	21—30	31—40	41—50	>50
	0	28	1	2	0	1	0	0
	1—10	0	2	1	0	7	5	11
% cover,	11—20	2	11	0	2	4	3	7
Species	21—30	1	1	0	3	5	4	4
B	31—40	0	2	4	1	0	2	3
	41—50	0	8	12	2	0	0	2
	>50	0	10	7	0	1	0	1

On the basis of presence and absence only, this gives

28	4
3	115

showing a marked positive association.

If the data are split at 40 % cover for each species, the 2×2 table is

68	39
40	3

showing a *negative* association, which is also marked if quadrats in which either species is absent are omitted:

40	35
37	3

On the other hand, if the whole data are split at 10 % cover, the 2×2 table becomes

31	27
25	67

which shows a *positive* association.

These different results are not mutually contradictory – they merely reflect a non-linear relation between the quantities of the two species, or indicate that their presence or absence may depend on factors differing from those which affect their cover when they are present.

While one recognizes that the association observed will depend on the arbitrary break-point selected for each species, its significance may be determined by any methods appropriate to presence and absence data. If a significant association is demonstrated by this rough-and-ready method, its reality need not be doubted, though its interpretation may call for more study.

Akin to the conversion of quantitative data into a 2×2 table is RAUNKIAER's (1928) "degree of codominance", which is the proportion of the area studied in which both species have frequencies in excess of 80 % Rather similar is IVERSON's (1936) "Affinitäts-koeffizient", which averages the frequency of species B in areas where species A is dominant, and that of species A where species B is dominant.

Other measures of correlation between species make direct use of the quantities of each of them in replicate samples. ELLENBERG (1956) proposed as a measure of quantitative association between species the "spezifische Massen-Gemeinschaftskoeffizient".

$$\sum_{i \varepsilon T} (x_i + y_i) \ / \ [\sum_{i \varepsilon T} (x_i + y_i) + 2(\sum_{i \varepsilon U} x_i + \sum_{i \varepsilon V} y_i)] \qquad (22)$$

where x_i and y_i are measures (biomasses) of species A and B in the ith sample, T is the subset of samples containing both species, and U and V are the subsets containing respectively species A or species B but not both.

WHITTAKER & FAIRBANKS (1958) used a related measure as "percentage similarity of distribution" (cf. (51)),

$$\sum_i \min (x_i, y_i) = 1 - 0.5 \sum_i |x_i - y_i| \qquad (23)$$

in which x_i and y_i are measures for species x and y in the i th sample, when those measures have been divivided by the row sums of a community table (i.e. for each species they are the decimal proportions of the sum of that species' measures in all samples of the set). MONK (1968) derives a measure of correlation between species, not from their joint occurrence in the same samples, but from the similarity of the environmental factors in the samples where they occur; in other words, it is an *ecological* correlation in the narrow sense. For each environmental factor, the mean is calculated for those subsets of samples in which each of the species occurs. These means are then used as the variables x_i, y_i in calculation of a similarity index (expression (50), section 5.3.3.2 below.)

A very commonly used measure of association making good use of measurement data, but depending on the assumption of linear relations between species, is the correlation coefficient:

$$r = \frac{\sum_i (x_i - \bar{x})(y_i - \bar{y})}{\sqrt{\sum_i (x_i - \bar{x})^2 \sum_i (y_i - \bar{y})^2}} \qquad (24)$$

where x_i y_i are the quantities of the two species, $\bar{x}$, $\bar{y}$ their means, and summation is over all samples. If presence of the species is scored as 1, and absence is 0, this reduces to the point correlation coefficient (13). Where there is no association between the two species, and the quantity measures of each are normally distributed, the expectation of r is zero, and its sampling distribution is well known. Ecological data are often very far from normally distributed — in fact, a distribution curve with a mode substantially below the mean is usual — and in this case, though the correlation coefficient may provide an acceptable measure of association, its distribution is unknown, and to rely on published tables may lead one far astray. Moreover, non-linear relations such as those in Table II are probably far from uncommon, and will not be revealed by a correlation coefficient. For this purpose, tests of quadratic or higher-order regression may be required.

To avoid the drawbacks of unknown but certainly non-normal distributions, the use of rank-order correlation has sometimes been suggested. The actual measures are replaced by the ranking of the sample in respect of each species, and then the SPEARMAN (or other) rank correlation coefficient is calculated without reference to the measurements (FAGER 1957, NUMATA & MITSUDERA 1960). Alternatively, the ranks may be used as the basis for a normalizing ("rankit") transformation, and a correlation coefficient calculated for the transformed data in the usual way — a coefficient which is now assured of having the properties and distribution expected for the correlation coefficient if the null hypothesis is true.

Information measures offer another way of avoiding distribution problems. MORISITA (1959) has proposed an index of interspecific overlapping calculated as follows: let the number of individuals of species x in the i th sample be n_{xi}, and the total individuals of this species in all samples be N_x. Then one defines

$$\delta_x = \frac{\sum_i n_{xi}(n_{xi}-1)}{N_x(N_x-1)} \tag{25}$$

and the index of interspecific overlapping between species x and y is

$$C_\delta = \frac{2\sum_i n_{xi}n_{yi}}{(\delta_x+\delta_y)\,N_x N_y} \tag{26}$$

taking values of approximately unity where the two species vary in proportion in the different samples, and zero where they do not occur together. This index ignores those samples where neither species is present, and MORISITA was able to prove by empirical trials that it is almost independent of N_x and N_y.

In order to get an equivalent index of interspecific correlation (i.e. taking quadrats from which both species were absent into account), MORISITA defined

$$R'_\delta = C_\delta - \frac{2}{(\delta_x+\delta_y)M} \tag{27}$$

where M is the number of samples. Then the correlation index is given by

$$\begin{aligned} R_\delta &= R'_\delta & R'_\delta > 0 \\ R_\delta &= \frac{M\sum_i n_{xi}n_{yi}}{N_x N_y} - 1, & R'_\delta < 0 \end{aligned} \tag{28}$$

which allows it to range from —1 to +1. Like C_δ, MORISITA found it almost independent of N_x and N_y; HURLBERT (1969), however, was able to show that for certain types of vegetation pattern it may give very different results with quadrats of different size, and that some of these results may be seriously at variance with common-sense views.

MORISITA extended this treatment from density to cover data, so that indices of interspecific overlapping and correlation for cover were

$$C_{\delta(p)} = \frac{M \sum_i p_{xi} p_{yi}}{(\delta_{x(p)} + \delta_{y(p)}) M^2 \bar{p}_x \bar{p}_y} \tag{29}$$

and

$$R'_{\delta(p)} = C_{\delta(p)} - \frac{2}{(\delta_{x(p)} + \delta_{y(p)}) M} \tag{30}$$

where p_{xi} is the cover of the x th species in the i th sample; $\bar{p}_x$ is the mean cover of the x th species; and

$$\delta_{x(p)} = \frac{\sum_i p^2_{xi}}{M^2 \bar{p}^2_x} \tag{31}$$

ESTABROOK (1967) also suggested an information measure for associations where the values observed could be expressed as a frequency table — e.g. the numbers of quadrats for which the quantities of the two species in question fell into each of a limited number of categories. If the probability with which the quantity of species j in a quadrat falls into the i th category is P_{ij}, then the information content of the species j data is

$$H_j = - \sum_i P_{ij} \log P_{ij} \tag{32}$$

By finding the information content for species j within each category of species k (or vice versa) one can then determine the information content of j which is not shared with k, $H(j/k)$. If then we define

$$D(j,k) = \frac{H(j/k) + H(k/j)}{H(j\,.\,k)} \tag{33}$$

where $H(j.k)$ is the total information in the two species considered together, we have a measure of the independence of the two species. This may readily be converted into a measure of association:

$$S(j,k) = \sqrt{1 - D^2(j,k)} \tag{34}$$

which will range between 0 and 1. As ORLOCI (1968) suggests, this measure could be applied directly to count or frequency data in ecology or (by arbitrary categorization) to continuous measures. It should be remembered, though, that any information measure of this sort takes no account of the fact that the categories are ordered.

5.2.1.4 *Species Correlation from Distance Measures*

Association between species can also be studied on the basis of distance between individuals. The first to do this was DICE (1952). He took the less abundant of two species, and measured the distance from each individual of this species to the nearest individual of the other species in each sextant around it. He showed that the square roots of these distances would be distributed approximately normally if the two species were uncorrelated, and that the skewness parameter g would be positive if they tended to occur together, negative if they tended to be separated.

PIELOU (1961) recorded random individuals together with their nearest neighbours, and studied the specific composition of the pairs thus formed. For every two species a 2×2 table is set up, partitioning the pairs according to whether species A was or was not the random individual, and whether B was or was not the nearest neighbour. The significance of the relation between the species is tested by the usual methods for 2×2 tables, and the measure

$$S = 1 - \frac{(b+c)M}{(a+b)(b+d) + (a+c)(c+d)} \qquad (35)$$

is used as a coefficient of segregation, showing the extent to which individuals tend to have as nearest neighbours others of the same species rather than a different species.

GOODALL (1965) proposed two methods using distance measurements, one based on the recognition of mutually nearest neighbours, the other on comparison of distances between individuals of the two species with distances from a random point. These techniques were primarily intended for significance tests; but the second could also, if desired, provide a measure of association – for instance, by the ratio of the random-point distance to the nearest-neighbour distance.

A rather similar method has been proposed by YARRANTON (1966) with special reference to lichen communities, in which a random point defines a pair of species — that of the individual on which the point falls, and the nearest other species in contact with it. A species x species matrix showing the number of times each possible pair-wise record occurs can then form the basis for association measures, just like the presence of species in other types of sample. It should be noted, however, that the observational matrix will not be symmetrical.

5.2.2 Relations Among More than Two Species

Sometimes the relations studied are not between species in the narrow sense at all, but rather between groups of species with similar biological characteristics, which may behave in a largely vicarious way. It was shown, for instance (Goodall 1965), that there was a close correlation in the mulga vegetation of Australia between a group of species of taller shrubs on the one hand, and another group of much lower-growing shrubs, which seemed to need protection at some stages of growth. Similarly, McDonough (1963) showed a relationship between some cactus and shrub species in the Sonoran Desert. In neither of these cases, though, was there any specific relation between pairs of species to be found.

One may wish to know, not whether particular pairs of species are correlated, but whether a whole group of species tend to occur together. For instance, correlations among groups of species provide an objective basis for the recognition of "ecological groups" or "sociological groups". This concept, originated by Duvigneaud (1946, 1953) and developed further by Ellenberg (1950) and Gounot (1961) among others (article 2.2.3.1), has affinities with the "characteristic species" of associations recognized by the Zürich-Montpellier School (article 20). A group of species with similar environmental requirements tend to occur together, and a given vegetation type is identifiable from the presence of one or more such ecological groups. A numerical method for recognizing correlated groups of species may accordingly be useful in vegetation classification.

The recognition of groups of correlated species is the counterpart of the problem of classifying stands (article 19), and many of the same methods can be applied. For instance the technique known as "inverse association analysis" (Williams & Lambert 1961, Lambert & Dale 1964) separates groups of species acording to whether or not they occur in certain key stands, whereas "inverse information analysis" (Williams et al. 1966) separates them according to the independence of their information content.

Another suitable approach is that of multivariate factor analysis (Dagnelie 1960a, b, 1965, article 8). Principal component analysis, for instance, discussed in connexion with stand ordination in article 10, is a procedure for finding orthogonal linear functions of the different variables (quantities of different species) which account for the maximum possible proportion of their variation. The species with large coefficients (of the same sign) in one of these functions then constitute a correlated group, whose structure may often be clarified by a rotation of axes. With presence and absence data, the principal component method may also be used, if presence and absence are recorded as "1" and "0".

Pair-wise correlations among species may serve as a basis for recognizing larger groups. WELTEN & HEGG (1958) in fact regarded correlations between single pairs of species as likely to be fortuitous unless they were supported by mutual correlations with other pairs of species, which thus constitute an ecological group. BEALS (1965b) used measures of "distance" (i.e., inverse correlation) between species pairs to ordinate a set of species in two dimensions, and then divided them into seven ecological groups from inspection of this ordination. KLEINER & HARPER (1966) distinguished groups of species on the basis of the number of significant positive correlations between pairs within the group.

Various ways have been proposed to express the degree of correlation among a number of species forming a group. HOPKINS (1957) and MC INTOSH (1962), for instance, used the expression

$$2A/\{s(s-1)\} \qquad (36)$$

where the group consists of s species, and A of the pair-wise correlations among them reach significance at a selected level. FAGER (1957) suggested the use of KENDALL's (1955) measure of "concordance"—analogous with rank-order correlation. MORISITA (1959) presented extensions of his C_δ, R_δ indices (expressions (26) and (28) above) to cases where several species are involved, each with the same total number of individuals. MARGALEF (1958) generalized the expression (9) he used for association between two species, in the form

$$a^{-1}(m_1 \,.\, m_2 \,.\, m_3 \ldots\ldots m_s)^{s-1} \qquad (37)$$

as a measure of association between s species present respectively in $m_1, m_2, \ldots\ldots m_s$ samples, and jointly present in a samples.

Questions regarding the significance of association among a group of species may be framed in different ways. Considering first presence and absence data, in the $2 \times 2 \times 2$ table

		Species A present		Species A absent	
Species B:		+	—	+	—
Species C	+	a	b	a'	b'
	—	c	d	c'	d'

one may wish to know whether the number of samples, a, in which all three species occurred is greater than was to be expected given the frequency of the species singly and in pairs. Or one could ask whether the eight cell frequencies are consistent with an assumption of complete independence of the species. Or, as a third possibility, one could consider the three species as a group which may occur in any combinations, and ask whether the four cell values *a*, *b*, *c*, and *a'* collectively are larger than expected, given the frequencies of the three species separately. All three questions are legitimate, and the answers will differ.

Taking GREIG-SMITH's (1952, 1964) figures for *Amaioua corymbosa*, *Lacistema aggregatum* and *Alibertia acuminata* in Trinidad as A, B, and C above, one may tabulate them:

23	6	6	7
14	15	11	18

GREIG-SMITH chooses to answer the second question raised above, and obtains a value for χ^2 of 16.69, with four degrees of freedom. He points out that this result may be ascribed to the large excess over expectation of *a* — in other words, that the negative answer to the second question depends on a positive answer to the first, though this is not explicitly answered. If, on the other hand, we ask the third question, we find that the four cells with two or more of the species contain *fewer* quadrats than would have been expected from their individual frequencies (49, as against 52.06, giving a non-significant χ^2 value of 0.38 with one degree of freedom); the low figures for quadrats with two of the three species counterbalance the high figure for quadrats with all three.

PIELOU & PIELOU (1967) proposed a significance test for "coexistence" among a group of species based on the number of samples without any of the group. Combinatorial expressions are given for the probability of an observed figure, taking the frequency of species into account, but will not be reproduced here. The test does not distinguish between cases where all of the group tend to occur together, and those in which the positive coexistence is confined to a subset. A normal approximation is also given for cases where the number of species in the group is small, the number of samples large.

5.2.3 PARTIAL CORRELATIONS

Where more than two species are consistently associated, the relationships among them may often be elucidated by calculating partial correlation coefficients (e.g. DAWSON 1951, VASILEVICH

1970). This is a well-established statistic for describing interrelations among normally-distributed variables. It expresses and tests the linear interrelation between two variables, after eliminating that part of their correlation which is attributable to a mutual linear association with one or more other variables. In the simplest case, it is expressed as

$$r_{ab \cdot c} = \frac{r_{ab} - r_{ac}\, r_{bc}}{\sqrt{(1 - r_{ac}^2)(1 - r_{bc}^2)}} \cdot \tag{38}$$

where *a* and *b* are the variables under study and *c* is the variable the relations with which are to be eliminated, while r_{ab} is the product-moment correlation (expression (24)) between *a* and *b*. The significance of this partial correlation may be tested against tabulated values showing the distribution of sample values from an uncorrelated population — the same as that of the product-moment correlation, with one fewer degrees of freedom.

Cole (1957) proposed a similar test for partial association between species, where presence and absence only had been recorded. Apart from relations among three or more species, however, he pointed out that the third variable might well be a habitat factor, so that the test would then examine the relation between the two species when this environmental factor was held constant. In this connexion, tests of partial association or correlation seem to have more ecological relevance; to know whether two species are still correlated in the absence of differences in an environmental factor which may affect them is likely to be more valuable in interpretation than to find the effects of removing mutual correlation with a third species, relations with which are less likely to be directly causal.

5.2.4 Non-Linear Relations among Species

Most of the methods of identifying and measuring the correlation between species which have been described above are based on a linear model, or assume a substantial linear component in the relationship. It is by no means clear that this assumption is generally justified. It has been shown (Whittaker 1967, article 2) that, where observations cover a fairly wide range of ecological conditions, different species may occur in maximum quantity at different environmental values, and decrease both above and below this optimum. Where the modes of two species do not coincide, this means that a linear model may fail to reveal the relationship between them. If a species with a narrow range occurs with a broad-ranging one, there may be a real relationship between their quantities which could be shown only by studying higher-order terms.

Consider, for instance, the hypothetical set of data:

Species A, cover %:	0	3	6	15	30	42	60
Species B, cover %:	0	9.5	18	37.5	50	42	0

which could arise, for instance, if the environmental range of A was much wider than that of B. There is an exact mathematical relation between the two series of figures ($18B=60A-A^2$), but usual measures of association will not show this. The correlation coefficient r (expression (24), for instance, is +.169. SPEARMAN's rank correlation coefficient is —.229. If these two are the only species present in the quadrats, WHITTAKER's index (23) will be .33. In such cases, a regression approach may be appropriate; a test of quadratic regression would demonstrate the relationship, and its closeness could, for instance, be expressed by R^2, the proportion of the variance of B accounted for by the regression on A.

Where the relation between two species is studied, not for its own sake, but as part of an ordination procedure, the use of a linear model for essentially non-linear relations may lead to considerable obscurity. If principal components analysis is applied, for instance, to data for species each showing a separate mode along a single axis, each of these modes will be represented by a distinct principal component, and the simple underlying model of a single axis with non-linear relations of each species to it will not be revealed (cf. article 10). Such models have been the subject of some study by pyschologists, and attempts are currently being made to apply non-linear models for principal components analysis to ecological data.

5.3 Similarity Between Vegetation Samples

5.3.1 INTRODUCTION

A problem converse to interspecific correlation is that of establishing and expressing the degree of similarity or dissimilarity between samples of vegetation — or, indeed, between entire stands. A matrix of similarity is, for instance, often required as a preliminary to the classification or ordination of a set of vegetation samples. Assessments of similarity can be based on any observable features of the vegetation — or even of the biotope — according to the basic assumptions and intentions underlying the investigations. Growth-forms or structural features might well be so used, for instance. In practice, however, almost all investigators have used floristic criteria, and it is with these that we shall concern ourselves.

A distinction needs to be made between measures of similarity which are context-dependent, and those which are, in a sense, absolute. "Similarity" is sometimes regarded as a concept with a fixed value and a fixed expression, so that the degree of similarity between two objects can be stated without reference to any others. Sometimes, on the other hand, the degree of similarity is considered in reference to a particular range of variation, either concretely represented in a set of objects, or present conceptually in the mind assessing it; thus a black mouse and a white mouse may be regarded as highly dissimilar within a population of mice, very similar in a mixed population of rodent species. Examples of both types of similarity concept will be found below.

Though formally the problem of inter-sample similarity seems to be on all fours with that of interspecific correlation already discussed, there are important differences between the two situations, as may be made evident by comparing the significance tests appropriate in each of the two cases.

5.3.2 Significance Tests of Similarity

For species association, the null hypothesis adopted was that each of the two species had a probability of being recorded which remained uniform throughout the set of samples, and was independent of whether the other species occurred in the sample or not. These probabilities are estimated from the marginal totals. In comparing the set of species occurring in two quadrats with a view to judging their similarity, the appropriate null hypothesis is clearly that the two quadrats were samples of the same vegetation type, and that consequently, for each species, the probability of occurrence was the same for both samples. But records in two quadrats only of the presence or absence of a species provide no basis for estimating its probability of occurrence — beyond the trivial conclusion that the probability is non-zero if the species is recorded at all. Only where the two are part of a larger collection in which, for the purposes of the null hypothesis, the probability for each species can be regarded as uniform does the basis exist for estimating the probability, and hence for a test of the null hypothesis.

An insensitive significance test may however be applied to the records in two samples only, without the need for frequency estimates based on a larger set of samples, by comparing the number of species present in one but not in the other (GOODALL 1969). If they are from the same community, the expectation is that these numbers will be equal for the two samples; and — on the same assumption as above, of independence — the distribution of unequal numbers will follow the binomial distribution. Particularly where the two samples come from vegetation types differing markedly in floristic diversity, significant dissimilarity may often be revealed by this test.

An assumption of independence between species is almost inevitable in testing a hypothesis regarding similarity of samples based on species presence and absence. Where the vegetation sampled is uniform, and the sample size is sufficient to cover the largest scale of pattern, this assumption seems reasonable (GOODALL 1953), apart from occasional instances of obligate relations between particular pairs of species, such as parasite and host. In the absence of such an assumption, there seems no basis for significance tests of similarity based on qualitative criteria.

The fact that a 2×2 table for similarity between quadrats or stands combines the records for a large number of species each with its own frequency in the population means that methods (such as the χ^2 test) suitable for testing association are inapplicable. The model underlying the χ^2 test is of a number of events all alike, and falling with constant frequency into the two classes of each of two binary classifications. Were the frequencies of all species the same, the same technique could be applied to tests of sample similarity. In practice, this requirement is never even approximately met, and, if a χ^2 statistic were calculated, its distribution would differ widely from tabulated values even if the samples compared were replicates from the same community. Thus, a frequency curve for χ^2 between samples differing from the standard distribution (JUHÁSZ NAGY 1964) does not necessarily imply heterogeneity among the samples.

Where quantitative variables are used (such as density, cover, basal area, etc. for each species), the distribution of any index of similarity between samples would depend on the multivariate distribution of the variables included in the index. In GOODALL's (1966) probabilistic index this is taken into account, and the distribution is rectangular. For other indices an exact solution would be extremely difficult, and it might be best to approach the problem empirically by studying the distribution of calculated similarities in the whole matrix of sample comparisons, and finding where the index in question fell in the tail of the observed distribution.

It has so far been assumed that a particular pair of samples is of special interest, and that the similarity between this pair only is to be tested. This is not the usual situation. Usually all pair-wise similarities within a set of samples are studied simultaneously, and the question is rather whether the maximum or minimum among them is significant, or whether the whole set differ from the expected distribution. The latter may generally be tested by calculating χ^2 after arbitrary division into classes, or (better) by the KOLMOGOROV-SMIRNOV test. A test of the maximum or minimum necessitates an adjustment of significance levels. If the probability with which the null hypothesis may be falsely rejected when true is α, and n independent values have been tested, the maximum value calling for rejection of the null hypothesis is such as would be exceeded by individual values with a probability

$$1-(1-\alpha)^{1/n}$$

The fact that a set of similarity values may not be fully independent will usually make this test over-conservative.

5.3.3 MEASURES OF SAMPLE SIMILARITY

5.3.3.1 *Presence and Absence Data*

Where only presence or absence of species is recorded, a series of coefficients corresponding with the association indices mentioned above (Section 5.2.1.2) can be derived from the resulting 2×2 tables:

		Sample X +	Sample X −
Sample Y	+	a	b
	−	c	d

where a, b, c and d are the numbers of species present or absent in the samples in question. Many of these indices have also been used in numerical taxonomy, and are discussed in this connexion by SOKAL & SNEATH (1963).

The oldest similarity index – that used by JACCARD (1901) in comparisons of species lists from different geographical areas in the Alps — is

$$a/(a+b+c) \qquad (39)$$

which is analogous with expression (4) above for interspecific association, while SØRENSEN (1948) proposed the index

$$2a/(2a+b+c) \tag{40}$$

analogous with expression (5). SØRENSEN's index has been used very widely for expressing similarity, and nomograms and tables for simplifying its calculation have been published by FALIŃSKI (1958). Expressions (39) and (40) have been called "coefficients of community", and are referred to elsewhere in this volume as CC. The complement of either is a coefficient of community difference (CD).

KULCZYNSKI (1928) proposed two indices, which may be represented

$$a/(b+c) \tag{41}$$

and

$$\frac{1}{2}\left(\frac{a}{a+b}\right)\left(\frac{a}{a+c}\right) \tag{42}$$

while OCHIAI (1957) and BARKMAN (1958) used

$$\frac{a}{\sqrt{(a+b)(a+c)}} \tag{43}$$

MAAREL (1966) suggested an index which is the square of (43), but in fact used that of SØRENSEN because of its greater simplicity.

WILLIAMS (1944, 1947a, b) proposed as an index of similarity the ratio of the number of species present in two sample areas to the number expected from a species-area curve if they had been samples of the same homogeneous area.

WILLIAMS et al. (1966) used an index based on information theory (see (64) below) which, in the case of data limited to presence and absence of species in the two samples under comparison, reduces to:

$$2\ (b+c)\log 2$$

As FIELD (1969) points out, this tends automatically to give species-poor samples greater similarity than species-rich ones, which does not always correspond with intuitive views of the concept.

Similarity indices giving equal weight to presence and absence, such as the matching coefficient (12), have been little used in plant sociology. One example is the work of CRAWFORD et al. (1970), who applied the index of RUSSELL & RAO (1940) to comparison of samples of scrub vegetation in Peru. In our terms, this index was a/S, where S is the total species number, $a+b+c+d$; d is thus included in the denominator, but not in the numerator.

The limited use in plant sociology of indices taking d into account is understandable because of their great dependence on the less common species absent from both samples. It might equally be argued that the presence in both samples of ubiquitous species is also uninformative. One way of surmounting this difficulty is by the probabilistic similarity index (GOODALL 1966), which gives equal weight to presence in both the quadrats of a rare species, and absence from both of a ubiquitous one. In the calculation of this index (which can use binary, ranked or quantitative data), all possible pairs of values for each variable are placed in order of similarity, the probability of the observed or any greater degree of similarity is estimated for each variable, and these estimates are then combined.

It is clear that even replicate samples from the same homogeneous area of vegetation are unlikely to have exactly the same species complements, and consequently the values of any of these similarity indices will only exceptionally reach their theoretical maxima. The expectation and distribution of values between replicate samples have been little studied. The probabilistic index (GOODALL 1966) has of course a retangular distribution with an expectation of 0.5; and it has been shown (GOODALL 1967) that the matching coefficient has an expectation of

$$\frac{1}{S}\sum_{i=1}^{S}(1-2f_i+2f_i^2) \qquad (45)$$

where f_i is the proportional frequency in the samples of the i th species (out of a total of S species), and that its distribution may be approximated by the binomial. Likewise, the expectation of the JACCARD and SØRENSEN coefficients may be expressed respectively as

$$\sum_{i=1}^{S} f_i^2 \sum_{i=1}^{S}(2f_i-f_i^2) \qquad (46)$$

and

$$\sum_{i=1}^{S} f_i^2 \sum_{i=1}^{S} f_i \qquad (47)$$

LOOMAN & CAMPBELL (1960) attempted to set significance limits for the SØRENSEN coefficient by deriving them from χ^2 tests; they recognized, however, that it is necessary to take into account d, which is excluded from the SØRENSEN coefficient. Its distribution was studied empirically by GOUNOT & CALLEJA (1962) in samples from a *Brachypodium ramosum* grassland. They showed that the mean value increased and the variance decreased as the sample size increased, but that even with a 32 m^2 sample, for which the mean was .827, the confidence limits were .750 and .904. They accordingly recommended the use of multiple samples from each stand. The mean similarity coefficient decreased as the distance between samples increased — at least up to a distance greater than 4m — which fits in with observations on variance of quantities of individual species (GOODALL 1961).

WILLIAMS (1949, 1964) showed that the expectation of the JACCARD coefficient (39) between samples of equal size from homogeneous vegetation was simply related to the index of diversity a, on the assumption that the species densities followed the logarithmic series. If the mean number of species per quadrat is S, the expected value is

$$\frac{S - a \log_e 2}{S + a \log_e 2} \qquad (48)$$

This implies that the value of the index, within homogeneous vegetation, will increase as the sample size increases.

5.3.3.2 *Quantitative Data*

We turn now to similarity indices based on quantitative measures of the contribution of each species to the vegetation – cover, biomass, basal area, frequency, etc.

A similarity index proposed by GLEASON (1920), may be expressed as

$$\frac{\sum\limits_{i \varepsilon T} (x_i + y_i)}{\sum\limits_{i} (x_i + y_i)} \qquad (49)$$

where x_i, y_i are the quantities of the i th species in the two samples, and T is the sub-set of species present in both. This index, closely analogous with ELLENBERG'S "spezifische Massen-Gemeinschafts-koeffizient" for interspecific correlation (22 above), has the drawback that the index may be affected greatly by a minimal occurrence of a species. Consider the two sets of data in Table III:

Quadrat		Species	Quadrat	
X	Y		X	Y
99	1	1	99	0
10	0	2	10	0
0	5	3	0	5
80	80	4	80	80

The GLEASON index is 0.82 for the data on the left-hand side of the Table, 0.41 for those on the right; yet the only difference is the presence of 1 % of Species 1 in quadrat Y in the former case.

CZEKANOWSKI's (1909) 'percentage similarity'

$$PS = \frac{2 \sum_i \min(x_i, y_i)}{\sum_i (x_i + y_i)} \tag{50}$$

was first introduced into plant sociology by DAHL & HADAČ (1941)[1], and subsequently used extensively by the Wisconsin school (e.g. CULBERSON 1955). If x_i, y_i are expressed as proportions of the total quantities of all species in samples, X and Y, index (50) reduces to

$$PS' = \sum_i \min(x_i, y_i) = 1 - 0.5 \sum_i |x_i - y_i| \tag{51}$$

in which form it was used by RENKONEN (1938, 1944) and WHITTAKER (1952) for comparisons of insect samples. Among others who have made use of percentage similarity, with various measures of sample composition, are DAHL (1957), CLAUSEN (1957), BRAY & CURTIS (1957), WHITTAKER & FAIRBANKS (1958), WHITTAKER (1960) and RAMSAY (1964a, b). It is analogous with index (23) for interspecific correlation.

Complements of percentage similarity are "percentage difference", which ODUM (1950) used to compare bird communities:

$$PD = \sum |x_i - y_i| / \sum (x_i + y_i) \tag{52}$$

KULCZYNSKI's (1928) "floristic degree of resemblance" was

$$\tfrac{1}{2} \left(\frac{1}{\sum_i x_i} + \frac{1}{\sum_i y_i} \right) \sum_i \min(x_i, y_i) \tag{53}$$

[1]) Their description of the index is ambiguous, but examination of their tables shows that it was calculated by (50), x_i and y_i being the mean value in the DOMIN scale for the stands in each of the two association tables compared.

Ružička's (1958) index was

$$\frac{\sum_i \min(x_i, y_i)}{\sum_i x_i + \sum_i y_i - \sum_i \min(x_i, y_i)} \tag{54}$$

and Pandeya (1961) used one very similar to that of Gleason (49), viz.

$$\frac{\sum_{i\epsilon T} (x_i + y_i)}{\sum_i (x_i + y_i) + \sum_{i\epsilon T} |x_i - y_i|} \tag{55}$$

where T is the sub-set of species present in both samples.

Forms of the correlation coefficient (24) have also been used as indices of similarity. Some investigators (e.g. Motomura 1952, Fresco 1969) have calculated the coefficient from unstandardized values of the variable; apart from the inapplicability of ordinary significance tests, this method will have the effect of giving much greater weight to similar values for the abundant species than to those less abundant in the vegetation.

A rather similar method was adopted by Orloci (1966) as a preliminary to ordination. He took deviations from the species mean without standardizing and showed that the scalar product

$$\sum_i (x_i - \bar{z}_i)(y_i - \bar{z}_i) \tag{56}$$

where $\bar{z}_i$ is the mean over all quadrats for the i th species, is a suitable measure of the similarity between quadrats X and Y for use in principal components analysis.

In calculating any of these indices there is, of course, no obligation to use the same measure of quantity for each species. Indeed, Beals & Cottam (1960) included two measurements — density and dominance — for each tree species, and one — frequency — for each herb species. But the similarity values obtained will, of course, be markedly affected by such arbitrary weightings.

Normally, like Motomura's "correlation coefficient", all these indices give the various species weight in proportion to the quantities in which they occur. This may or may not be desirable in expressing similarity, but many, like Kontkanen (1950), have felt that some redressing of this inequality in weighting is desirable. One way of doing this is to rank the species in each sample, and calculate a rank correlation coefficient, as was done by Ghent (1963).

If the values for each variable are first standardized by converting to zero mean and unit variance, and these standardized values are used to calculate a "correlation coefficient" between stands as an index of similarity, values of the index will follow the tabulated distributions on the assumption of species independence and normal distribution. If the distributions deviate widely from normality, a rankit transformation may be preferable to standardization. Either of these procedures will give all species equal weight.

In the Wisconsin modification (BRAY & CURTIS, 1957) of expression (50), the observations for each species are expressed as a percentage of the maximum for that species over all samples; these figures for each sample are then adjusted to a total of 100 before calculation of the coefficient. This modification, like those of the previous paragraph, has the effect of giving equal weight to all species, and making the similarity depend on their relative rather than absolute contribution to the vegetation, so that two samples in which the totals differed by a large factor would have 100 % similarity if their quantitative floristic composition were identical. As WHITTAKER (1967) points out, this procedure has the further advantage that different types of measurement (e.g. basal area and density) may be selected for different species. BRAY & CURTIS claim that this coefficient differentiated pairs of samples in the middle range of similarity better than the correlation coefficient.

HALL (1970) considered that the contribution of different species to a similarity function should be weighted in accordance with their mean quantity or frequency, within the set in which the similarity is being assessed. He was concerned with the more general case of similarity among a number of samples (see below, Section 5.3.5), but if only two samples are compared his index becomes

$$\frac{m + m^2/n}{m + n} \tag{57}$$

where m and n are the smaller and larger of the two values for each species.

ELLENBERG (1956) used a "Massen-Gemeinschaftskoeffizient"– corresponding with the expression (22) already cited for interspecific correlation —

$$\frac{\sum_{i \varepsilon T} (x_i + y_i)}{2 \sum_{i \varepsilon U} x_i + 2 \sum_{i \varepsilon V} y_i + \sum_{i \varepsilon T} (x_i + y_i)} \tag{58}$$

where T is the subset of species occurring in both samples, U and V those occurring in one or the other only.

MORISITA (1959) proposed a measure of community similarity, C_λ, calculated in exactly the same way as his index of interspecific overlapping C_δ (expression (26) above), and having the same independence of the total numbers of individuals in the samples except when they are small. He also extended this expression to weight (or cover), on the assumptions that the weight per individual within a species can be treated as constant, and that the number of individuals is large. The appropriate expression is then

$$C_{\lambda(w)} = \frac{2 \sum_{i} w_{1j} w_{2j}}{(\lambda_{(w)1} + \lambda_{(w)2}) W_1 W_2} \tag{59}$$

where w_{ij} is the weight of the j th species in the i th quadrat, W_i the weight of all species, and

$$\lambda_{(w)i} = \sum_{j} w_{ij}^2 W_i^2 \tag{60}$$

BEALS (1965a) introduced a new consideration into similarity by taking growth-form into account. His modification of expression (50) for lichen and moss communities was

$$\frac{3 \sum_{i} \min (x_i, y_i) + \sum_{j} \min (a_j, b_j)}{2 \sum_{i} (x_i + y_i)} \tag{61}$$

where x_i, y_i were the frequencies of the i th species and a_j, b_j were the sums of species frequencies for the j th growth-form in the two samples respectively.

As with the similarity indices based on presence and absence, it is clearly relevant to consider, for those based on quantitative data, their expectation and distribution when the two samples are replicates from the same homogeneous community. WHITTAKER (1952) found that, in an insect community, values of expression (51) ranged from .5 to .8 in comparisons between replicate samples, while BRAY & CURTIS (1957) found that the Wisconsin coefficient averaged .82 between replicates. It is not, however, to be expected that this average value would be independent of the floristic composition of the vegetation. In fact, it is clearly desirable that every use of similarity coefficients should be accompanied by a study of the distribution of the same coefficient in replicate samples, for the cases where this distribution is known *a priori* are few.

5.3.3.3 *Expressions of Sample Distance*

An inverse measure of similarity between samples is provided by the compositional distance between them (compare "taxonomic distance" as discussed by SOKAL & SNEATH, 1963.) If we consider a multidimensional system in which each axis represents the quantity of a species, and a point represents the floristic composition of each vegetation sample, then the distance between the points representing two samples is their compositional distance. This may be calculated by the usual Euclidean distance expression:

$$ED = \sqrt{\sum_i (x_i - y_i)^2} \qquad (62)$$

where x_i, y_i are the quantities (using any appropriate measure) of the i th species in samples X and Y. This was used, for instance, by FIJALKOWSKI (1954) and VASILEVICH (1963), and by NEWBOULD (1960) under the designation Ω^2, without taking the square root.

ORLOCI (1967, 1968) suggested replacing the absolute Euclidean distance by a relative distance, each observation being standardized by dividing by the length of the stand vector. This means that the values of the distance measure will be between 0 and $\sqrt{2}$, and will be zero if all species occur in the same proportions in the two stands; thus, the effects of varying quantities of the vegetation as a whole — of varying productivity between stands — are eliminated from the measure. Another way to improve Euclidean distance as an index of inter-stand relationship would be to replace the original variables by an orthogonal system based on them, such as is provided by principal component analysis; one would thus avoid giving excessive weight to a highly correlated group of species just because they happen to be numerous. This is closely akin to the generalized measure of distance (MAHALANOBIS, 1936; see also RAO, 1952), which has been used in ecology by HUGHES (1954), HUGHES & LINDLEY (1955), and HARBERD (1962). It may be expressed as

$$D^2 = \sum_i \sum_j a_{ij}\, d_i d_j \qquad (63)$$

where d_i is the difference between the values for the i th species, $|x_i - y_i|$ in the notation of (62), and a_{ij} is an element of the dispersion matrix.

Distance measures may also be based on presence and absence of species, values of "0" or "1" replacing the usual quantitative measures (WILLIAMS et al. 1966). The Euclidean distance (62) is here simply the square root of the number of species not common to the two samples. Again, however, standardization or — better — replacement by an orthogonal uncorrelated system of variables is a desirable improvement.

5.3.4 SIMILARITY BETWEEN TWO SETS OF SAMPLES

To comparisons between two sets of samples – two association tables, for instance — much the same considerations apply as between a pair of samples. If the number of samples in the two sets differ, this may have marked effects where the comparisons are based on numbers of species, as in the JACCARD (39) and SØRENSEN (40) indices (DAHL 1957, MORISITA 1959). The effects of this have been reduced by limiting the comparison to species with > 10 % presence (BARKMAN 1958) or > 50 % constancy (NORDHAGEN 1943, DAHL 1957), or by using a fixed number of species (RAABE 1950). Quantitative measures are not subject to the same difficulties. Percentage similarity (50, 51) and related measures can be applied to presence or constancy values (DAHL 1957, FRYDMAN & WHITTAKER 1968, MAAREL 1969) or mean density values (WHITTAKER & FAIRBANKS 1958) of species in two sets of samples being compared. WILLIAMS et al. (1966) calculated the Euclidean distance between the centroids of the clusters of points representing the composition of the different samples in the set. BARKMAN (1958) used

$$A_2 = \frac{\sum_j \min(x_{1j}, x_{2j})}{\sqrt{\sum_{i \varepsilon T} x_{1i} \sum_{i \varepsilon U} x_{2i}}} \tag{64}$$

where x_{ij} is the mean % cover of the j th species in the i th set of samples, T is the set of species for which $x_{1j} > x_{2j}$, and U the set for which $x_{1j} < x_{2j}$.

ČEŠKA (1966) proposed an index based on SØRENSEN's presence similarity index (40), viz.

$$\frac{2 \sum_j C_{1j} \cdot C_{2j}}{\sum_j C_{1j} + \sum_j C_{2j}} \tag{65}$$

where C_{ij} is the proportion of samples within the i th set including species j.

In a subsequent paper (ČEŠKA 1968), he gave an alternative expression:

$$\frac{\sum_j C_{1j} \cdot C_{2j}}{\sum_j C_{1j} + \sum_j C_{2j} - \sum_j C_{1j} \cdot C_{2j}} \tag{66}$$

but gave no arguments in favour of it.

5.3.5 MULTI-SAMPLE SIMILARITY

In inter-sample comparisons, as in interspecific association, it is possible to generalize consideration to subsets larger than two.

But even less attention has been given to multi-sample comparisons than to multiple relations among species.

The problem here is of testing for clustering, or devising a cluster index — one which will express the similarity among a group of more than two vegetation samples. In numerical taxonomy (SOKAL & SNEATH 1963), considerable attention has been given to methods of recognizing clusters — groups of taxa which are more similar to one another than to others not within the cluster. Some of these techniques have also been applied in plant sociology; further information will be found in article 19. Most of the methods, however, are highly arbitrary and dependent upon the particular similarity index used. Significance of the clustering — whether it could be the result of random assortment of attributes — has rarely been considered.

ČEŠKA (1966, 1968) derived a measure of mean similarity within a set of samples from SØRENSEN's presence similarity index (40). It may be expressed as:

$$\frac{\sum a_i(a_i-1)}{(M-1)\sum a_i} \tag{67}$$

where the i th species occurs in a_i out of a total of M samples. This expression excludes self-comparisons.

HALL (1967a, b) proposed as a measure of heterogeneity (i.e. an inverse measure of similarity) among a group of samples

$$H_t = \frac{1}{S}\sum_{j=1}^{S}\frac{a_{jr}}{M} \tag{68}$$

where a_{jr} is the number of samples (out of a total of M) with the rarer state (presence or absence) for the jth out of S species. For quantitative data, this was replaced by

$$H_{qm} = \sum_{j=1}^{S}\frac{\sum_{i=1}^{M} x_{ij}}{\sum_{j=1}^{S}\sum_{i=1}^{M} x_{ij}}\left(1-\frac{\sigma_j}{\sigma_M}\right) \tag{69}$$

where x_{ij} is the value for the j th species in the i th sample, σ_j is the standard deviation of these data, and σ_M is the hypothetical maximum standard deviation if a group of M observations for this species were maximally heterogeneous — were divided as equally as possible between zero and the largest possible value (clearly this implies a type of measurement which has a natural upper bound).

MORISITA (1959) showed that his C_λ index of community similarity could be extended to comparisons among several community samples, provided the numbers of individuals were the same in each. WILLIAMS et al. (1966) used another information statistic as a measure of the differences among a group of quadrats, viz.

$$I = SM \log M - \sum_{j=1}^{S} [a_j \log a_j + (M - a_j) \log (M - a_j)] \tag{70}$$

where S species are recorded in M samples, and the j th species is present in a_j samples. As mentioned above, this reduces to (44) where only two samples are under comparison.

ORLOCI (1968) suggested that the overall disorder among c samples could be measured, for count or frequency data summarized in contingency tables, by the minimum discrimination information statistic:

$$I = \sum_i \sum_j x_{ij} \log x_{ij} + x_{..} \log x_{..} - \sum_i x_{i.} \log x_{i.} - \sum_j x_{.j} \log x_{.j} \tag{71}$$

where the cells of the table are x_{ij}, the row totals $x_{i.}$, the column totals $x_{.j}$, and the overall total $x_{..}$ Yet another information statistic:

$$\frac{1}{S} \sum_{j=1}^{S} \log_2 {}^{M}C_{a_j} \tag{72}$$

was proposed by GODRON (1966) as an inverse measure of similarity among samples. Where $M=2$, this is the complement of JACCARD'S similarity coefficient (39).

5.4 Conclusion

5.4.1 GENERAL PERSPECTIVES

In the first place, I repeat that there is nothing absolute about correlation between species — it is entirely context-dependent. Statements that a particular pair of species are positively correlated, negatively correlated, or uncorrelated may all be correct in different contexts.

Interspecific correlation means little or nothing if divorced from considerations of scale. When measured in small quadrats within a reasonably homogeneous community, it depends upon – and is indeed a good method of studying — the pattern of this community. If large quadrats are used in such a community, most interspecific correlations will disappear. This fact can be used at the same time for testing homogeneity and for determining the minimal area of the community, in a very useful sense of the term.

If one compares different communities by this size of quadrat or larger (or compares complete stands of different communities), interspecific correlations are generated by a different set of factors operating on a larger scale. They depend, not on the pattern within the community, but on the biotope differences leading to the establishment of the different communities, or resulting from them. These are the correlations which permit the recognition of ecological or sociological groups of species, so important for ordination and classification of vegetation and for indicating environmental conditions or successional status.

If small quadrats only are used in heterogeneous vegetation one risks confounding correlations from these two sources. The data will reflect both the pattern within each community, and the differences in composition between communities. These elements may be separated if correlations are computed separately for sets of quadrats within limited portions of the area covered and for comparisons between such sets, from which the effects of correlations within the sets can then be eliminated. But it may be simpler to use different sizes of quadrats for these two purposes.

The same remarks apply, *mutatis mutandis*, to correlations based on distance measurements — though in general these will be concerned with small-scale correlations. If one were interested in the larger-scale relationships, nearer neighbours would have to be ignored in order to avoid confounding these relationships with the effects of local pattern.

The preceding paragraphs assume a vegetational structure in which reasonably discrete communities are recognizable and transition zones can be ignored. Where variation in the vegetation is continuous, no "minimal area" can be found (GOODALL 1961), and satisfactory representation of interspecific relationships may then take the form of a continuous function expressing correlation in terms of quadrat size and interspace. For any given quadrat size, for instance, one might develop a transition probability function giving the probability of finding different quantities of the two species in one quadrat in terms of the known composition of another quadrat at a known distance from it. More simply, correlation in quadrats of a given size could be expressed as a continuous function of the interspace between them.

Apart from their use in ordination and classification, studies of interspecific correlation have often been concerned with significance tests. One wonders whether this emphasis may not be misplaced. Underlying any significance test is a null hypothesis -- which in this case is that the species in question are, at the scale tested, distributed independently of one another. But is this null hypothesis ever ecologically reasonable? One knows that the establishment and growth of every species are affected, to a greater or lesser extent, by every factor of the environment. Consequently, the null hypothesis is strictly incredible, and a test of it is pointless. Estimates of correlation should be the ecologist's aim rather than tests of their significance — though these estimates should be associated with intervals expressing their uncertainty due to sampling errors.

In the field, pair-wise correlations between species are usually artificial isolates from a more complex situation, better represented in terms of overlapping groups of ecologically related species ("ecological groups" or "sociological groups") or, better still, as a multidimensional continuum. If the main axes of variation within this continuum can be identified and the association of each species with each of the main axes (including possible non-linear relationships) can be defined, the correlations between pairs or groups of species follow directly. This type of analysis of vegetation is dealt with elsewhere (articles 2,6,8).

While correlation between species is a relationship between populations which can be sampled, and one can consider the correlation index obtained as an estimate of a true value obtaining between the populations, the situation is somewhat different for similarity indices between quadrats or stands. One has two actual, concrete quadrats existing at a fixed time and place, and the calculated similarity between them is not subject to sampling error. The quadrats, on the other hand, may themselves be regarded as samples of larger areas of vegetation, and in this case one may be interested in the way in which the error-free similarity indices between pairs may vary from pair to pair. This variation will in part be due to heterogeneity (at the scale of quadrat size used) within each of the vegetation areas sampled, in part to differences which would arise even if a completely homogeneous area were sampled. There have been some tests of the variation of similarity indices within actual areas of vegetation assumed homogeneous. But it might well be easier and more satisfactory to study this in artificial populations, the properties of which can be defined exactly, for in real-life vegetation complete homogeneity is probably never attained.

Similarity between stands need not be context-dependent as is interspecific correlation. It can be, and often is, defined in absolute terms. This is possible because the variables on which it is based – the species present – are a definite and enumerable set, at least when species absent from both stands are ignored. As pointed out above, though, to treat similarity as an absolute relationship between two stands does not square well with a common-sense approach, according to which two objects which may be regarded as similar within a very broad context are considered quite dissimilar when the focus narrows.

Unless similarity measurements have some ulterior purpose (such as ordination or classification), the weights given to the different species are purely arbitrary. As we have seen, many indices take account of species presence or absence only, which amounts to giving all species equal weight. This means that the chance presence of the rarer species in one or both samples will play a part in determining similarity which could be reasonably regarded as excessive. The same is true of indices using quantitative data standardized for each species.

More in accordance with ecological thinking are indices in which each species is weighted according to some quantitative measure of its contributions to the vegetation – cover, basal area, biomass, etc. Which of these quantitative measures is used will affect the value of the indices obtained, and is an arbitrary decision. Even more arbitrary is a decision, say, to give trees double weight compared with herbs. Weighting by some measure of species contribution which is of biological importance at least gives an index which can claim to be natural and meaningful. It could be argued, however, that this over-weights any group of highly-correlated species, the information from which includes a large share of redundancy. Such an argument could be answered by using as an inverse measure of similarity the Euclidean distances between points in an orthogonal system provided by principal components analysis – *ex hypothesi* uncorrelated.

Discussion of pair-wise similarity of stands leads naturally to the question of similarities between and within groups of stands. Some of the indices mentioned cannot be extended to cover similarities among a group of stands larger than two; for others, this extension would appear quite natural. For comparison *between* two groups of stands, though, any of the similarity indices available for pair-wise comparisons among stands can be used. If they are based on presence and absence of species, different numbers and sizes of samples in the two groups will affect the results, just as two stands of different size cannot very satisfactorily be compared by these indices. If, however, the index is based on quantitative measures of the species present, there is no reason why the mean of these values within each group should not be used for computation of inter-group similarity. These questions, however, are closely associated with the use of quantitative methods for classification of vegetation, and will be discussed further in article 19.

5.4.2 Choice of Indices

One of the purposes of this article is to display to the interested reader the wide range of indices available in the literature. There has at present, however been far too little informed comparison of their merits, and consequently choice among them is still largely a matter of taste. The different indices, however, mean different things. It is quite possible to find populations where the true values of the matching coefficient (12), say, are identical, while those of the IVERSEN coefficient (4) differ, and *vice versa*.

In choosing among the range of indices that have been proposed and used, the following guidelines may be suggested:

1. One is not usually concerned with statistical tests of the significance of correlation between species, or differences between samples, so much as with quantitative expression of these relationships.
2. Other things being equal, a simpler index is to be preferred.
3. Appreciation of the meaning of an index is easier if it has known upper and lower limits (e.g. +1 and −1, or +1 and 0); for interspecific association and correlation, it is also helpful if it has a known central value (for instance zero, where the limits are ±1) when the species are independently distributed.
4. Since the index will have to be estimated from samples its sampling properties are of some importance. It is desirable, in particular, that a sample estimate should be unbiased, and should be independent of sample size once a 'minimal area' is exceeded. Few sampling studies of association and correlation indices for use in ecology have been performed. The example in Table I shows that some indices – the matching coefficient (12), for instance – may be markedly biased in small samples, while others are not. Among the indices for which estimates are unbiased, choice would often depend on sensitivity in relation to sampling variance.

None of the indices of species association (those based on presence and absence) is, in general, independent of sample size. Consider two situations where samples of given size have expected proportional frequencies as follows:

		Species A				Species A	
		+	−			+	−
Species B	+	.2	.2	Species B	+	.7	.7
	−	.3	.3		−	.1	.1

If the sample size is doubled (the two halves of the sample being assumed independent of one another in respect of species occurrence) the frequencies become respectively

		Species A				Species A	
		+	−			+	−
Species B	+	.48	.16	Species B	+	.93	.03
	−	.27	.09		−	.03	.01

In the first case (where the species are unassociated) the expectation of the point correlation coefficient (15) is zero in each case, but the expectation of the IVERSEN index (4) is .29 and .53 respectively. In the second case, where there is a strong positive association, both indices differ with sample size, the expected values of the point correlation coefficient being .38 and .22 respectively, and those for the IVERSEN coefficient .78 and .94.

These considerations argue strongly for the use of quantitative data if interspecific indices independent of sample size are required. If, however, one is content to work within the framework of a single fixed sample size, the great simplicity of the indices based on presence and absence, and the ease of collecting the field data needed, make them attractive. This applies particularly to such an index as that of IVERSEN, but the fact that the point correlation coefficient has a well-defined expectation where the species are uncorrelated seems to make the slight extra effort in computing it worth while.

Where quantitative data are being used, expression (23) has the advantage of simplicity, but the product-moment correlation (24) may again be preferable because of its known zero expectation if the species are uncorrelated. (Further note: Product-moment correlation is most appropriate when positive and negative values express direction of slope of monotonic distributions of species in relation to the environmental gradients represented in a set of samples. The wider the range of environments and communities sampled, the more the curvilinear and non-monotonic features of species distributions may affect the measurement, and the less interpretable may be 'positive,' 'neutral,' and 'negative' correlations. The centred measures of species association and correlation should in general be most useful for narrow ranges of sample difference; for wide ranges of samples they may express relative distributional overlap of species less effectively than the non-centred measures — the Editor).

In selecting an index of similarity between stands similar considerations apply. Presence-and-absence indices (coefficients of community) like those of JACCARD (39) and SØRENSEN (40) are less affected by sample size than the interspecific indices mentioned above. The influence of sample size depends here on the form of the species-area curve in the two samples compared; if the proportion of common species is the same in different parts of the curve, then the value of the similarity index will be almost independent of sample size. Broadly, however, presence-and-absence indices should be chosen only if samples of the same size are to be compared. Since no special intuitive significance attaches to a null similarity (as it does to a null correlation), there seems little reason to prefer a centred index of similarity, and the SØRENSEN index has the merit of simplicity.

Where similarity is considered inadequately represented by species presence and absence, or where samples of different size are to be compared, an index based on measures of the quantities of the species present is to be preferred. Percentage similarity is an appropriate and simple index, either in its absolute form (50), or in the relative form (51) preferred by some authors – the choice depending on whether the total quantity of plant material present is regarded as a relevant consideration in assessing the similarity of two samples. Euclidean distance (62) is an alternative appropriate to the assumptions of some ordination techniques. Sample similarity measures are further discussed elsewhere in this volume – in their mathematical aspects in article 9, their effects on ordination performance in article 10.

If, for some special reason, significance tests are required, the choice of index will be different. For presence-and-absence data, interspecific association may be examined by the tests for 2×2 tables, either exact (3) or approximate (1); for quantitative data, the rank correlation coefficient seems best. For similarity, the probabilistic index (GOODALL 1966) may be used with either presence-and-absence or quantitative data.

It is admitted that indices not mentioned is this sub-section may have advantages, at least in some circumstances. No clear advantages have been demonstrated, though, and until this has happened the choice should go to the simplest index which appears to meet the particular need.

5.5 Summary

Similarity of distribution between species, and similarity in composition between samples, have much in common; they are in a sense opposite faces of the same coin. Both are of use in the ordination and classification of vegetation, and also for testing homogeneity of samples or sets of samples. Statistical tests for significance of differences can be applied to species distributions and composition of samples, but such tests are not usually appropriate. The problem is usually one of measuring relative similarity, rather than testing a null hypothesis.

A range of expressions are reviewed which espress relative similarity of species distributions. Some of these measure species association in terms of presence or absence in samples, with or without taking into account samples from which both species are absent. Other expressions of species correlation are based on quantitative measures of the contributions of two species to the different samples in which they occur. Measures of both types may be either non-centred (ranging from 1.0 to 0), or centred (ranging from +1.0 through 0 to −1.0 for positive, neutral, and negative distributional relationships). The measures can also be applied to groups of more than two species, and they then link with the concept of 'ecological groups'.

Various expressions of relative similarity of vegetation samples are also discussed. Some of these are based on presence and absence of species in samples; others are based on quantitative measures of the contribution of each species to the samples being compared. In both types of index, a distinction is drawn between measures based wholly on characteristics of a given pair of samples being compared, and measures which relate a given pair to one another within the context of a broader set of samples. Complements of sample similarity express relative difference of samples from one another in respect of floristic composition. The concept of similarity can be applied to sets of samples as well as to individual samples, and can be extended to relations between more than two entities, whether these entities are individual samples or sets of samples. Here similarity measurement impinges directly on the classification of vegetation.

Various merits and limitations affect the different similarity measures. Choices of measures may differ with research purposes and kinds of samples, but the following are suggestions of indices of wide utility, with abbreviations used in other articles and the numbers of the formulas in this article:

i) Species distributional similarity
- a) Species association (SA), from presence and absence data
  - Percentage co-occurrence (4), non-centred
  - Coefficient of point correlation (15), centred
- b) Species correlation (SC), from quantitative data
  - Percentage similarity of distribution (23), non-centred
  - Coefficient of correlation (24) or rank correlation, centred

ii) Sample similarity of composition
- a) Presence and absence data
  - Coefficient of community (CC) of Sørensen (40) and its complement, coefficient of difference (CD)

b) Quantitative data

Percentage similarity (PS), (50, 51) and its complement percentage difference (PD)

Euclidean distance (ED),(62).

REFERENCES

AGRELL, I., – 1945 – An objective method for characterization of animal and plant communities. *K. fisiogr. Sällsk. Lund Förh.* 15: 63—67.

BARKMAN, J. J., – 1958 – Phytosociology and Ecology of Cryptogamic Epiphytes, Including a Taxonomic Survey and Description of their Vegetation Units in Europe. Van Gorcum, Assen. 628 pp.

BARNARD, J. L., – 1964 – Marine Amphipoda of Bahia de San Quintin, Baja California. *Pacif. Nat.* 4: 55—139.

BEALS, E. W., – 1965a – Ordination of some corticolous cryptogamic communities in south-central Wisconsin. (Russ. summ.) *Oikos* 16: 1—8.

BEALS, E. W., – 1965b – Species patterns in a Lebanese Poterietum. *Vegetatio* 13: 69—87.

BEALS, E. W. & G. COTTAM, – 1960 – The forest vegetation of the Apostle Islands, Wisconsin. *Ecology* 41: 785—790.

BOURDEAU, P., – 1961 – L'outil statistique en écologie et sociologie végétales *Biométr.-Praxim.* 2: 193—216.

BRAY, J. R., – 1956 – A study of mutual occurrence of plant species. *Ecology* 37: 21—28.

BRAY, J. R. & J. T. CURTIS, – 1957 – An ordination of the upland forest communities of southern Wisconsin. *Ecol. Monogr.* 27: 325—349.

BYER, M. D., – 1970 – Interspecific plant association as influenced by three variables. *Ecology* 51: 103—112.

CANCELA DA FONSECA, J. P., – 1966 – L'outil statistique en biologie du sol. III. Indices d'intérêt écologique. *Revue Écol. Biol. Sol* 3: 381—407.

CASSAGNAU, P., – 1961 – Écologie du sol dans les Pyrénées centrales: Les biocénoses de Collemboles. Hermann, Paris. 235 pp.

ČEŠKA, A., – 1966 – Estimation of the mean floristic similarity between and within sets of vegetation ıelevés. *Folia geobot. phytotax. bohemoslovaca* 1: 93—100.

ČEŠKA, A., – 1968 – Application of association coefficients for estimating the mean similarity between sets of vegetational relevés. *Folia geobot. phytotax. bohemoslovaca* 3: 57—64.

CLAUSEN, J. J., – 1957 – A phytosociological ordination of the conifer swamps of Wisconsin. *Ecology* 38: 638—646.

COLE, L. C., – 1949 – The measurement of interspecific association. *Ecology.* 39: 411—424.

COLE, L. C., – 1957 – The measurement of partial interspecific association. *Ecology* 38: 226—233.

CRAWFORD, R. M. M., D. WISHART, & R. M. CAMPBELL, – 1970 – A numerical analysis of high altitude scrub vegetation in relation to soil erosion in the Eastern Cordillera of Peru. *J. Ecol.* 58: 173—191.

CULBERSON, W. L., – 1955 – The corticolous communities of lichens and bryophytes in the upland forests of northern Wisconsin. *Ecol. Monogr.* 25: 215—231.

CZEKANOWSKI, J., – 1909 – Zur differential Diagnose der Neandertalgruppe. *Korrespbl. dt. Ges. Anthrop.* 40: 44—47.

DAGNELIE, P., – 1960a – Contribution à l'étude des communautés végétales par l'analyse factorielle. (Engl. summ.) *Bull. Serv. Carte phytogéogr.* Sér. B. 5: 7—71, 93—195.

DAGNELIE, P., – 1960b – Quelques problèmes statistiques posés par l'utilisation de l'analyse factorielle en phytosociologie. *Gembloux Inst. Agron. Stat. Rech., Bull. Hors Sér.* 1: 430—437.

DAGNELIE, P., – 1962a – Étude statistique d'une pelouse à *Brachypodium ramosum.* III. les liaisons interspécifiques. Première partie. *Bull. Serv. Carte phytogéogr.* Sér. B, 7: 87—97.

DAGNELIE, P., – 1962b – Étude statistique d'une pelouse à Brachypodium ramosum. V. Les liaisons interspécifiques. Deuxième partie. *Bull. Serv. Carte phytogéogr.* Sér. B, 7: 149—160.

DAGNELIE, P., – 1965 – L'étude des communautés végétales par l'analyse statistique des liaisons entre les espèces et les variables écologiques: principes fondamentaux. *Biometrics* 21: 345—361.

DAHL, E., – 1957 – Rondane. Mountain vegetation in South Norway and its relation to the environment. *Skr. norske Videns-Akad.*, Mat.-naturw. Kl. 1956 (3): 1—374.

DAHL, E. & E. HADAČ – 1941 – Strandgesellschaften der Insel Ostøy im Oslofjord. *Nytt Mag. Naturv.* 82: 251—312.

DAWSON, G. W. P., – 1951 – A method for investigating the relationship between the distribution of individuals of different species in a plant community. *Ecology* 32: 332—334.

DICE, L. R., – 1945 – Measures of the amount of ecologic association between species. *Ecology* 26: 297—302.

DICE, L. R., – 1952 – Measure of the spacing between individuals within a population. *Contrib. Lab. vertebr. Biol. Univ. Mich.* 55: 1—23.

DUVIGNEAUD, P., – 1946 – La variabilité des associations végétales. *Bull. Soc. Bot. Belg.* 78: 107—134.

DUVIGNEAUD, P., – 1953 – Les savanes du Bas-Congo. Essai de phytosociologie topographique. *Lejeunia* 10: 1—92.

ELLENBERG, H., – 1950 – Landwirtschaftliche Pflanzensoziologie. I. Unkrautgemeinschaften als Zeiger für Klima und Boden. Ulmer, Stuttgart. 141 pp.

ELLENBERG, H., – 1956 – In, 'Einführung in die Phytologie,' by H. WALTER, Vol. 4, Part I. Aufgaben und Methoden der Vegetationskunde. Ulmer, Stuttgart. 136 pp.

ESTABROOK, F. F., – 1967 – An information theory model for character analysis. *Taxon* 16: 86—97.

FAGER, E. W., – 1957 – Determination and analysis of recurrent groups. *Ecology* 38: 586—595.

FAGER, E. W., – 1963 – Communities of organisms. In, 'The Sea', ed. M. N. HILL. Vol. 2: 415—437. Interscience, New York & London.

FALINSKI, J. B., – 1958 – Nomogramy i tablice współczynników podobieństwa między zdjęciami fitosocjologicznymi według wzoru Jaccarda i Steinhaus (Engl. summ.: Nomograms and tables for similarity coefficients). *Acta Soc. bot. pol.* 27: 115—130.

FIELD, G., – 1969 – The use of the information statistic in the numerical classification of heterogeneous systems. *J. Ecol.* 57: 565—569.

FIJALKOWSKI, D., – 1954 – Szata roślinna wawozów okolie Lublina na tle niektorych warunków siedliskowych. (Russ. & Engl. summs.: Vegetation of loess ravines near Lublin on the background of some environmental conditions) *Annls Univ. Mariae Curie-Skłodowska* Sect. B, 9: 125—215.

Forbes, S. A., – 1907 – On the local distribution of certain Illinois fishes: an essay in statistical ecology. *Bull. Ill. St. Lab. nat. Hist.* 7: 273—303.
Forbes, S. A., – 1925 – Method of determining and measuring the associative relations of species. *Science, N. Y.* 61: 524.
Fresco, L. F. M., – 1969 – Factor analysis as a method in synecological research. *Acta bot. neerl.* 18: 477—482.
Frey, T. E. A., – 1963 – On the application of the correlation coefficient of the species interrelations (In Russian). *Bot. Zhur. SSSR* 48: 235—239.
Frydman, I. & R. H. Whittaker, – 1968 – Forest associations of southeast Lublin Province, Poland. (Germ. summ.) *Ecology* 49: 896—908.
Gardner, J. L., – 1951 – Vegetation of the creosotebush area of the Rio Grande Valley in New Mexico. *Ecol. Monogr.* 21: 379—403.
Ghent, A. W., – 1963 – Kendall's 'Tau' coefficient as an index of similarity in comparisons of plant or animal communities. *Can. Ent.* 95: 568—574.
Gilbert, N., – 1963 – Non-additive combining abilities. *Genet. Res.* 4: 65—78.
Gilbert, N. & T. C. E. Wells, – 1966 – Analysis of quadrat data. *J. Ecol.* 54: 675—685.
Gleason, H. A., – 1920 – Some applications of the quadrat method. *Bull. Torrey bot. Club* 47: 21—33.
Godron. M., – 1966 – Application de la théorie de l'information à l'étude de l'homogénéité et de la structure de la végétation. *Oecol. Pl.* 1: 187—197.
Goodall, D. W., – 1952 – Quantitative aspects of plant distribution. *Biol. Rev.* 27: 194—245.
Goodall, D. W., – 1953 – Objective methods for the classification of vegetation. I. The use of positive interspecific correlation. *Aust. J. Bot.* 1: 39—63.
Goodall, D. W., – 1961 – Objective methods for the classification of vegetation IV. Pattern and minimal area. *Aust. J. Bot.* 9: 162—196.
Goodall, D. W., – 1965 – Plot-less tests of interspecific association. *J. Ecol.* 53: 197—210.
Goodall, D. W., – 1966 – A new similarity index based on probability. *Biometrics* 22: 883—907.
Goodall, D. W., – 1967 – The distribution of the matching coefficient. (French summ.) *Biometrics* 23: 647—656.
Goodall, D. W., – 1969 – Procedure for recognition of uncommon species combinations in sets of vegetation samples. *Vegetatio* 18: 19—35.
Goodman, L. A. & W. H. Kruskal, – 1954 – Measures of association for cross-classifications. *J. Am. statist. Ass.* 49: 732—764.
Goodman, L. A. & W. H. Kruskal, – 1959 – Measures of association for cross-classifications. II. Further discussion and references. *J. Am. statist. Ass.* 54: 123—163.
Gounot, M., – 1961 – Les méthodes d'inventaire de la végétation. *Bull. Serv. Carte phytogéogr.*, Sér. B, 6: 7—73.
Gounot, M., – 1969 – Méthodes d'étude quantitative de la végétation. Masson, Paris. 314 pp.
Gounot, M. & M. Calleja, – 1962 – Coefficient de communauté, homogénéité et aire minimale. *Bull. Serv. Carte phytogéogr.* Sér. B, 7: 181—200.
Greig-Smith, P., – 1952 – Ecological observations on degraded and secondary forest in Trinidad, British West Indies. II. Structure of the communities. *J. Ecol.* 40: 316—330.
Greig-Smith, P., – 1964 – Quantitative Plant Ecology. 2nd ed. Butterworths, London. 256 pp.
Hacker, H. P., – 1921 – On the distribution of anophelines in the Brickfields

Road area of Kuala Lumpur, Novembei, 1917. *Rep. Fed. Malay States Malaria Bur.* 2: 3—28.
HALL, A. V., – 1967a – Methods for demonstrating resemblance in taxonomy and ecology. *Nature, Lond.* 214: 830—831.
HALL, A. V., – 1967b – Studies in recently developed group-forming procedures in taxonomy and ecology. *Jl S. Afr. Bot.* 33: 185—196.
HALL, A. V., – 1970 – A computer-based method for showing continua and communities in ecology. *J. Ecol.* 58: 591—602.
HARBERD, D. J., – 1962 – Application of a multivariate technique to ecological survey. *J. Ecol.* 50: 1—17.
HOPKINS, B., – 1957 – Pattern in the plant community. *J. Ecol.* 45: 451—463.
HUGHES, R. E., – 1954 – The application of multivariate analysis to (a) problems of classification in ecology; (b) the study of the interrelationships of the plant community and environment. *8th Int. bot. Congr., Paris, Rapp. Commis.* Sect. 7 & 8: 16—18.
HUGHES, R. E. & D. V. LINDLEY, – 1955 – Application of biometric methods to problems of classification in ecology. *Nature, Lond.* 175: 806—807.
HURLBERT, S. H., – 1969 – A coefficient of interspecific association. *Ecology* 50: 1—9.
IVERSEN, J., – 1936 – Biologische Pflanzentypen als Hilfsmittel in der Vegetationforschung. Ein Beitrag zur ökologischen Charakterisierung und Anordnung der Pflanzengesellschaften. *Medd. Skalling-Lab.* 4: 1—224.
IVERSEN, J., – 1954 – Über die Korrelationen zwischen den Pflanzenarten in einem grönländischen Talgebiet. *Vegetatio* 5—6: 238—246.
JACCARD, P., – 1901 – Distribution de la flore alpine dans le Bassin des Dranses et dans quelques régions voisines. *Bull. Soc. vaud. Sci. nat.* 37: 241—272.
JUHÁSZ NAGY, P., – 1964 – Continuum studies on meadow vegetation. (Russ. summ.) *Acta bot. hung.* 10: 159—173.
KASSAS, M. & M. S. EL-ABYAD, – 1962/3 – On the phytosociology of the desert vegetation of Egypt. *Ann. arid Zone* 1: 54—83.
KENDALL, M. G., – 1955 – Rank Correlation Methods. 2nd ed. C. Griffin, London. 196 pp.
KERSHAW, K. A., – 1961 – Association and co-variance analysis of plant communities. *J. Ecol.* 49: 643—654.
KLEINER, E. F. & K. T. HARPER, – 1966 – An investigation of association patterns of prevalent grassland species in Red Butte Canyon, Salt Lake County, Utah. *Proc. Utah Acad. Sci.* 43: 29—36.
KOLBE, W., – 1956 – Korrelationsstatistische Probleme der Grünlandsoziologie. *Inaug-Diss, Landw. Fak., Rheinischen Friedrich-Wilhelms-Universität, Bonn.* 91 pp.
KONTKANEN, P., – 1950 – Quantitative and seasonal studies on the leafhopper fauna of the field stratum on open areas in North Karelia. (Finn. summ.) *Annls zool., Soc. Zool.-bot. fenn. Vanamo* (*Suomal. eläin-ja kasvit. Seur. van. eläin. Julk.*) 13 (8): 1—91.
KULCZYŃSKI, St., – 1928 – Die Pflanzenassoziationen der Pieninen. *Bull. int. Acad. pol. Sci. Lett.*, Cl. Sci. Math. Nat., sér. B. 1927 (Suppl. 2): 57—203.
LAMBERT, J. M. & M. B. DALE, – 1964 – The use of statistics in phytosociology. *Adv. ecol. Res.* 2: 59—99.
LOOMAN, J. & J. B. CAMPBELL, – 1960 – Adaptation of Sorensen's K (1948) for estimating unit affinities in prairie vegetation. *Ecology* 41: 409—416.
McDONOUGH, W. T., – 1963 – Interspecific association among desert plants. *Amer. Midl. Nat.* 70: 291—299.
McINTOSH, R. P., – 1962 – Pattern in a forest community. *Ecology* 43: 25—33.

MAAREL, E. VAN DER, – 1966 – Over vegetatiestructuren, -relaties en -systemen in het bijzonder in de duingraslanden van Voorne. *Thesis, Univ. Utrecht.* 170 pp.
MAAREL, E. VAN DER, – 1969 – On the use of ordination models in phytosociology. (Germ. summ.) *Vegetatio* 19: 21—46.
MAHALANOBIS, P. C., – 1936 – On the generalized distance in statistics. *Proc. natn. Inst. Sci. India* 2: 49—55.
MARGALEF, T., – 1958 – Distribución de los crustaceos en las aguas continentales españolas. Grado de associación entre las especies en relación con factores ecologicos e historicos. *Publnes Inst. Biol. apl., Barcelona* 27: 17—31.
MICHAEL, E. L., – 1920 – Marine ecology and the coefficient of association: a plea in behalf of quantitative biology. *Ecology* 8: 54—59.
MIYATA, I., – 1966 – Ecological studies on the vegetation of Akiyoshi-dai limestone plateau (2) Structure of the reserved forest developed on the upper slopes of Akiyoshi-do cave. *Jap. J. Ecol.* 16: 241—247.
MONK, C. D., – 1968 – Successional and environmental relationships of the forest vegetation of North Central Florida. *Amer. Midl. Nat.* 79: 441—457.
MORISITA, M., – 1959 – Measuring of interspecific association and similarity between communities. *Mem. Fac. Sci. Kyushu Univ.*, Ser. E (Biol) 3: 65—80.
MOTOMURA, I., – 1952 – Comparison of communities based on correlation coefficients (In Japanese). *Ecol. Rev.*, Sendai 13: 67—71.
MOZLEY, A., – 1936 – The statistical analysis of the distribution of pond molluscs in Western Canada. *Am. Nat.* 70: 237—244.
NEWBOULD, P. J., – 1960 – The ecology of Cranesmoor, a New Forest valley bog. I. The present vegetation. *J. Ecol.* 48: 361—383.
NORDHAGEN, R., – 1943 – Sikilsdalen og Norges fjellbeiter. En plantesosiologisk monografi. *Bergens Mus. Skr.* 22: 1—607.
NUMATA, M. & M. MITSUDERA, – 1960 – Discrimination of the rank of species in a plant community. *Biol. J. Nara Women's Univ.* 19: 57—62.
OCHIAI, A., – 1957 – Zoogeographical studies on the soleoid fishes found in Japan and its neighbouring regions. *Bull. Jap. Soc. scient. Fish.* 22: 526—530.
ODUM, E. P., – 1950 – Bird populations of the Highlands (North Carolina) Plateau in relation to plant succession and avian invasion. *Ecology* 31: 587—605.
ORLOCI, L., – 1966 – Geometric models in ecology. I. The theory and application of some ordination methods. *J. Ecol.* 54: 193—215.
ORLOCI, L., – 1967 – An agglomerative method for classification of plant communities. *J. Ecol.* 55: 193—206.
ORLOCI, L., – 1968 – Definitions of structure in multivariate phytosociological samples. *Vegetatio* 15: 281—291.
PANDEYA, S. A., – 1961 – On some new concepts in phytosociological studies of grassland. II. Community coefficient (F×C) ICC. *J. Indian bot. Soc.* 40: 267—270.
PIELOU, D. P. & E. C. PIELOU, – 1967 – The detection of different degrees of coexistence. *J. theoret. Biol.* 16: 427—437.
PIELOU, E. C., – 1961 – Segregation and symmetry in two-species populations as studied by nearest-neighbour relationships. *J. Ecol.* 49: 255—269.
PIELOU, E. C., – 1969a – Association tests versus homogeneity tests: their use in subdividing quadrats into groups. *Vegetatio* 18: 4—18.
PIELOU, E. C., – 1969b – An Introduction to Mathematical Ecology. Wiley-Interscience, New York. 286 pp.
RAABE, E. W., – 1950 – Über die 'Charakteristische Arten-Kombination' in der Pflanzensoziologie. *Schr. naturw. Ver. Schlesw.-Holst.* 24: 8—14.

Ramsay, D. M., – 1964a – An analysis of Nigerian savanna. II. An alternative method of analysis and its application to the Gombe sandstone vegetation. *J. Ecol.* 52: 457—466.

Ramsay, D. M., – 1964b – Some recent work on savanna vegetation-soil relationships. *Proc. 1st Nigerian For. Conf., Kaduna,* 156—159.

Rao, C. R., – 1952 – Advanced Statistical Methods in Biometric Research. John Wiley, New York; Chapman & Hall, London. 390 pp.

Raunkiaer, C., – 1928 – Dominansareal, Artstaethed og Formationsdominanter. *Biol. Meddr.*, K. dansk Vidensk. Selsk. 7 (1) :1—47.

Renkonen, O., – 1938 – Statistisch-ökologische Untersuchungen über die terrestrische Käferwelt der finnischen Bruchmoore. (Finn. summ.) *Annls. zool., Soc. Zool. -bot. Fenn. Vanamo* (*Suomal. eläin-ja kasvit. Seur. van. eläin. Julk.*) 6 (1): 1—231.

Renkonen, O., – 1944 – Die Carabiden- und Staphylinidenbestände eines Seeufers in SW Finnland. Ein Beitrag zur Theorie der statistischen Insektensynökologie. *Ann. ent. fenn.* (*Suom. hyont. Aikak.*) 10: 33—104.

Russell, P. F. & T. R. Rao, – 1940 – On habitat and association of species of anopheline larvae in south-eastern Madras. *J. Malar. Inst. India* 3: 153—178.

Ružička, M., – 1958 – Anwendung mathematisch-statistischer Methoden in der Geobotanik (Synthetische Bearbeitung von Aufnahmen). *Biológia, Bratisl.* 13: 647—661.

Sokal, R. R. & C. D. Michener, – 1958 – A statistical method for evaluating systematic relationships. *Kans. Univ. Sci. Bull.* 38: 1409—1437.

Sokal, R. R. & P. H. A. Sneath, – 1963 – Principles of Numerical Taxonomy. Freeman, San Francisco & London. 359 pp.

Sørensen, T., – 1948 – A method of establishing groups of equal amplitude in plant sociology based on similarity of species content. *Biol. Skr..*, K. danske Vidensk. Selsk. 5 (4): 1—34.

Tarwid, K., – 1960 – Szacowanie zbierzności nisz ekologicznych gatunków droga oceny prawdobieństwa spotykania sie ich w połowach (Engl. summ.: Assessment of the coincidence of ecological niches of species by estimating the probability of their concurrence in samples). *Ekol. pol.*, Ser. B, 6: 115—130.

Vasilevich, V. I., – 1963 – A tentative morphological analysis of the meadow continuum (In Russian), *Bot. Zhur. SSSR.* 48: 1653—1659.

Vasilevich, V. I., – 1969 – Statistical Methods in Geobotany (In Russian). Nauka, Leningrad. 231 pp.

Vasilevich, V. I., – 1970 – Partial linkages among species in a meadow community (In Russian). *Dokl. Akad. Nauk SSSR* 190: 1246—1249. *Dokl. Bot. Sci.* 190—192: 5—7.

Vries, D. M. de, – 1953 – Objective combinations of species. *Acta bot. neerl.* 1: 497—499.

Vries, D. M. de, – 1954 – Ecological results obtained by the use of interspecific correlation. *European Grassl. Conf.*, Paris, paper 224/53, 11 pp.

Vries, D. M. de, J. P. Baretta, & G. Hamming, – 1954 – Constellation of frequent herbage plants based on their correlation in occurrence. *Vegetatio* 5: 105—111.

Welten, M. & O. Hegg, – 1958 – Vorschlag für die statistiche Erfassung pflanzensoziologischer Zusammenhange. *Verh. schweiz. naturf. Ges.* 1958: 127—129.

Whittaker, R. H., – 1952 – A study of summer foliage insect communities in the Great Smoky Mountains. *Ecol. Monogr.* 22: 1—44.

Whittaker, R. H., – 1960 – Vegetation of the Siskiyou Mountains, Oregon and California. *Ecol. Monogr.* 30: 279—338.

WHITTAKER, R. H., – 1967 – Gradient analysis of vegetation. *Biol. Rev.* 42: 207—264.
WHITTAKER, R. H. & C. W. FAIRBANKS, – 1958 – A study of plankton copepod communities in the Columbia Basin, southeastern Washington. *Ecology* 39: 46—65.
WILLIAMS, C. B., – 1944 – Some applications of the logarithmic series and the index of diversity to ecological problems. *J. Ecol.* 32: 1—44.
WILLIAMS, C. B., – 1947a – The logarithmic series and the comparison of island floras. *Proc. Linn. Soc. Lond.* 158: 104—108.
WILLIAMS, C. B., – 1947b – The logarithmic series and its application to biological problems. *J. Ecol.* 34: 253—272.
WILLIAMS, C. B., – 1949 – Jaccard's generic coefficient and coefficient of floral community in relation to the logarithmic series and the index of diversity. *Ann. Bot.*, N. S. 13: 53—58.
WILLIAMS, C. B., – 1964 – Patterns in the Balance of Nature and Related problems in Quantitative Ecology. Academic Press, London & New York. 324 pp.
WILLIAMS, W. T. & J. M. LAMBERT, – 1961 – Multivariate methods in plant ecology. III. Inverse association-analysis. *J. Ecol.* 49: 717—729.
WILLIAMS, W. T., J. M. LAMBERT, & G. N. LANCE, – 1966 – Multivariate methods in plant ecology. V. Similarity analyses and information-analysis. *J. Ecol.* 54, 427—445.
YARRANTON, G. A., – 1966 – A plotless method of sampling vegetation. *J. Ecol.* 54: 229—237.
YATES, F., – 1934 – Contingency tables involving small numbers and the χ^2 test. *Jl. R. statist. Soc., Suppl.* 1: 217—235.

6 MATRIX AND PLEXUS TECHNIQUES

Robert P. Mc Intosh

Contents

6 MATRIX AND PLEXUS TECHNIQUES

6.1 Introduction

For several decades ecologists have concerned themselves with devising intelligible and mutually compatible methods of describing, analyzing and synthesizing the often bewildering complexity of plant and animal communities. GAMS (1961) and WHITTAKER (1967) consider many of these. Similar techniques have been developed or adapted, independently of ecologists, by others concerned with the analysis of character variation in diverse objects of study. CROVELLO (1970) has provided an enlightening and useful compendium and codification of the many methods of character variation analysis which have been used in ecology and systematics. This paper is concerned with some of the relatively simple techniques of arrangement and graphic representation which may reveal relationships between community samples (and their species).

6.2 Primary and Secondary Matrices

6.2.1 COMMUNITY TABLES

Phytosociologists traditionally have worked with a two-dimensional table consisting of lists of species, of measured quantities of species, or of environmental measurements, for community samples. The table may include data collected in a series of sample quadrats in one stand of a community, or in different stands of the same or of different community-types. The table thus represents some range of variation in characteristics of the samples; the range may be narrow or wide. Each column (object) of the table is a sample plot list or stand record (the relevé of the BRAUN-BLANQUET school of phytosociology). The rows (characters) correspond either to species or to environmental characteristics. The phytosociological table forms a primary matrix from which the ecologist hopes to synthesize a description of the pattern of variation in nature and, subsequently, to explain it. Description of pattern may be approached through recognition of groups of similar samples (classification), or directions (gradients or axes) of variation interrelating the samples (ordination).

The traditional uses of such tables in various schools of ecology are reviewed by WHITTAKER (1962), while BECKING (1957) and ELLENBERG (1956) give accounts of the methods of the most widely known of these, the BRAUN-BLANQUET school (cf. BRAUN-BLANQUET 1951, POORE 1955, DAGNELIE 1961, and article 20 in this volume). The basic approach (used also by other disciplines with multiple lists to order) is to rearrange the sequence of the plot lists or relevés, both horizontally and vertically, to form blocks of characteristic or differentiating species (or combinations of species and site characteristics) which best distinguish vegetational units by their occurrence in many or all of the samples representing a unit. In the process, some plots may be excluded. The final arrangement commonly is subjective, based upon the experience of the phytosociologist. Two major difficulties of such approaches to interpreting the primary matrix are that the arrangements of the matrix may, and commonly do, differ from botanist to botanist and that, in some instances, the stands sampled or relevés included in the primary matrix are subjectively selected by some predetermined criterion. Visual rearrangements of a primary matrix of taxonomic characters have been used by systematists in similar ways (SOKAL & SNEATH 1963, CARMICHAEL & SNEATH 1969). Whether such tables are used casually, as many ecologists have done, or in the more formalized manner of the BRAUN-BLANQUET school, a major problem is the specification of the characters (characteristic, constant, diagnostic, or dominant species) or other attributes of the objects (plot lists or relevés) which are to be used to differentiate the community-types in the primary matrix (WHITTAKER 1962, POORE 1955). MOORE et al. (1967) developed a computer method for objectively arranging phytosociological tables of presence-absence data which was designed to minimize the presence-absence exchanges and to provide an objective sorting of the relevés.

6.2.2 SIMILARITY INDICES

The difficulty in comparing visually with one another numerous lists representing different sample plots or communities has led to use of numerical indices. There are several such numerical coefficients or indices which attempt to measure objectively the relative similarity (affinity, association, correlation, or conversely, dissimilarity, difference, distance) between pairs of samples or communities as columns of the primary matrix, or between species or environmental characteristics as rows of the primary matrix. These numerical coefficients are not unique to ecology, and similar coefficients based on relative overall similarity have been developed and used in systematics, psychology and other disciplines. SOKAL & SNEATH (1963) and BOYCE (1969) reviewed various coefficients used for assessing phenetic similarity in numerical taxonomy and those (some identical) used in ecology are discussed by COLE (1949, 1957), CLAUSEN (1957a), BECKING (1957), DAHL (1957), MORISITA (1959), DAGNELIE (1960), GREIG-SMITH (1964), WHITTAKER (1967), ČEŠKA (1968), HURLBERT (1969), and GHENT (1972).

ORLOCI (1972) compares distance, probability, and information coefficients and provides arithmetic examples of their calculation. Among the numerous coefficients which have been used are coefficients of community (CC), percentage similarity (PS), Euclidean distance (ED), species association using presence and absence data (SA), and species correlation using quantitative data (SC). GOODALL reviews these in the preceeding article of this volume so they are not discussed or evaluated here. It is important to observe, however, that different coefficients differently assess the relationships of samples or of species to one another, and that choice of the coefficient may influence the results of the study. There is often advantage in seeing how more than one coefficient expresses the relationships; in particular it may be desirable to compare results with coefficients using quantitative data or presence and absence only (PS or ED vs. CC, and SC vs. SA).

6.2.3 SECONDARY MATRICES

The similarity coefficient, in effect, reduces the comparison between two samples or species to a single numerical value or to a point in a multidimensional character space. Similarity coefficients (or their complements, distance measures) may be calculated between all pairs of objects (sample plots, relevés, community-types), or between all pairs of characters (species, environmental characteristics) in the primary data matrix. The similarity coefficients are arranged in new data arrays in tables we may term here secondary matrices. Two secondary matrices (a 'Q' matrix of similarities between samples as objects, and an 'R' matrix of similarities between species or environmental characters) may be derived from a given community table as primary matrix, and similar kinds of analyses can be applied to the two secondary matrices.

The matrix may be used in various ways to elucidate the relations between the objects (or characters), since it contains a similarity value for each object of the original matrix with every other object. The problem of the ecologist is to find order (structure, pattern) in the matrix representative of the order in the natural system from which it was abstracted. He seeks to identify orders or clusters of the samples, communities, species or environmental attributes which reveal a pattern of relationships among them. The secondary matrix of similarity coefficients is the basic substrate from which many different techniques attempt to extract relationships (Mc Intosh 1967b, Whittaker 1967, Crovello 1970). In this paper we shall consider further only techniques based on rearrangements of the secondary matrix, and 'plexus' methods using graphical representation of relationships in the secondary matrix by points and lines in a two- or three-dimensional space.

6.3 **Matrix Arrangement**

Unless the object or characters of the primary data matrix were ordered in some way, there will be no evident pattern in the secondary similarity matrix.

In some cases the matrix may be simply inspected and the incidence and degree of significant similarities, or their absence, noted (Greig-Smith 1952, Whittaker 1952, Shanks 1953, Jones 1955, Préscényi 1958, Kershaw 1959, 1960, Ono 1961, Dagnelie 1962, Cook & Hurst 1963, Mc Donough 1963, Beaman & Andresen 1966). In the effort to clarify relationships the ecologist may choose one of several additional ways of examining or reordering the secondary matrix.

6.3.1 Sample and Species Ordering

Sanders & Hessler (1969) ordered a matrix of faunal similarity coefficients (CC) on an external criterion, latitude, (Fig. 1) using a common method of differential shading of similarity classes to make for easy visualization of the pattern. They subjectively interpreted the diagonal pattern from upper left to lower right as an indication that high faunal indices are always with neighboring stations and that there is a gradual and continuous change along the transect.

A more familiar method is to rearrange the secondary matrix on internal criteria, by trial and error, to bring the highest similarity values nearest the diagonal. This method, essentially an ordering of the matrix, was apparently originated independently by Polish

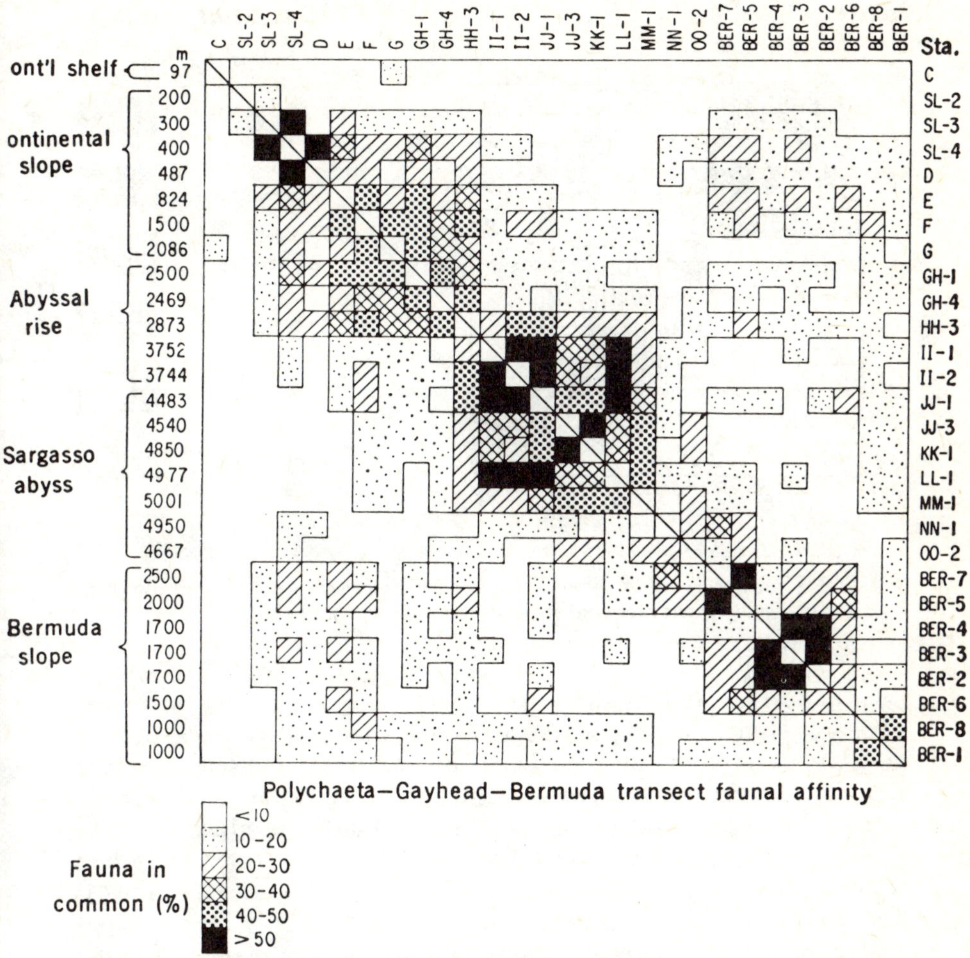

Fig. 1. A matrix diagram showing faunal similarity among sampling stations for marine bottom communities at different latitudes on a north-south axis (SANDERS & HESSLER 1969, Fig. 3).

and Danish ecologists and has been widely used by ecologists, systematists and psychologists. An extreme individual sample of a matrix is located on one end of the sequence and the sample having the highest similarity measure with it is located next to it, the third having maximum similarity to the second is then located next and the process continued up to the last sample which, presumably, has the minimum similarity value with the first. The numbers in each element of the resulting matrix may be replaced with a coded scale of shadings the highest generally being darkest (SOKAL & SNEATH

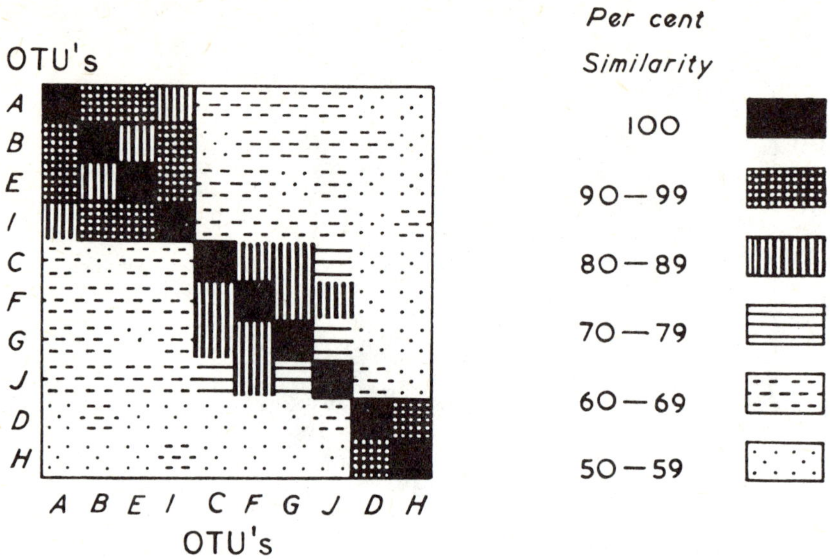

Fig. 2. A symmetric matrix diagram showing high similarity classes on the principal diagonal (SOKAL & SNEATH 1963, Fig. 7—4).

1963). By rearranging the numbers or shaded squares and trying different choices of the first sample, the matrix is rearranged until the highest values, or darkest shadings, are moved toward the diagonal placing the plots with high similarity values close to each other. Fig. 2 is an idealized symmetrical matrix produced by this method. Matrices of this type may be constructed ordering species as well as samples or stands (GREIG-SMITH 1952, GILBERT & CURTIS 1953, BRAY 1956, WHITTAKER & FAIRBANKS 1958). The ordered matrix or 'trellis diagram' has been widely used by plant ecologists; (KULCZYŃSKI 1928, MATUSZKIEWICZ 1948, SØRENSEN 1948, MOTYKA et al. 1950, GREIG-SMITH 1952, GILBERT & CURTIS 1953, GUINOCHET 1955, HALE 1955, BRAY 1956, CLAUSEN 1957a, DAHL 1957, GUINOCHET & CASAL 1957, DIX & BUTLER 1960, BRODO 1961, RUŽICKA 1961, STRUIK & CURTIS 1962, IZDEBSKI 1962, JUHÁSZ-NAGY 1963, YARRANTON 1967, YARRANTON & BEASLEIGH 1968, VASILEVICH 1969); animal ecologists, particularly entomologists, (RENKONEN 1938, 1949, MACFADYEN 1954, 1963, KONTKANEN 1957, FAGER 1957, WHITTAKER & FAIRBANKS 1958, BALOGH 1958, AGRELL 1963, HAGEMEIER 1966) and systematists (SOKAL & SNEATH 1963, MOSS 1967, TAYLOR & CAMPBELL 1969). Sometimes, especially in psychology, the matrix is arranged following a factor analysis by grouping the elements which are highly correlated with each of the principal factors in sequence. This places the high correlations along the diagonal (T. Crovello, pers. comm.)

Commonly, the matrix is rearranged by inspection to produce orders or clusters. Visual rearrangement of the matrix, however, is always subject to the subjective interpretation of the arranger and large matrices become very unwieldy, so a number of methods of evaluating similarity matrices in objective and repeatable fashion have been proposed (Sørensen 1948, Renkonen 1949, Clausen 1957a, Fager 1957).

A simplified method, devised by J. R. Bray, described by Clausen (1957a, b) used a partial matrix of the similarity values of each of the samples (or species) of the complete matrix with the two extremes (the samples related by the lowest similarity value in the full matrix). The extreme or end-point samples are placed at the ends of a line, and each of the other elements is placed along this line by the formula

$$X = \frac{ICC + CRCC}{2}$$

Where: $X =$ number of units along the line from the first extreme community.

$ICC =$ the similarity coefficient value of the element with the second extreme community.

$CRCC =$ maximum similarity coefficient of the incomplete matrix minus RCC.

$RCC =$ the similarity coefficient value of the element with the first extreme community.

The samples are arranged in sequence by values of X to form an ordered matrix in which, again, high similarity values should lie on or near the diagonal.

This method suffered from the difficulty of later methods of ordination using extreme end stands that, if these stands were very different, the remainder clustered toward the center. The positions of the communities were also weighted only by their similarity with the extremes and not with each other. Despite the difficulties, this technique (or that of placing most similar samples next to one another) will in some cases produce an effective ordination of the samples along a major environmental complex-gradient. Such is most likely to be the case if one environmental gradient overshadows all others in determining variation among the samples. Whether or not the arrangement is an effective ordination, it may be of interest to observe the groups of closely related samples that may be revealed in it.

6.3.2 Sample and Species Grouping

Sørensen (1948) arranged the elements of the secondary matrix into groups such that each member resembled all others of the group at an arbitrarily selected value of the similarity coefficient (CC). The smallest (most similar) groups had coefficient values > 50. Groups so defined could be assembled into larger, less restrictive, groups with lower values of the similarity coefficient, forming a hierarchy (Sørensen 1948, Hosakawa et al. 1957). Clausen (1957a) compared the results of Bray's and Sørensen's methods, preferring the latter, and Kontkanen (1957) compared Sørensen's method with Renkonen's (1938) and also prefered Sørensen's. Looman & Campbell (1960) criticized Sørenson's method because of his choice of an arbitrary value of the coefficient for establishing groups and developed a new coefficient by combining Sørenson's with Kendall's rank correlation coefficient (*Tau*). This is used to calculate an expected value of the similarity coefficient, based on the smallest number of species in common, which would be required if two sites are to be considered similar at the .01 level of significance. They use a grouping based on this coefficient to validate the results of a conventional association-table analysis. Ramsey (1964) and Greig-Smith (1964) questioned their detailed results because they referred to a chi-square table with two degrees of freedom rather than to the one appropriate to a 2×2 table, but the method has the merit of providing a statistical measure of significance.

Fager (1957) developed a method of producing 'recurrent groups' of species from a similarity matrix (SA). He defined a recurrent group as one in which:

1. The coefficient of similarity is significant at the .05 level for all pairs of species within the group.
2. The group includes the greatest possible number of species.
3. If several groups with the same number of members are possible, those are selected which will give the greatest number of groups without members in common.
4. If two or more groups with the same number of species and with members in common are possible, the one which occurs as a unit in the greatest number of samples is chosen.

Fager & Mc Gowan (1963) identified recurrent groups of zooplankton species; Davies (1971) identified groups of algae in beach sands using Fager's method.

Dahl (1957) used matrices and clusters based on the Sørensen coefficient (CC) to verify the results of a subjective analysis of association tables according to the methods of Continental ecological schools. Results may be subject to dispute as RENKONEN (1949) points out. Different methods may agree in broad features, which suggests some basic regularity. RENKONEN also warns against blind dependence on a statistical method and urges that the whole primary data base be available so that it can be used for comparisons as new and, hopefully, better methods of analysis become available (*cf.* CROVELLO 1970). Given the expense it may be impossible to publish entire primary data matrices, but these should be deposited in appropriate repositories for the use of later investigators. SOKAL & SNEATH (1963) discussed other methods of identifying clusters in a matrix which have not been used in ecology. MOSS (1967) used a shaded matrix (or 'trellis diagram') approach to compare results of matrices derived from correlation coefficients (r) and the Euclidean distance measure at three levels of similarity. There have been few comparative studies of the several methods described and these are only the simplest and most graphic ways in which patterns may be sought in a matrix of similarity coefficients. One difficulty is that these similarity or distance coefficients may be influenced by the diversity of the communities being compared. It is possible for two stands that share all or many of their species to have a greater interstand distance than two other stands having no species in common (MC INTOSH 1967a, NEWSOME & DIX 1968, BANNISTER 1968, GHENT 1972, ORLOCI 1972).

Objective comparisons of clusters or ordinations of ecological data produced by matrix methods have not usually been attempted. CROVELLO (1969) reviews several summary statistics for objective comparison of different analyses used in numerical taxonomy. The cophenetic correlation coefficient (SOKAL & SNEATH 1963) compares the similarity coefficients and the cophenetic values which are a matrix of coded similarity values extracted from the phenogram (diagram of the clusters) constructed from the similarity matrix. A perfect correlation shows that the original matrix is faithfully reproduced by the phenogram. The correlation between two or more similarity matrices may be calculated and the correlation between two or more cophenetic value matrices may also be useful, as CROVELLO suggests, in providing additional support for visual methods of comparing similarity matrices or patterns of relationship abstracted from them.

6.4 Plexus Techniques

It is natural to use a matrix of similarity coefficients to construct a graphic representation of the relations between the objects or characters. Samples or species then become points in a spatial pattern representing relative similarity by the arrangement of these points. The points may or may not be connected by lines. Quantitative relationships of the samples may be indicated by the spacing of the points, or by making the length or thickness of the lines proportional to the degree of similarity or dissimilarity (distance). Similarity coefficients are usually transformed because their values are inverse to measures of distance; these similarities are often subtracted from 100 %, or 85 %, or some other maximum value. The diagrams are variously known in ecology as dendrites (Faliński 1960, Izdebski 1963, Vasilevich 1969); pleiades or constellations (Vries 1953, Vries et al. 1954); correlata (Juhász-Nagy 1963); plexuses (Whittaker & Fairbanks 1958, Whittaker 1967) or, more generally, graphs.

6.4.1 Sample and Species Grouping

A plexus diagram is a logical extension of the traditional phytosociological concept of groups or 'associations' of species (or stands) with strong ecological ties with one another or to a common environment. As ecologically related species are expected to be spatially proximate on the ground, and related stands are expected to occupy similar environments, so they should be represented close together in abstract vegetation space. Plexus diagrams may be used to identify groups of associated or correlated species, or clusters of similar samples or stands, or to relate to one another groups of species and stands (Hopkins 1957, Whittaker & Fairbanks 1958). The latter approach to classification resembles Lambert & Williams, (1962) identification of noda from a matrix of chi-square values as species as species-in-habitat units.

The existence of natural groupings of species (recurrent arrays, imbricated ecological groups, iterative groups, associations) is a common premise which has permeated both plant and animal ecology. But there has been little agreement on the identity, or the methods of definition and delimitation, of these natural groups. A clear natural group would be one in which each member has greater affinity for all other members of the group than for any nonmember, producing a marked discontinuity between groups. Commonly, biological groups are not so clearly circumscribed, and manifestly, if the criteria for recognition of the group are broadened the scope and discreteness of the group will change (Sørensen 1948, Wirth et al. 1966, Gould & Garwood 1969). In some uses a simple linkage by any criterion is considered

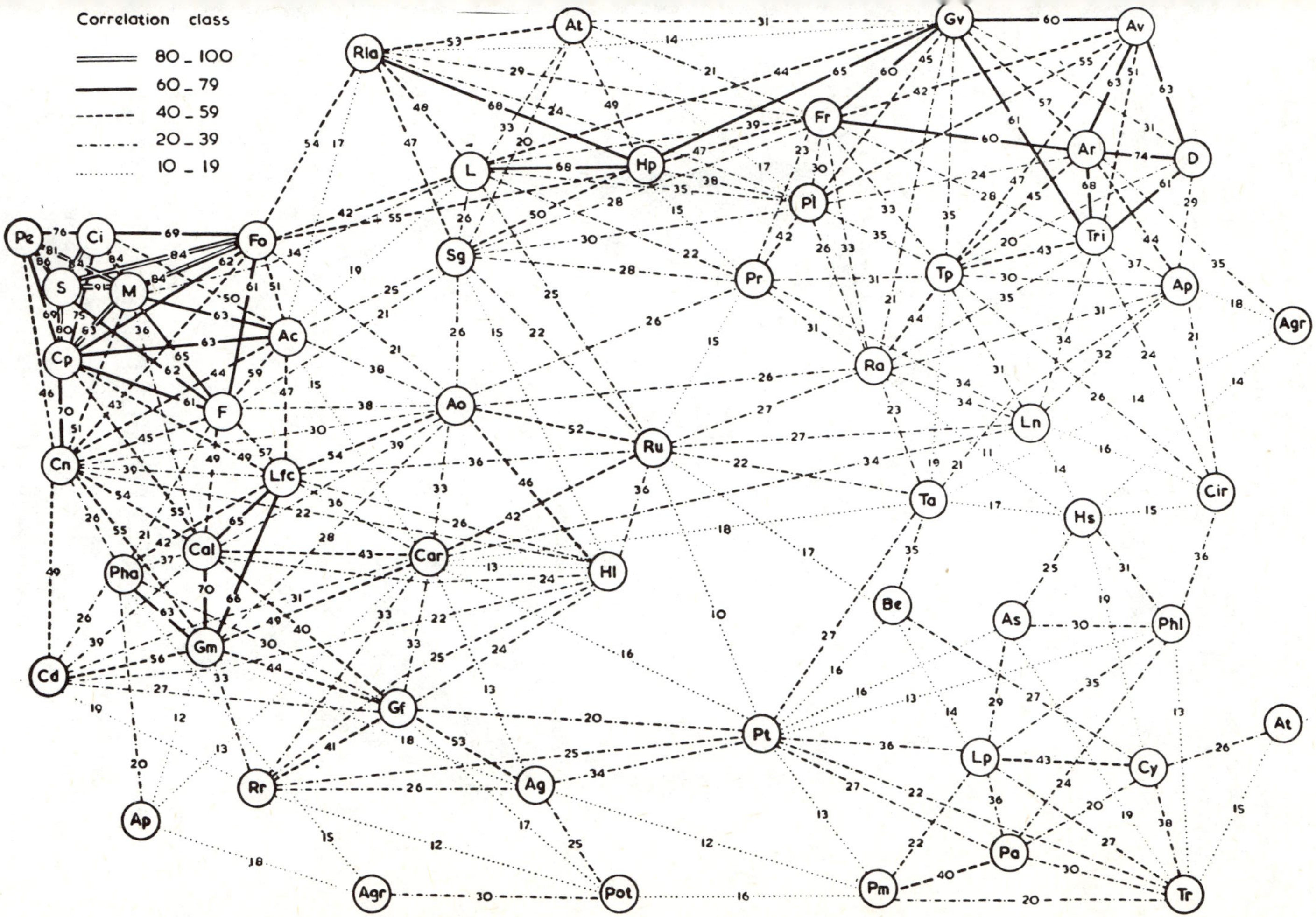

Fig. 3. A constellation or plexus diagram of species in Dutch grasslands based on correlation coefficients × 100 for species co-occurrence (VRIES et al. 1954, Fig. 2). Drought-tolerant species appear in the upper part of the figure, with calciphile species to the right and acidiphile species to the left. Mesic grassland species are on the lower right, and wet-soil species on the left of the figure.

sufficient to include a species as a member of a group while a more restrictive usage (complete linkage) requires that it be linked with each member of the group (SOKAL & SNEATH 1963). In some single linkage groups the extreme members may be less similar to each

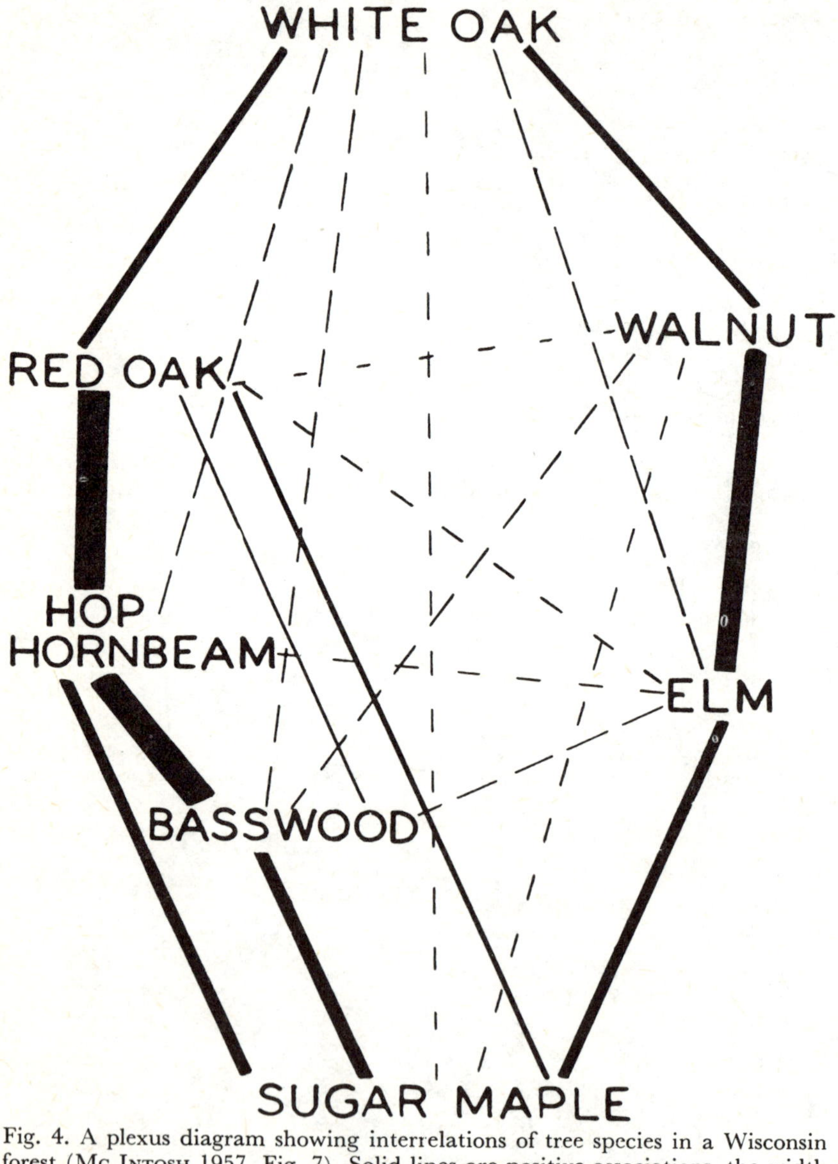

Fig. 4. A plexus diagram showing interrelations of tree species in a Wisconsin forest (MC INTOSH 1957, Fig. 7). Solid lines are positive associations, the width proportional to the magnitude, dashed lines are negative correlations, the space proportional to the magnitude.

other than either is to a nonmember of the group, whereas completely linked groups are more compact.

VRIES (1953) and VRIES et al. (1954) produced a plexus or constellation figure from a matrix of similarity coefficients (SA) by subjectively placing species with high positive correlation close together, and negatively correlated species far apart (Fig. 3). Because it was not possible to place the species correctly in two dimensions, classes of correlation were indicated by different types of lines. VRIES et al. recognized various areas of the diagram and clusters of species as being related to different site conditions. Similar methods relating species or plots have been used by (MACAN 1954, MC INTOSH 1957, HOSOKAWA et al. 1957, BARKMAN 1958, 1965, AGNEW 1961, QUARTERMAN & KEEVER 1962, JUHÁSZ-NAGY 1963, LOOMAN 1963, RAMSEY & DE LEEUW 1964, RAMSEY 1964, BEALS 1965, KLEINER & HARPER 1966, GODRON 1967, YARRANTON 1967, FRYDMAN & WHITTAKER 1968). Commonly only positive coefficients were used, but in some (Fig. 4) an attempt was made to illustrate negative values as well (MC INTOSH 1957).

The plexus diagrams most commonly used showed a pattern of species linked by positive similarity coefficients (WELCH 1960, AGNEW 1961, JUHÁSZ-NAGY 1963). The diagrams were usually interpreted as suggesting groups or clusters of related species as represented by positions in a single plane (Fig. 5). LOOMAN (1963) arranged species (by SA) in three dimensions and photographed the plexus to show relationships that could not be represented on a plane (Fig. 6). HEGG (1965) arranged species (by SA) in a plexus on which linkage-groups of species (ecological groups) on two levels are encircled (Fig. 7). The linkage-groups of the outer ring characterize lower-level community-types, those of the inner ring higher-level community-types, and each of the inner groups is linked with a number of the outer groups. Many species did not fit into the linkage-groups but could be given positions in the pattern of relationships the plexus represents. HEGG's treatment made possible both effective relation of species association to community classification, and indication of a number of directions of community relationships in two dimensions by the radial arrangement. Patterns of relationships of the plexus diagram are commonly interpreted, and 'groups' recognized subjectively, by occurrence of more links or higher coefficients among members of a group (AGNEW 1961, LOOMAN 1963, RAMSEY & DE LEEUW 1964). The subjectivity of recognizing groups in a maze of links of different degrees is a basic limitation on the use of plexuses for classification.

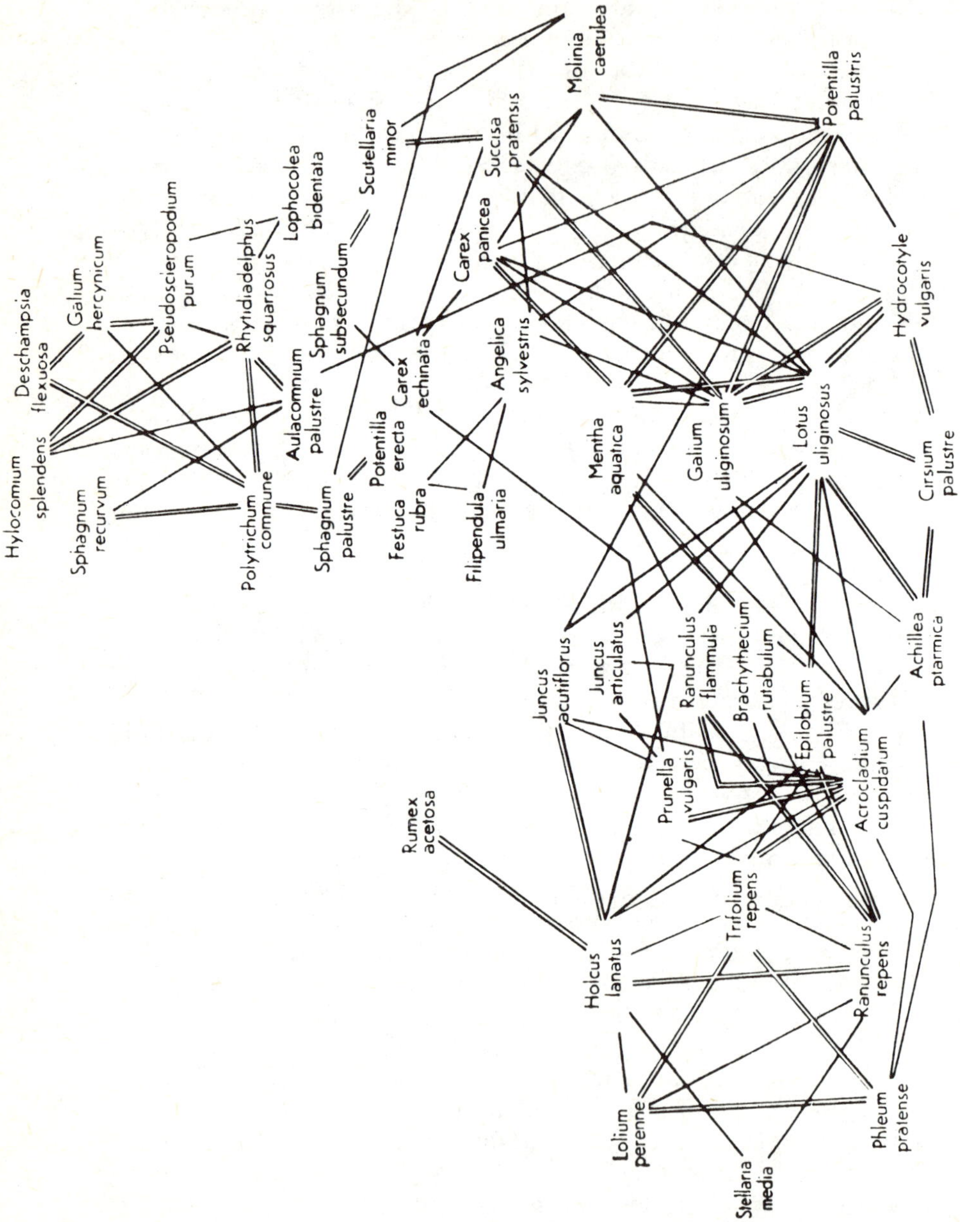

Fig. 5. A typical plexus diagram of positive correlations between species in pastures with *Juncus effusus* (Agnew

This evident subjectivity may be a reason plexuses have not been more widely used as an aid to the classification of community samples, though such use is feasible in some cases. The Polish phytosociologists, who were among the early users of graphical methods,

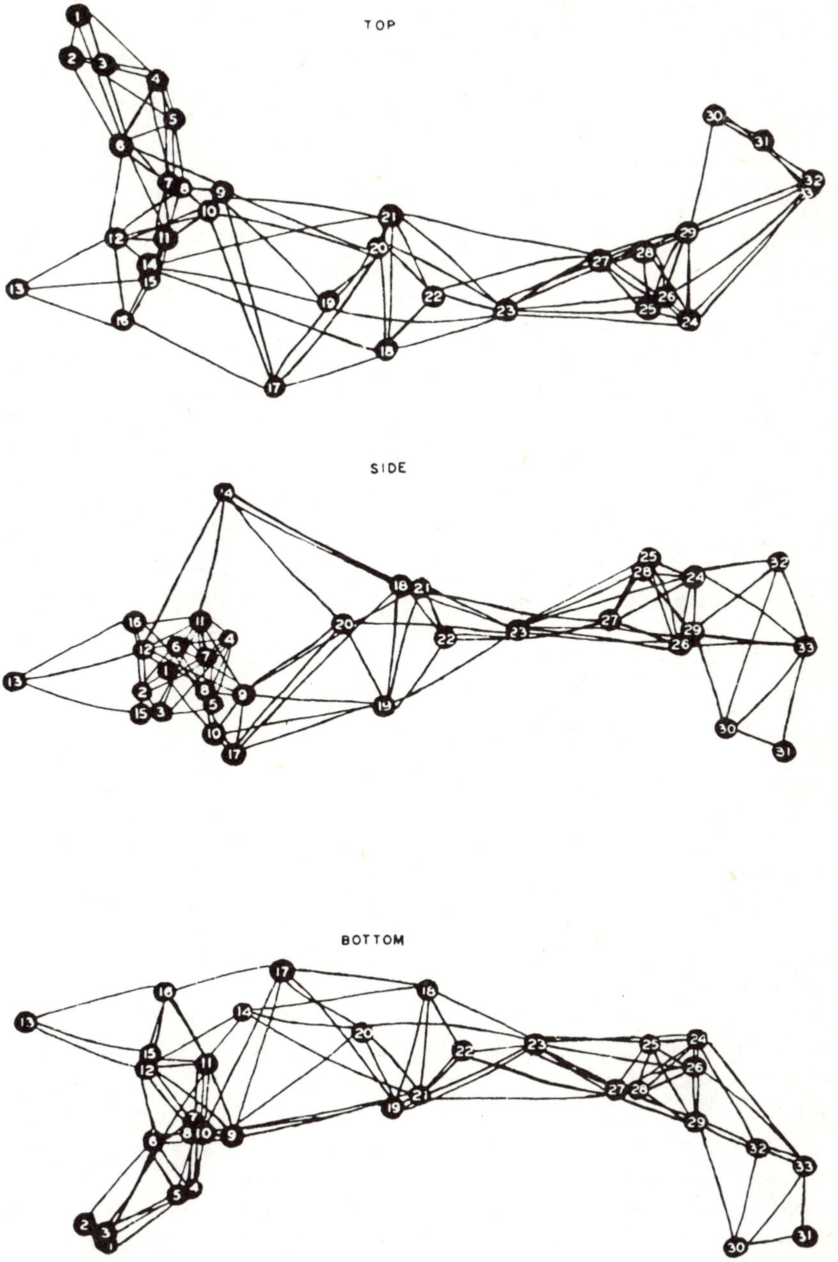

Fig. 6. A plexus or constellation of grassland species in Saskatchewan, a three dimension model photographed from three sides (LOOMAN 1963, Fig. 1).

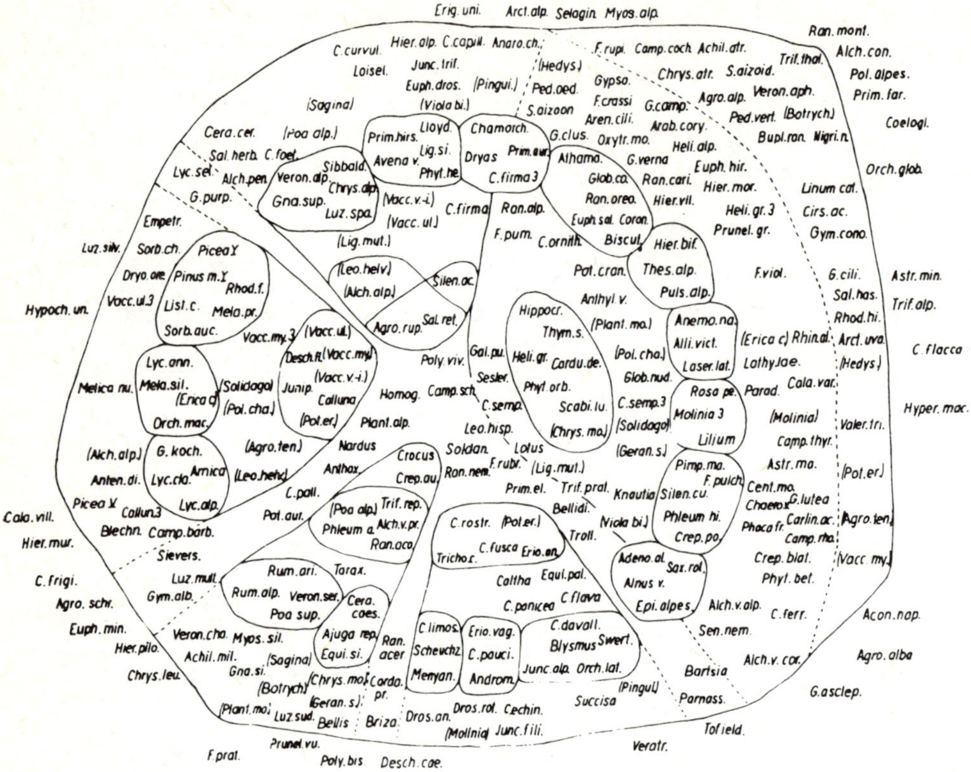

Fig. 7. An arrangement of species by association measurements for Swiss mountain vegetation (HEGG 1965). Linkage-groups of species, by which community-types are characterized, are circled. A number of directions of community variation in relation to environment are indicated: acid soils (left), alpine vegetation (upper left), calcareous soils (right), wet soils (down), and nitrogen-rich soils (lower left).

used the method of dendrites along with the matrix method to show relations between phytosociological units (Fig. 8). Coefficients of similarity (CC) were shown on the links between units, although the graphical distance was not proportional to the magnitude of the coefficient (FALIŃSKI 1960, IZDEBSKI 1963). RAMSEY (1964) used the SØRENSEN coefficient (CC), with a minimum value of 70, to delimit initial group noda or central clusters of plots. These form the foci of groups in a plexus diagram and plots are added with progressively lower similarity values (*cf.* RAMSEY & DE LEEUW 1964).

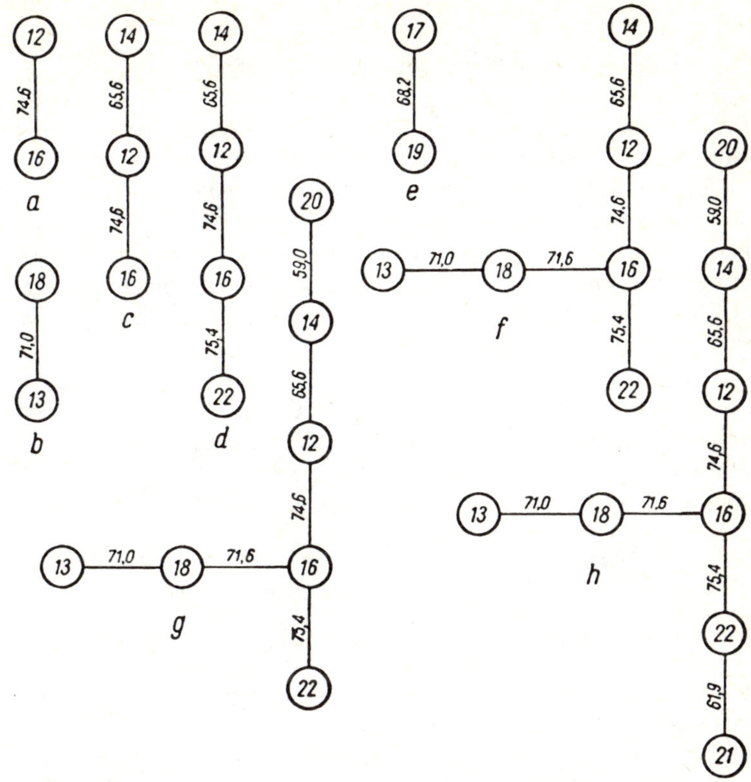

	12	13	14	16	17	18	19	20	21	22
12		61,3	**65,6**	**74,6**	55,2	64,7	51,8	58,5	61,0	65,7
13	62 19		62,1	59,0	65,4	**71,0**	58,3	50,8	52,6	50,0
14	64 21	58 18		57,1	51,8	62,5	48,0	**59,0**	56,4	63,6
16	67 25	61 18	63		59,6	**71,6**	56,6	46,9	59,2	**75,4**
17	58 16	52 17	54 14	57 17		62,1	**68,2**	50,9	50,0	43,3
18	68 22	62 22	64 20	67 24	58 18		55,5	55,4	58,5	62,8
19	54 14	48 14	50 12	53 15	44 15	54 15		47,1	47,0	50,0
20	65 19	59 15	61 18	64 15	55 14	65 18	51 12		48,1	47,8
21	82 25	76 20	78 22	81 24	72 18	82 24	68 16	79 19		**61,9**
22	70 23	64 16	66 21	69 26	60 13	70 22	56 14	67 16	84 26	

Fig. 8. A table of similarity coefficients and dendrite (FALIŃSKI 1960, Fig. 1,2). The matrix of KULCZYŃSKI (1928) coefficients (below) is the basis for a cumulative series of plexus diagrams *a* through *h* (above).

6.4.2 Definition and Representation of Groups

Groups or clusters may be objectively recognized when all pairs of the original matrix are at least as similar as some agreed-upon value of the similarity index, *i.e.* when the elements of the comparison matrix exceed a given value. All pairs of objects which are linked at the given level with any member of the group may be considered members of the group or it may be required, as Sørensen (1948) did, that all members be linked with each of the others above the selected similarity values.

The cohesiveness of a cluster may be expressed simply as a proportion of the possible links between the members of any group which is realized (*i.e.* 4 objects would have 6 possible links but could be linked with a minimum of 3). Wirth et al. (1966) proposed that cohesiveness of clusters may be expressed by the ratio:

$$\frac{\text{Total number of existing connections minus necessary connections.}}{\text{Total possible connections minus necessary connections.}}$$

If the components are simply linked linearly the connectedness is zero; when the ratio is one the cluster is maximally connected. Thus, patterns of clusters can be quantitatively compared.

Hopkins (1957) devised an objective method of producing 'basic units' or groups of significantly ($P > 0.001$ for SA) positively associated species within a site of a plant community. He suggested a method of identifying these as follows:

1. List all species with negative associations in order of decreasing number of these associations.
2. To any of these species add species which are positively associated with them.
3. When two or more groups contain a common species combine them.
4. Add extra groups of any positively related species to the list of groups.
5. Repeat 3.

The remaining groups are the basic units, and Hopkins shows a plexus diagram of such groups. Since by definition no member of a group has a significant association with any member of another group, there is no ready way of showing relations among groups. Hopkins secondarily assigned each of the sample areas to one or more basic units and estimated the proportion of the study area occupied by each basic unit. Araoz et al. (1971) form groups or noda by grouping sample quadrats according to their distances from randomly selected points. The interrelations between these are shown by linking noda with high association and low dissociation values to form a plexus diagram.

Most of the work using plexus diagrams has been concerned with identification of species groups or community-types. Numerical taxonomists have been doing much the same kind of thing (WIRTH et al. 1966, MOSS 1967, GOULD & GARWOOD 1969). WIRTH et al. show a series of successive plexus graphs formed as the strictness of the criterion for linking their objects (taxa) is sequentially relaxed. This is similar to SØRENSEN's method of forming clusters and is useful in showing the sequence of links formed. GOULD & GARWOOD, similarly, form a series of increasingly encompassing (hierarchial) groups by cumulating patterns of correlation coefficients among dentition measurements of animals at decreasing levels of *r*. The representation of points (vertices) and lines (edges) in abstract space is the subject matter of graph theory which ecologists have been empirically, and largely unwittingly, dabbling in for four decades. Numerical taxonomists have been, more recently and self-consciously, doing the same thing, and have incorporated some of the more sophisticated aspects of graph theory into their analyses. BUSACKER & SAATY (1965) provide an introduction to graph theory which may help ecologists to improve their largely ad hoc graphic procedures. The plexus graphs commonly used by ecologists are usually 'undirected' — the lines connecting the points not having a direction. If these lines are given direction (by arrows) the graph is a 'directed' graph. This may be useful in graphic representations of a plexus to show nearest neighbor or 'most similar' relations not possible to show in the ordinary representation. JACKSON & CROVELLO (1971) show a plexus imbedded in the first three dimensions or components of a factor analysis with the nearest neighbors or most similar points connected by directed lines. As ecologists have often noted, it is not usually possible to show the full matrix of similarity relations in two or three dimensions hence the plexus is always somewhat subjective and the actual relations distorted. In a directed graph the relations can be objectively shown by a factor analysis technique and the nearest neighbor relations superimposed on the graph illustrating where the full relationships are inadequately represented by the analysis.

6.4.3 ORDINATION

Plexus treatment may be effective also as an approach to ordination. SUKATSCHEW (1928, 1932) and other Russian authors have,

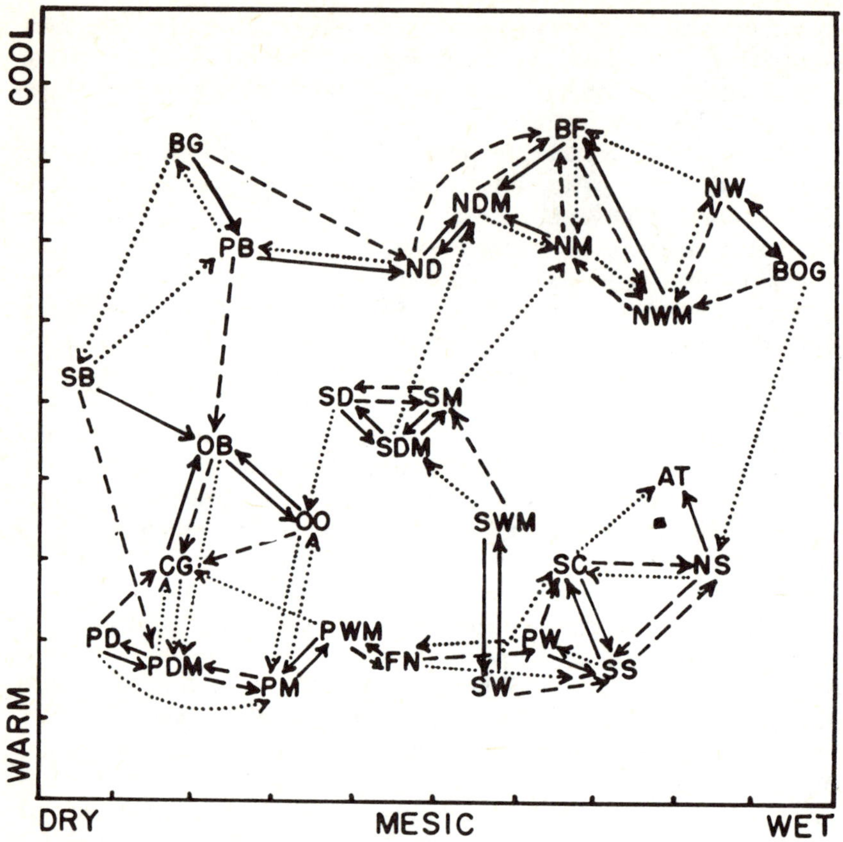

Fig. 9. A plexus ordination of community-types of Wisconsin in relation to moisture and temperature axes (CURTIS 1959, Fig. 35). Solid arrows point to the most closely related community, dashed arrows to the second, and dotted arrows to the third most closely related community. Community-types ordinated: AT alder ticket, BF boreal forest, BG bracken-grassland, BOG open bog, CG cedar glade, FN fen, ND northern dry forest, NDM northern dry-mesic forest, NM northern mesic forest, NWM northern wet-mesic forest, NW northern wet forest, OB oak barrens, OO oak opening, PB pine barrens, PD dry prairie, PDM dry-mesic prairie, PM mesic prairie, PWM wet-mesic prairie, PW wet prairie, SB sand barrens, SC shrub-carr, SD southern dry forest, SDM southern dry-mesic forest, SM southern mesic forest, SS southern sedge meadow, SWM southern wet-mesic forest, SW southern wet forest.

without using similarity measurements, arranged community-types into ecological series intersecting one another and representing major directions of environmental and vegetational variation (see article 17, Fig. 1, and other figs. in articles 15 and 17). WAGNER (1958) arranged middle European forest community-types into a plexus by their relations to elevation, moisture, and aridity. Comparable representations appear in the long tradition of succession diagrams in classical ecology showing the seral relationships of communities (or species) in successions. These diagrams were entirely subjective, and the essential difference of the plexus from them is the addition of quantitative methods of collecting data, the technique of condensing the primary data matrix into a matrix of similarity values, and the use of these values to identify clusters or directions of variation in the sample material.

CLAUSEN (1957a) described a graphic method of representing relationships by subjectively placing the elements of the matrix on a plane surface so that the distance from the nearest neighbors approximated the similarity values although they were not connected by lines. CURTIS (1959) arranged a wide range of different plant communities of Wisconsin into a plexus of community-types, comparing composite samples of each type by percentage presence of the more important undergrowth species and a PS calculation. A community-type central to the range of variation was placed in the center of a cubical space, and the other types were located with respect to it by triangulation procedures (Fig. 9). Two-dimensional views of this arrangement were used to plot environmental measurements for the community-types and to show how moisture, temperature, and internal light conditions in particular varied along the three axes of the plexus. The plexus was consequently an ordination of the community-types in relation to these three complex-gradients. The plexus was used also to show the distribution of species in relation to the community pattern and one another (Fig. 10), and the patterns of such broader community characteristics as the proportions in the vegetation of evergreen species, of shrubs and woody vines, and of major taxonomic groupings (Fig. 11).

In other studies the species or community plexus has been displayed on the main axes of environmental variables (WHITTAKER & FAIRBANKS 1958, OMURA & HOSOKAWA 1959, MARTIN 1960, AGNEW 1961, QUARTERMAN & KEEVER 1962, RAMSAY 1964, RAMSAY & DE LEEUW 1964). GIMINGHAM (1961, 1969) arranged a plexus diagram of heath stands from a broad geographic area of western Europe, considering SØRENSEN's CC values, to show major directions of variation (north-south and oceanic-continental climates). CURTIS' (1959) may be most significant among these applications of plexus technique for two reasons. First his plexus, rather than being a substitute for more formal ordination of a limited range of material, was the means of ordinating a range of samples so wide as to make difficult any other kind of ordination. Second his study, more effectively than any other, used the plexus to relate

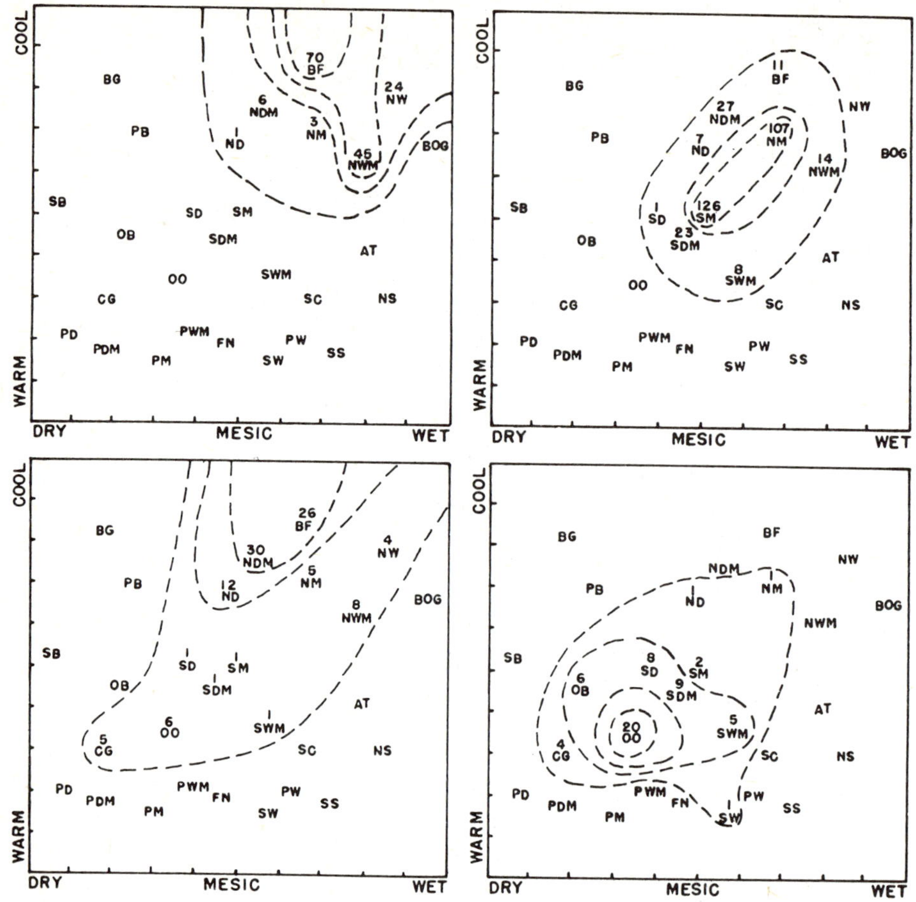

Fig. 10. Population distributions of four tree species in the ordination field of Fig. 9 (CURTIS 1959, Fig. 48; cf. WHITTAKER 1956 and Fig. 6, article 2 in this volume). Isodems are drawn for Wisconsin synthetic importance values, for *Abies balsamea* (upper left), *Acer saccharum* (upper right), *Betula papyrifera* (lower left), and *Carya ovata* (lower right).

environmental factors, species distributions, community variables, and community-types to one another.

In some cases species, or samples, or composite samples for community-types, can be arranged into a pattern showing their relationships to one another and recognized directions of environmental variation. In other cases the arrangement itself may reveal main directions of community variation that can be related to major environmental complex-gradients. The plexus may thus, in favorable cases, serve the purposes of direct and indirect gradient

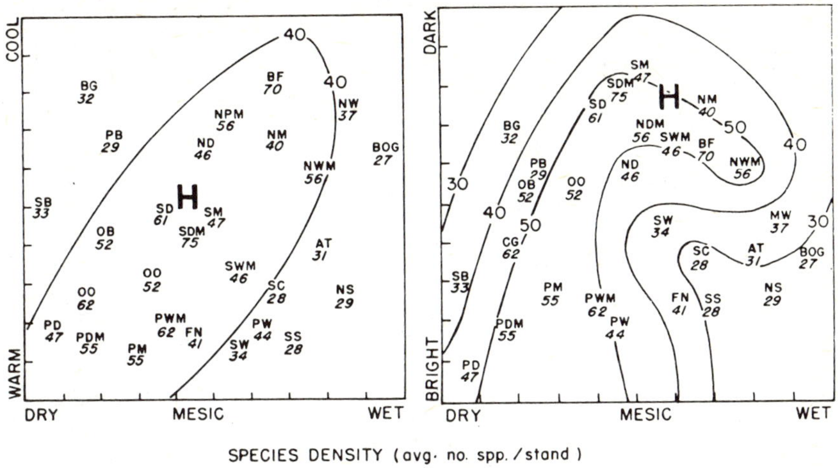

Fig. 11. Patterns of species diversity (average numbers of vascular plant species per sample) in the ordination field of Fig. 9, and the comparable ordination in relation to moisture and light axes (prepared from CURTIS 1959, Table IV—2 by R. H. WHITTAKER).

analysis, respectively. It is also possible to ordinate samples into a plexus by similarity measurements applied to environmental characteristics. HUGHES & LINDLEY (1955) calculated a matrix of similarity coefficients from a table of soil characteristics for four soil types and illustrated the relations between these as a plexus. HOLE & HIRONAKA (1960) calculated matrices of similarity values for Miami Family soil characteristics separately from laboratory and field data and a third matrix of representative soils of Great Soil Groups of the world. They represented the results as a three-dimensional model relating the soils types by perspective projection (Fig. 12).

There may be further interest in the construction of the different plexuses (of species, samples, and biotope measurements) possible from the same set of samples (or community types). No study was found comparing all three of these plexuses. Fig. 13 illustrates a case in which species distributions in samples (SA) and similarities (PS) of composite samples representing community-types were used to prepare plexuses which showed the same environmental axes. In this case the two plexuses independently revealed major directions of environmental effects on communities that had not been clearly recognized before the plexus treatment. BEALS & COPE (1964) calculated three separate matrices of similarity coefficients using the trees, spring herbs, and summer herbs found on different forest soil types and compared plexuses for the soil types based on

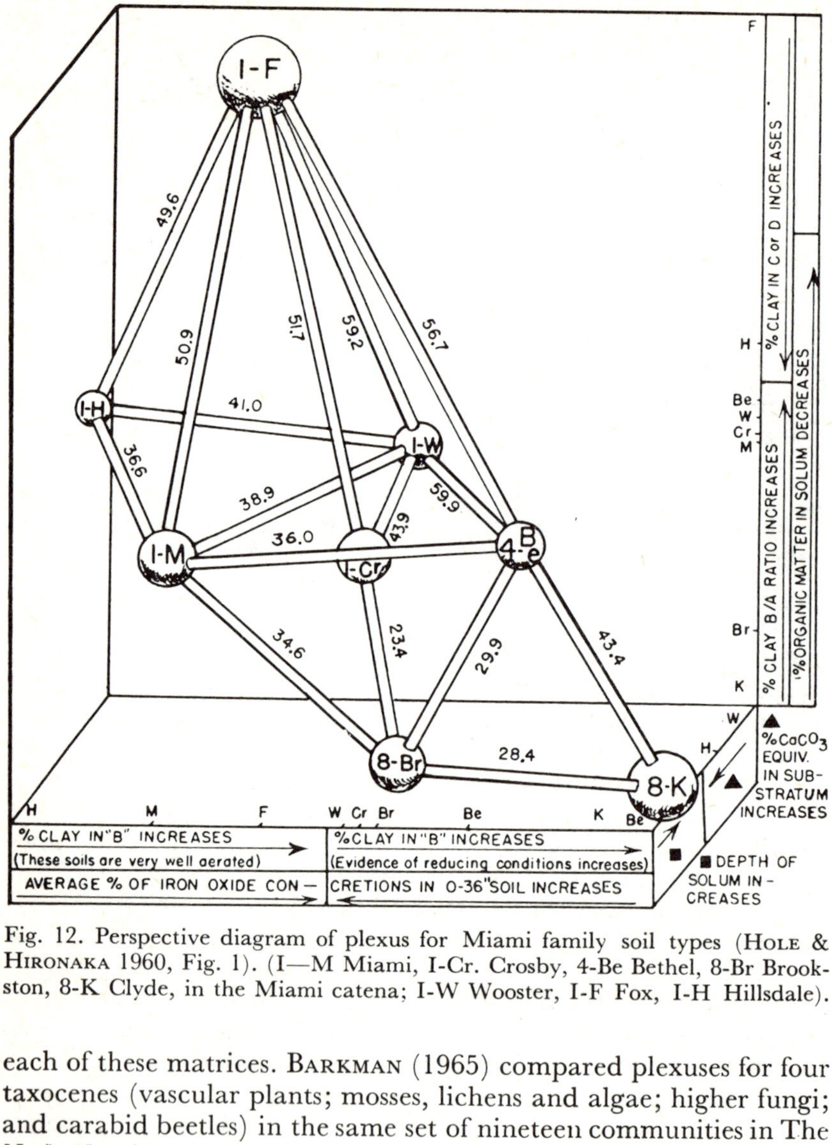

Fig. 12. Perspective diagram of plexus for Miami family soil types (Hole & Hironaka 1960, Fig. 1). (I—M Miami, I-Cr. Crosby, 4-Be Bethel, 8-Br Brookston, 8-K Clyde, in the Miami catena; I-W Wooster, I-F Fox, I-H Hillsdale).

each of these matrices. Barkman (1965) compared plexuses for four taxocenes (vascular plants; mosses, lichens and algae; higher fungi; and carabid beetles) in the same set of nineteen communities in The Netherlands.

6.5 **Conclusion**

The versatility of the techniques we have been discussing deserves emphasis. The secondary matrix and the plexus may be used, separately or in combination, to seek or to display relationships of ob-

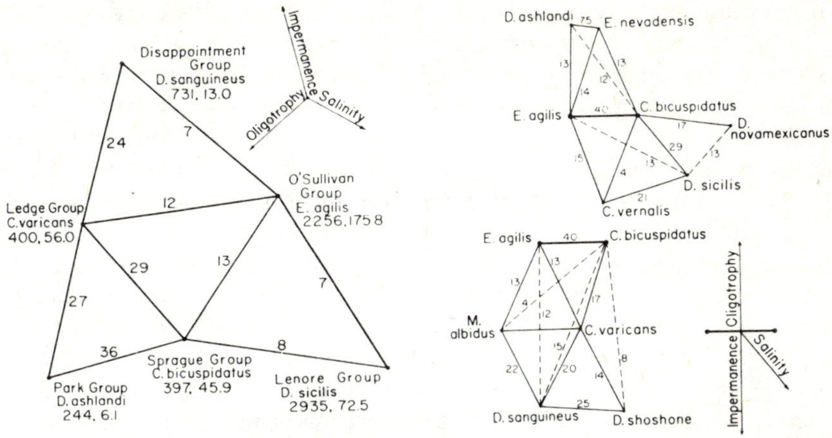

Fig. 13. A community-type plexus (left) and species-association plexus (right) for copepod samples from lakes and ponds in southeastern Washington (WHITTAKER & FAIRBANKS 1958, Figs. 3 and 5). Community-types are arranged by percentage similarities (numbers on lines) for composite samples; for each type are entered a characteristic species, mean total salt content of the water in parts per million (first number), and mean microsrustacean standing crop in individuals per liter (second number). Copepod species are arranged by percentage association values (numbers on lines); the solid lines are scaled for these values. The heavy line connecting *Eucyclops agilis* and *Cyclops bicuspidatus* is a center from which the remaining species radiate on the three environmental axes of oligotrophy (*Diaptomus ashlandi* and *Epischura nevadensis*), salinity (*C. vernalis*, *D. sicilis*, and *D. novemexicanus*), and seasonal impermance (*D. sanguineus* and *D. shoshone*).

jects to one another and environments. Species, community samples and biotopes may be the objects of such treatment, and there may be interest in comparing the results for these different objects in a given study. The levels on which the matrix and plexus can be applied range from use of replicate samples from a single stand to study species association and intra-community differentiation, to the use of individual samples from a narrower or wider range of community variation, to use of composite samples summarizing data for many individual samples representing a range of community-types. The techniques may be used to group or classify the objects or to ordinate them, or for both purposes; and the ordination makes possible relating to one another variables of environment, species distribution, and community characteristics.

The most general rewards of matrix and plexus approaches lie in the ordering of ecological material, and in revealing or representing relationships that are not otherwise evident. The plexus in particular may reveal relations of samples and species to one another in relation to environmental gradients. The plexus may then, like other ordinations, serve as a means of abstracting and understanding a community pattern in relation to a range of environmental variation. In some simple cases the matrix and plexus treatment may serve these purposes as well as any other, and there is only limited gain from the use of other, more formal techniques of ordination. Matrix and plexus techniques may become increasingly ineffective, however, with increase in number of objects, range of community variation, and number of directions of environmental and community interrelation. Though other means of ordination are affected by these same characteristics of the sample material, Wisconsin comparative ordination and factor analysis techniques will in some cases offer much more effective ordination than a plexus. Recent studies of distortions inherent in these other techniques imply that they too must be interpreted with caution (La France 1972, Austin & Noy-Meir 1972, article 10).

The matrix and plexus approach is thus not a generally sufficient answer to the problems of classification and ordination. However, the matrix and plexus techniques are more direct, and their implications are in some cases clearer to the investigator, than is the case with mathematically more complex procedures. These relatively simple approaches to community variation have continued value –as an alternative means of treating either small sample sets or very wide ranges of community variation, and as a preliminary step toward or a complement and aid to interpretation of other ordinations – that should not be overlooked.

6.6 Summary

Data from natural communities are often summarized as a community table or primary matrix, in which the columns represent samples and the rows species or environmental measurements. A variety of measurements can be applied to such tables to express the relative similarity to one another of samples, or of species, or of environmental variables. These similarity measurements of a given type may be compiled into a table (secondary matrix) in which the measurements indicate relationships of each sample (or species, or variable) to every other.

The secondary matrices can be rearranged that high similarity values are concentrated along the diagonal of the table. The ordered matrix may then be used either as: (i) an aid to classification of samples or grouping of species, when clusters of high similarity values are observed in the matrix, or (ii) an approach to ordination when the sequence of samples or species expresses their relation to a major environmental complex-gradient.

Values from the secondary matrix may also be used to arrange samples, or species, as points in a diagram or plexus in which the closeness of points to one another expresses their relative similarity. The plexus also may be used as an aid to classification or a means of ordination. With some sample sets, the plexus is an effective means of ordination in relation to two or three axes of environmental and community variation. The plexus may then be used to relate environmental measurements, species distributions, and community characteristics to one another, and as an abstract representation of a community pattern.

Though their effectiveness is more limited than that of some other ordination techniques, matrix and plexus approaches have continued value.

REFERENCES

AGNEW, A. D. Q., – 1961 – The ecology of *Juncus effusus* L. in North Wales. *J. Ecol.* 49: 83—102.

AGRELL, I., – 1963 – A sociological analysis of soil Collembola. *Oikos* 14: 237—247.

ARAOZ, J., G. SARMIENTO, & M. MONASTERIO, – 1971 – An essay in the use of association and dissociation measures in phytosociological classification. *J. Ecol.* 59: 39—50.

AUSTIN, M. P. & I. NOY-MEIR, – 1972 – The problem of non-linearity in ordination: experiments with two-gradient models. *J. Ecol.* 59: 763—773.

BALOGH, J., – 1958 – Lebensgemeinschaften der Landtiere. Akademie-Verlag, Berlin. 560 p.

BANNISTER, P., – 1968 – An evaluation of some procedures used in simple ordinations. *J. Ecol.* 56: 27—34.

BARKMAN, J. J., – 1958 – Phytosociology and Ecology of Cryptogamic Epiphytes. Van Gorcum, Assen. 628 pp.

BARKMAN, J. J., – 1965 – Die Kryptogamenflora einiger Vegetationstypen in Drente und ihr Zusammenhang mit Boden und Mikroklima. (Engl. summ.) In 'Biosoziologie,' ed R. TÜXEN, *Ber. Symp. Int. Ver. Vegetskunde.*, Stolzenau/Weser, 1960: 157—171.

BEALS, E. W., – 1965 – Species patterns in a Lebanese Poterietum. *Vegetatio* 13: 69—87.

BEALS, E. W. & J. B. COPE, – 1964 – Vegetation and soils in an eastern Indiana woods. *Ecology* 45: 777—792.

BEAMAN, J. H. & J. W. ANDRESEN, – 1966 – The vegetation, floristics and phytogeography of the summit of Cerro Potosi, Mexico. *Am. Midl. Nat.* 75: 1—33.

BECKING, R. W., – 1957 – The Zurich-Montpellier School of phytosociology. *Bot. Rev.* 23: 411—488.

BOYCE, A. J., – 1969 – Mapping diversity: A comparative study of some numerical methods. In 'Numerical Taxonomy', ed. A. J. COLE, pp. 1—30. Academic Press, New York.

Braun-Blanquet, J., – 1951 – Pflanzensoziologie: Grundzüge der Vegetationskunde. 2nd ed., Wien, 631p. 3rd ed., 1964, 865p.

Bray, J. R., – 1956 – A study of mutual occurrence of plant species. *Ecology* 37: 21—28.

Brodo, I. M., – 1961 – A study of lichen ecology in central Long Island, New York. *Am. Midl. Nat.* 65: 290—310.

Busacker, R. G. & T. L. Saaty, – 1965 – Finite Graphs and Networks. McGraw-Hill. New York. 294p.

Carmichael, J. W. & P. H. A. Sneath, – 1969 – Taxometric maps. *Syst. Zool.* 18: 402—415.

Češka, A., – 1968 – Application of association coefficients for estimating the mean similarity between sets of vegetational relevés. *Folia geobot. phytotax., bohemoslov.*, Praha. 3: 57—64.

Clausen, J. J., – 1957a – A comparison of some methods of establishing plant community patterns. *Bot. Tidsskr.* 53: 253—278.

Clausen, J. J., – 1957b – A phytosociological ordination of the conifer swamps of Wisconsin. *Ecology* 38: 638—646.

Cole, L.C., –1949 –The measure of interspecific association. *Ecology* 30: 411—424.

Cole, L. C., – 1957 – The measurement of partial interspecific association. *Ecology* 38: 226—233.

Cook, C. W., & R. Hurst, – 1963 – A quantitative measure of plant association on range in good and poor condition. *J. Range Mgmt.* 15: 266—273.

Crovello, T. J., – 1969 – Effects of change of characters and of number of characters in numerical taxonomy. *Am. Midl. Nat.* 81: 68—86.

Crovello, T. J., – 1970 – Analysis of character variation in ecology and systematics. *Ann. Rev. Ecol. Syst.* 1: 55—98.

Curtis, J. T., – 1959 – The Vegetation of Wisconsin: An Ordination of Plant Communities. Univ. Wisconsin, Madison. 657 pp.

Dagnelie, P., – 1960 – Contribution à l'étude des communautés végétale par l'analyse factorielle. (Engl. summ.) *Bull. Serv. Carte phytogéogr.* Sér. B, 5: 7—71, 93—195.

Dagnelie, P., – 1961 – L'application de l'analyse multivariable à l'étude des communautés végétales. *Bull. Inst. int. Statist.*, Rome, 33c Sess. p. 1—12.

Dagnelie, P., – 1962 – Étude statistique d'une pelouse à *Brachypodium ramosum.* III. Les liaisons interspécifiques. Presence et absence d'espèces. *Bull. Serv. Carte phytogéogr.* Sér. B, 7: 85—97.

Dahl, E., – 1957 – Rondane: Mountain vegetation in South Norway and its relation to the environment. *Skr. norske Vidensk-Akad.* Mat- naturv. Kl. 1956 (3): 1—374.

Davies, W., – 1971 – The phytopsammon of a sandy beach transect. *Am. Midl. Nat.* 86: 292—308.

Dix, R. L. & J. E. Butler, – 1960 – A phytosociological study of a small prairie in Wisconsin. *Ecology* 41: 316—327.

Ellenberg, H., – 1956 – Aufgaben und Methoden der Vegetationskunde. In 'Einführung in die Phytologie' by H. Walter, vol. IV. Grundlagen der Vegetationsgliederung, pt. I. Ulmer, Stuttgart. 136p.

Fager, E., – 1957 – Determination and analysis of recurrent groups. *Ecology* 38: 586—595.

Fager, F. W. & J. A. Mc Gowan, – 1963 – Zooplankton species groups in the North Pacific. *Science, N. Y.* 140: 453—460.

Faliński, J., – 1960 – Zastosowanie taksonomii wrocławskiej do fitosocjologii- (Germ. summ.: Anwendung der sog. 'Breslauer Taxonomie' in der Pflanzensoziologie). *Acta Soc. bot. Pol.* 29: 333—361.

FRYDMAN, I. & R. H. WHITTAKER, – 1968 – Forest associations of southeast Lublin Province, Poland. (Germ. summ.) *Ecology* 49: 896—908.

GAMS, H., – 1961 – Erfassung und Darstellung mehrdimensionaler Verwandtschaftsbeziehungen von Sippen und Lebensgemeinschaften. *Ber. geobot. Inst. ETH, Stiftg. Rübel*, Zurich, 1960, 32:96—115.

GHENT, A. W., – 1972 – A simplified computation procedure for Kendall's Tau suited to extensive species - density comparisons. *Am. Midl. Nat.* (in press).

GILBERT, M. & J. T. CURTIS, – 1953 – Relation of the understory to the upland forest in the prairie-forest border region of Wisconsin. *Trans. Wis. Acad. Sci. Arts Lett.* 42: 183—195.

GIMINGHAM, C. H., – 1961 – North European heath communities: a 'network of variation'. *J. Ecol.* 49: 655—694.

GIMINGHAM, C. H., – 1969 – The interpretation of variation in north-European dwarf-shrub heath communities. (Germ. summ.) *Vegetatio* 17: 89—108.

GODRON, M., – 1967 – Les groupes écologiques imbriqués en éceailles. *Oecol. Plant.*, Paris 2: 217—226.

GOULD, S. J. & R. A. GARWOOD, – 1969 – Levels of integration in mammalian dentitions: An analysis of correlations in *Nesophontes micrus* (Insectivora) and *Oryzomys couesi* (Rodentia). *Evolution* 23: 276—300.

GREIG-SMITH, P., – 1952 – Ecological observations on degraded and secondary forest in Trinidad, British West Indies. II. Structure of the communities. *J. Ecol.* 40: 316—330.

GREIG-SMITH, P., – 1964 – Quantitative Plant Ecology. 2nd ed. Butterworths, London. 256 pp.

GUINOCHET, M., – 1955 – Logique et dynamique du peuplement végétale. Masson, Paris. 143 pp.

GUINOCHET, M. & P. CASAL, – 1957 – Sur l'analyse différentielle de Czekanowski et son application à la phytosociologie. *Bull. Serv. Carte phytogéogr.* Sér. B, 2: 25—33.

HAGEMEIER, E. M., – 1966 – A numerical analysis of the distributional patterns of North American mammals. II. Reevaluation of the provinces. *Syst. Zool.* 15: 279—299.

HALE, M. E., – 1955 – Phytosociology of corticolous cryptogams in the upland forests of southern Wisconsin. *Ecology* 36: 45—63.

HEGG, O., – 1965 – Untersuchungen zur Pflanzensoziologie und Ökologie im Naturschutzgebiet Hohgant (Berner Voralpen). (French & Engl. summs.) *Beitr. geobot, Landesaufn. Schweiz* 46: 1—188.

HOLE, F. D. & M. HIRONAKA, – 1960 – An experiment in ordination of some soil profiles. *Proc. Soil Sci. Soc. Am.* 24: 309—312.

HOPKINS, B., – 1957 – Pattern in the plant community. *J. Ecol.* 45: 451—463.

HOSOKAWA, T., M. OMURA & Y. NISHIHARA, – 1957 – Grading and integration of epiphyte communities. *Jap. J. Ecol.* 7: 93—98.

HUGHES, R. E. & D. V. LINDLEY, – 1955 – Application of biometric methods to problems of classification in ecology. *Nature, Lond.* 175: 806—807.

HURLBERT, S. H., – 1969 – A coefficient of interspecific association. *Ecology* 50: 1—9.

IZDEBSKI, K., – 1963 – Bory na Roztoczu Środkowym. (Russ. & Engl. summs: Pineto-Vaccinietum uliginosi, Pineto-Vaccinietum myrtilli and Abietum polonicum in Central Roztoze). *Annls Univ. Mariae Curie-Skłodowska*, Lublin, Sec. C. 1962, 17: 313—362.

JACKSON, R. C. & T. J. CROVELLO, – 1971 – A comparison of numerical and biosystematic studies of *Haplopappus*. *Brittonia* 23: 54—70.

JONES, E. W., – 1955 – Ecological studies on the rain forest of southern Nigeria. IV. The plateau forest of the Okumo Forest Reserve. *J. Ecol.* 43: 564—594.

JUHÁSZ-NAGY, P., – 1963 – Investigations on the Bulgarian vegetation. Some hygrophilous plant communities. (I-III) *Acta biol. debrecena* (*Acta Univ. debrecen*, Budapest, Ser. 2) 2: 47—70.

KERSHAW, K. A., – 1959 – An investigation of the structure of a grassland community. II. The pattern of *Dactylis glomerata*, *Lolium perenne* and *Trifolium repens*. *J. Ecol.* 47: 31—43.

KERSHAW, K. A., – 1960 – The detection of pattern and association. *J. Ecol.* 48: 233—242.

KLEINER, E. F. & K. T. HARPER, – 1966 – An investigation of association patterns of prevalent grassland species in Red Butte Canyon, Salt Lake County, Utah. *Proc. Utah Acad. Sci.* 43 (2) : 29—36.

KONTKANEN, P., – 1957 – On the delimitation of communities in research on animal biocoenotics. *Cold Spring Harb. Symp. quant. Biol.* 22: 373—378.

KULCZYŃSKI, S. M., – 1928 – Die Pflanzenassoziationen der Pieninen. *Bull. int. Acad. pol. Sci. Lett.*, Cl. Sci. Math. Nat., Sér. B., 1927 (Suppl 2): 57—203.

LAFRANCE, C. R., – 1972 – Sampling and ordination characteristics of computer-simulated individualistic communities. *Ecology* 53: 387–397.

LAMBERT, J. M. & W. T. WILLIAMS, – 1962 – Multivariate methods in plant ecology. IV. Nodal analysis. *J. Ecol.* 50: 775—802.

LOOMAN, J., – 1963 – Preliminary classification of grasslands in Saskatchewan. *Ecology* 44: 15—29.

LOOMAN, J. & J. B. CAMPBELL, – 1960 – Adaption of Sørensen's K. (1948) for estimating unit affinities in prairie vegetation. *Ecology* 41: 409—416.

MACAN, T. T., – 1954 – A contribution to the study of the Corixidae (Hemipt). *J. Anim. Ecol.* 23: 115—141.

MAC FADYEN, A., – 1954 – The invertebrate fauna of Jan Mayen Island (East Greenland). *J. Anim. Ecol.* 23: 261—297.

MAC FADYEN, A., – 1963 – Animal Ecology, Aims and Methods. 2nd ed. Pitman, New York 344p.

MARTIN, R. H., – 1960 – The ecology of Groenvlei, a South African fen. *J. Ecol.* 48: 307—329.

MATUSZKIEWICZ, W., – 1948 – Roślinność lasów okolik Lwowa (Engl. summ.: The vegetation of the forests of the environs of Lvov.) *Annls Univ. Mariae Curie-Skłodowska*, Lublin, C, 3: 119—193.

MC DONOUGH, W. T., – 1963 – Interspecific associations among desert plants. *Am. Midl. Nat.* 70: 291—299.

MC INTOSH, R. P., – 1957 – The York Woods, a case history of forest succession in southern Wisconsin. *Ecology* 38: 29—37.

MC INTOSH, R. P., – 1962 – Pattern in a forest community. *Ecology* 43: 25—33.

MC INTOSH, R. P., – 1967a – An index of diversity and the relation of certain concepts to diversity. *Ecology* 48: 392—404.

MC INTOSH, R. P., – 1967b – The continuum concept of vegetation. *Bot. Rev.* 33: 130—187.

MOORE, G. W., W. S. BENNINGHOFF & P. S. DWYER, – 1967 – A computer method for the arrangement of phytosociological tables. In 'Proc. Ass. for Computing Machinery National Meeting, 1967,' pp. 297—299.

MORISITA, M., – 1959 – Measuring of interspecific association and similarity between communities. *Mem. Fac. Sci. Kyushu Univ.*, Ser. E. (Biol.) 3: 65—80.

MOSS, W. W., – 1967 – Some new analytic and graphic approaches to numerical taxonomy with an example from the Dermanyssidae (Acari). *Syst. Zool.* 16: 177—207.

Motyka, J., B. Dobrzański, & S. Zawadzki, – 1950 – Wstępne badania nad łąkami poludiniowo-wschodniej Lubelszczyzny. (Russ. & Engl. summs.: Preliminary studies on meadows in the southeast of the Province of Lublin). *Annls Univ. Mariae Curie-Skłodowska*, Lublin, Sec. E, 5: 367—447.

Newsome, R. D. & R. L. Dix, – 1968 – The forests of the Cypress Hills Alberta and Saskatchewan, Canada. *Am. Midl. Nat.* 80: 118—185.

Omura, M. & T. Hosokawa, – 1959 – On the detailed structure of a corticolous community analysed on the basis of interspecific association. *Mem. Fac. Sci. Kyushu Univ.*, Ser. E. (Biol.) , 3: 51—63.

Ono, Y., – 1961 – An ecological study of the brachyuran community on Tomoika Bay, Amakusa, Kyushu. *Rec. oceanogr. Wks Japan*, spec. no. 5: 199—210.

Orloci, L. – 1972 – On objective measures of phytosociological resemblance. *Am. Midl. Nat.* 88: 28–55.

Poore, M. E. D., – 1955 – The use of phytosociological methods in ecological investigations. I. The Braun-Blanquet system. *J. Ecol.* 43: 226—244. II. Practical issues involved in an attempt to apply the Braun-Blanquet system. *J. Ecol.* 43: 245—269. III. Practical application. *J. Ecol.* 43: 606—651.

Précsényi, O., – 1958 – Über die interspezifische Korrelation. *Acta bot. hung.* 4: 155—158.

Quarterman, E. & C. Keever, – 1962 – Southern mixed hardwood forest: Climax in the southeastern Coastal Plain: U. S. A. *Ecol. Monogr.* 32: 167—185.

Ramsay, D. Mc C., – 1964 – An analysis of Nigerian savanna II. An alternative method of analysis and its application to the Gombe sandstone vegetation. *J. Ecol.* 52: 457—466.

Ramsay, D. Mc C. & P. N. De Leeuw, – 1964 – An analysis of Nigerian savanna I. The survey area and the vegetation developed over Bima sandstone. *J. Ecol.* 52: 233—254.

Renkonen, O., – 1938 – Statistisch- ökologische Untersuchungen über die terrestriche Käferwelt der finnishchen Bruchmoore. (Finnish summ.) *Annls zool., Soc. zool. - bot. fenn. Vanamo* (*Suomal. eläin-ja kasvit Seur. van. eläin. Julk.*) 6: 1—231.

Renkonen, O., – 1949 – Discussion on the ways of insect synecology. *Oikos* 1: 122—126.

Ružička, M., – 1961 – Flechten - Kieferwald auf den Flugsanden der Tiefebene Záhorská nížina (Cladonio - Pinetum zahoricum). (Russ. & Slovak summs.) *Biológia Bratisl.* 16: 881—894.

Sanders, H. L. & R. R. Hessler, – 1969 – Ecology of the deep-sea benthos. *Science, N. Y.* 163: 1419—1424.

Shanks, R. E., – 1953 – Forest composition and species association in the beech-maple forest region of western Ohio. *Ecology* 34: 455—466.

Sokal, R. R. & P. H. A. Sneath, – 1963 – Principles of Numerical Taxonomy. Freeman, San Francisco. 359pp.

Sørensen, T. A., – 1948 – A method of establishing groups of equal amplitude in plant sociology, based on similarity of species content, and its application to analysis of the vegetation on Danish commons. *Biol. Skr.*, K. Danske Vidensk Selsk. 5 (4) : 1—34.

Struik, G. J. & J. T. Curtis, – 1962 – Herb distribution in an *Acer saccharum* forest. *Am. Midl. Nat.* 68: 285—296.

Sukachev, V. N., – 1928 - Principles of classification of the spruce communities of European Russia. *J. Ecol.* 16: 1—18.

Sukatschew, W. N., – 1932 – Die Untersuchung der Waldtypen des osteuropäischen Flachlandes. *Handb. biol. ArbMeth.* 11, 6: 191—250.

Taylor, R. J. & D. Campbell, – 1969 – Biochemical systematics and phylogenetic interpretations in the genus *Aquilegia*. *Evolution* 23: 153—162.

Vasilevich, V. I., – 1969 – Statistical Methods in Geobotany. (In Russian). USSR Academy of Sciences, Leningrad. 232pp.

Vries, D. M. de, – 1953 – Objective combinations of species. *Acta bot neerl.* 1: 497—499.

Vries, D. M. de, J. P. Baretta & G. Hamming, – 1954 – Constellation of frequent herbage plants, based on their correlation in occurrence. *Vegetatio*: 5/6: 105—111.

Wagner, H., – 1958 – Grundfragen der Systematik der Waldgesellschaften. *Veröff. geobot. Inst. Rübel*, Zürich 33: 241—252.

Welch, J. R., – 1960 – Observations on deciduous woodland in the eastern province of Tanganyika. *J. Ecol.* 48: 557—573.

Whittaker, R. H., – 1952 – A study of summer foliage insect communities in the Great Smoky Mountains. *Ecol. Monogr.* 22: 1—44.

Whittaker, R. H., – 1956 – Vegetation of the Great Smoky Mountains. *Ecol. Monogr.* 26: 1—80.

Whittaker, R. H., – 1962 – Classification of communities. *Bot. Rev.* 28: 1—239.

Whittaker, R. H., – 1967 – Gradient analysis of vegetation. *Biol. Rev.* 42: 207—264.

Whittaker, R. H. & C. W. Fairbanks, – 1958 – A study of plankton copepod communities in the Columbia Basin, southeastern Washington. *Ecology* 39: 46—65.

Wirth, M., G. F. Estabrook & D. J. Rogers, – 1966 – A graph theory model for systematic biology, with an example for the Oncidiinae (Orchidaceae) *Syst. Zool.* 15: 59—69.

Yarranton, G. A., – 1967 – A quantitative study of the bryophyte and macrolichen vegetation of the Dartmoor granite. *Lichenologist* 3: 392—408.

Yarranton, G. A. & W. J. Beasleigh, – 1968 – Towards a mathematical model of limestone pavement vegetation I. Vegetation and microtopography. *Can. J. Bot.* 46: 1591—1599.

7 WISCONSIN COMPARATIVE ORDINATION

Grant Cottam, F. Glenn Goff, and R. H. Whittaker

Contents

7 WISCONSIN COMPARATIVE ORDINATION

7.1 Introduction

Some of the earliest work in vegetational ordination was done at the University of Wisconsin more than twenty years ago. Since that time there has been a continuous increase in the use of various types of ordination procedures in vegetation studies. With wider usage of the approach have come better understanding and more formal mathematical techniques. The increasing capacity and availability of computers have promoted use of multivariate analytic methods in ecological research.

Because of the early development of certain methods of vegetation analysis at Wisconsin and the extensive application of these methods to the vegetation of the region, the University of Wisconsin has come to be known as one of the centers of 'ordination ecology.' Some authors (not, in general, those from Wisconsin) speak of a 'Wisconsin School' of plant ecology. While it is true that several techniques were independently developed by Wisconsin workers, most of these can be thought of as special cases of more general methods of multivariate analysis (e.g. SEAL 1964, KENDALL 1965) that are now coming into wider usage. The distinctive contribution from Wisconsin is discussed also by CURTIS (1959), MCINTOSH (1958, 1967), GREIG-SMITH (1964), and WHITTAKER (1967).

The purpose of this paper is (i) to present a brief review of the development of ordination at the University of Wisconsin, (ii) to illustrate the technique of polar ordination and its effectiveness for vegetation research, and (iii) to indicate how the compositional gradient analysis and polar ordination techniques developed at Wisconsin relate to other techniques now in use.

7.2 Compositional Gradient Analysis

CURTIS & MCINTOSH's (1951) compositional gradient analysis was based on use of species 'adaptation values' for sample ordination. To obtain these values, the quantity of each species within each stand or sample was first expressed as a synthetic importance value — the sum of relative density, relative basal area, and relative frequency (CURTIS & MCINTOSH 1951).

Assigning adaptation values to species was actually a two-stage process. Samples were first grouped into subsets, members of each of which were similar in their dominant species. These subsets were then arranged in a preliminary order along a compositional gradient by visual inspection of the patterns of species importance. An arrangement was sought which produced smooth patterns of change in the importance values of dominant species (either a monotonic taper from one extreme of the gradient toward the other, or tapers on each side of a peak within the gradient). This approach to ordination through the distributions of major species was termed the 'method of leading dominants.' Species distributions in this preliminary ordering of samples were observed, and each species was assigned an approximate value for relative position along the gradient on a scale from 1 to 10. These preliminary values were multiplied by the synthetic importance values of species to calculate a compositional index (cf. the weighted average technique, article 2) for each sample. Since the sum of the synthetic importance values for species in a given sample is 300 %, the compositional indices with species weighted 1 to 10 range between 300 and 3000. By plotting the importance values of a particular species in samples against the compositional indices of the samples, a scatter diagram such as that shown in Fig. 1 was produced.

The peak position of the species along the compositional gradient (x-axis) was determined by use of a moving-average curve

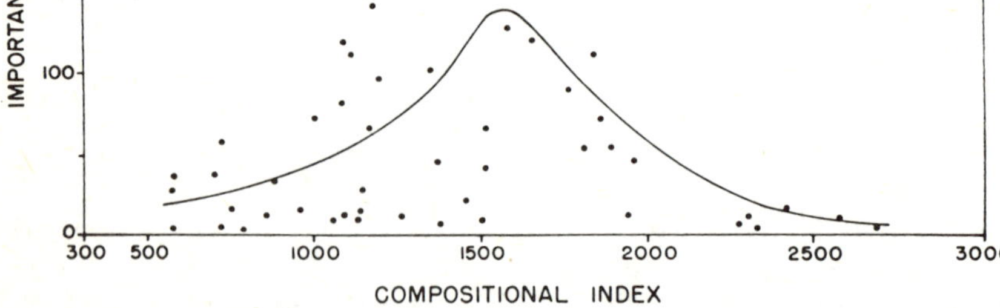

Fig. 1. A scatter diagram showing importance values of *Pinus strobus* in samples along the northern Wisconsin compositional gradient for forests. The solid line is a moving mean through the scatter of points; data are from Brown & Curtis (1952) records in the Wisconsin Plant Ecology laboratory files.

fitting technique. The adaptation value corresponding to this peak position (i.e., the peak position along the compositional gradient divided by 300) was the final adaptation value or weight assigned to the species. The final adaptation values were quite similar to the preliminary values. In some of the early studies this was taken as a validation of the preliminary ordering of samples. It is not, for the first assignment of values tends to fix the portion of the ordination transect in which a given species occurs. Studies independently relating species populations to environmental gradients are to be desired as a check on the assignment of species adaptation values.

The final adaptation values were used to arrange and group the samples into a composite transect, the data for which could be summarized in the form of Table I, article 2. Such transects effectively showed the rise and fall of species populations, and the change in community characteristics, along the compositional gradient. Characteristics of soils were found to vary along the gradient (Brown & Curtis 1952, Maycock & Curtis 1960), and it was shown that

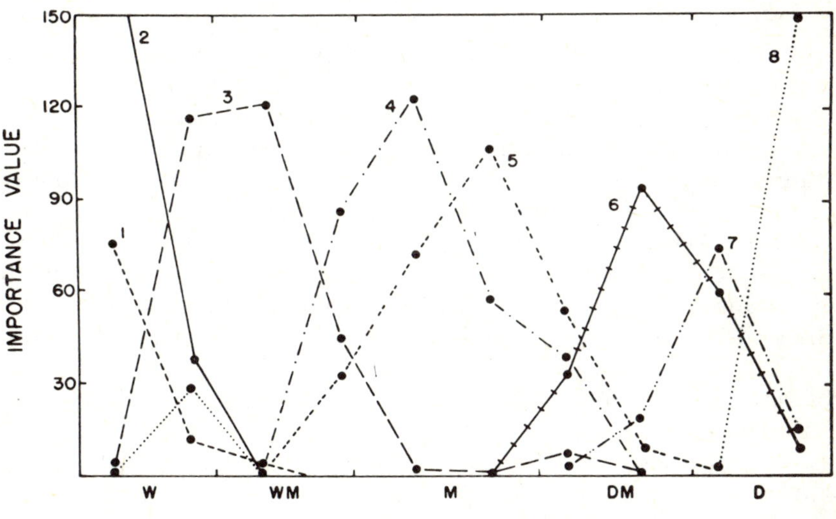

Fig. 2. Behavior of major tree species on the combined ordination of upland and lowland forests in northern Wisconsin: 1 tamarack (*Larix laricina*), 2 black spruce (*Picea mariana*), 3 white cedar (*Thuja occidentalis*), 4 eastern hemlock (*Tsuga canadensis*), 5 sugar maple (*Acer saccharum*), 6 white pine (*Pinus strobus*), 7 red pine (*P. resinosa*), and 8 jack pine (*P. banksiana*). Points are average Wisconsin synthetic importance values for species in sets of samples representing positions along the compositional gradient from wet (W, left), through mesic (M, middle), to dry (D, right) sites. From '*The Vegetation of Wisconsin*' by J. T. Curtis (1959) with permission of the University of Wisconsin Press.

other groups of organisms formed compositional gradients paralleling those for trees — undergrowth plants (GILBERT & CURTIS 1953), soil fungi (TRESNER et al. 1954, CHRISTENSEN et al. 1962), mosses and lichens on bark (HALE 1955, CULBERSON 1955), and birds (BOND 1957). One popular diagram resulting from these early studies showed the importance curves of several species superimposed as Fig. 2. Although studies of upland and lowland vegetation were conducted separately, the series of overlapping species curves, including both upland and lowland, was often shown on a single 'continuum' graph. Often in diagrams of this sort the curves were hand-smoothed.

GILBERT & CURTIS (1953) were the first workers at Wisconsin to use a matrix of interspecific association to order species. They used intensity of shading in the cells of the matrix to represent degree of interspecific association and simply rearranged the species to produce an ordering with heavily shaded cells along the main diagonal and lighter cells away from the diagonal (see article 6). Although they did not claim quantitative rigor, the ordering of species that resulted should be very similar if not identical to the ordering that would be produced by extracting the first eigenvector (principal component) from the same matrix of interspecific association. A similar approach has been applied by CLAUSEN (1957), SWINDALE & CURTIS (1957), and others to order species.

Following the work of CURTIS & MCINTOSH, several other studies used essentially the same techniques. BROWN & CURTIS (1952) applied the method of leading dominants to the forests of northern Wisconsin, CURTIS (1955) used a modification of the technique based on 'joint occurrence groups' to obtain gradient values for prairie species. Some studies applying the approach other than those mentioned include CHRISTENSEN et al. (1959), BRAY (1960), LINDSEY et al. (1961), ANDERSON (1963), LARSEN (1965), BUELL et al. (1966), ANDERSON et al. (1966), GEHLBACH 1967, and GEMBORYS & HODGKINS (1971).

All these studies had two characteristics in common. First, index values were assigned as weights for the species; values for samples could be found indirectly by use of the species values, but the samples were not ordered directly. Secondly, the species values were inferred from sociological data. While the patterns of community variation presumably resulted from the operation of underlying environmental variables, the indices were assigned primarily on the basis of species association or correlation and not by observation of species distributions along particular environmental gradients. It is in this respect that most of the Wisconsin work differs from the direct gradient analysis of WHITTAKER (1951, 1956, 1967), ROWE

(1956), WARING & MAJOR (1964) and others who have assigned species indices along specified environmental gradients (article 2).

7.3 Polar Ordination

7.3.1 TECHNIQUE

Wisconsin comparative or polar ordination was introduced by BRAY & CURTIS in 1957. The essential techniques of this, which may be the most distinctive contribution of Wisconsin plant ecology, can best be explained with a simple example. (The example and discussion relating to it were prepared by the third author on the basis of experience with the technique at Cornell University, and using 12 New Jersey saltmarsh samples from a larger set made available by the courtesy of L. F. M. FRESCO.)

Table I

Primary matrix of species coverage percents in

New Jersey saltmarsh samples.

	Sample number	1	2	3	4	5	6	7	8	9	10	11	12
	Species												
1	Spartina alterniflora	75	30	5	20	5	1		10	1	2		
2	Salicornia europaea	5	10	2	1	1		2			2		
3	Atriplex patula var. hastata	1	10	2	1	1	2	5		1		5	2
4	Distichlis spicata		15	80	2	10	15	30	1	10	10	20	
5	Suaeda maritima				20	10							
6	Salicornia virginica				5	10							
7	Juncus gerardii			1			40	1					
8	Scirpus olneyi						5	20				1	
9	Spartina patens							20	10	50		2	5
10	Iva frutescens							5	1	2	1	20	10
11	Phargmites communis								1	10	20	5	30
12	Solidago sempervirens									1	5	1	2
	Species cover total	81	65	90	49	37	63	83	23	75	40	54	49

The samples are first compiled into a standard community-table or primary matrix, as illustrated in Table I. Species that occur in a single sample have been omitted. Further analysis is based on quantitative comparisons among these samples. A number of similarity measurements can be used (articles 5, 10) but many applications of Wisconsin comparative ordination since BRAY & CURTIS (1957) have used a double standardization before computing percentage similarities. The rows of the primary matrix are first standardized, with the importance values of each species converted into percentages of the maximum value in the row for that species. Second, these standardized percentages are summed in each column, and the values in each column are converted to percentages of the total for that column. Each sample is now represented by a column of relative importance values (totaling 100 % or 1.0). This double standardization is in no way essential to the BRAY & CURTIS technique, but it may be advantageous (see 7.3.3). Table II is the same primary matrix as Table I after such double standardization.

Sample similarities are now computed. When the columns in the table are relative importance values, percentage similarity

Table II

Primary matrix of species coverages in New Jersey saltmarsh samples after double standardization.

	Sample number Species	1	2	3	4	5	6	7	8	9	10	11	12
1	Spartina alterniflora	62.5	15.5	4.5	13.4	3.5	0.8		31.1	0.7	1.3		
2	Salicornia europaea	31.3	38.6	13.4	5.0	5.3		7.3			9.7		
3	Atriplex patula var. hastata	6.2	38.6	13.4	5.0	5.3	12.1	18.2		5.3		22.7	9.1
4	Distichlis spicata		7.2	67.0	1.3	6.6	11.4	13.6	2.9	6.7	6.0	11.3	
5	Suaeda maritima				50.2	26.4							
6	Salicornia virginica				25.1	52.9							
7	Juncus gerardii			1.7			60.6	0.9					
8	Scirpus olneyi						15.1	36.4				2.3	
9	Spartina patens							14.5	46.6	53.4		1.8	4.5
10	Iva frutescens							9.1	11.7	5.3	2.4	45.3	22.7
11	Phragmites communis								7.8	17.8	32.2	7.6	45.5
12	Solidago sempervirens									10.7	48.3	9.1	18.2

computation simplifies to $PS = \Sigma \min(x,y)$, in which min (x,y) is the smaller of the two values for a given species in samples X and Y (article 5, formula 51). These values, computed in all possible directions between samples, are compiled into the secondary, sample-similarity matrix given as Table III. The problem is now to extract from these numbers major directions of variation in community composition, and to arrange the samples in relation to axes representing these major directions.

If the ordination is indirect, two samples which are most dissimilar may be chosen as the first end-points. The sample that has the lowest mean similarity with all other samples may be located first, in our example sample 6 (15.5 % bottom row of Table III). The sample with the lowest similarity value with this (sample 8, 3.7 %, Table III) may become the other end-point .These samples are now the poles of an axis along which all other samples are to be arranged, or ordinated.

Table III

Secondary matrix for sample similarities, New Jersey saltmarshes, percentage similarities of samples after double standardization of species coverage data.

Sample	1	2	3	4	5	6	7	8	9	10	11
1		53.0	24.1	23.4	14.1	7.1	13.5	31.1	6.1	11.0	6.2
2	53.0		38.5	24.7	20.7	20.2	32.7	18.4	12.7	17.0	29.9
3	24.1	38.5		15.8	20.7	26.0	35.2	7.4	12.7	17.0	24.7
4	23.4	24.7	15.8		66.4	7.1	11.3	14.6	7.0	7.6	6.3
5	14.1	20.7	20.7	66.4		12.7	17.2	6.4	12.6	12.6	11.9
6	7.1	20.2	26.0	7.1	12.7		39.5	3.7	12.7	6.9	25.7
7	13.5	32.7	35.2	11.3	17.2	39.5		26.5	31.9	15.7	42.7
8	31.1	18.4	7.4	14.6	6.4	3.7	26.5		63.3	14.4	23.9
9	6.1	12.7	12.7	7.0	12.6	12.7	31.9	63.3		37.7	35.8
10	11.0	17.0	17.0	7.6	12.6	6.9	15.7	14.4	37.7		25.1
11	6.2	29.9	24.7	6.3	11.9	25.7	42.7	23.9	35.8	25.1	
12	6.2	9.1	9.1	5.0	5.3	9.1	22.7	24.0	43.7	52.8	50.2
Mean	17.8	25.2	21.0	17.2	18.2	15.5	26.3	21.3	25.1	19.7	25.6

A distance between these end-point samples must be obtained; this distance becomes the length of the first axis. The complement of the similarity value for the end-point samples, their percentage difference $PD = 100\% - PS'$, is an expression of that distance, but not necessarily the best value. (i) If replicate samples from the same community are compared, their percentage similarities are not 100 % but some lower value; replicate similarities of 70 to 90 % are often observed (article 3). Distance between two samples is better expressed by this replicate similarity minus PS', than by 100 minus PS'. In many cases similarity measurements for replicate samples are not available ; the replicate similarity may be estimated as somewhat higher than the highest value in a secondary matrix of non-replicate samples, but usually less than 100 %. Bray & Curtis (1957) obtained a mean similarity (PS') of 82 % for seven replications; Beals (1960) used 85 % as a reasonable value for ordinating a matrix in which no similarity exceeds 85 %. (ii) If the similarity value for the two end-points is zero, the distance between them may be not smaller than 100 but larger. Removal of one or both samples may be considered. It may also be possible to estimate the distance by summing the distances from an intermediate sample to the two end-points. There are no zero values in Table III. Since some, single-species saltmarshes in the larger set had 100 % similarities, 100 % has been used as the replicate similarity in the example; and the first axis length is $L = 100 - 3.7 = 96.3$. (The ordination itself is not much affected by the replicate similarity estimate, or by whether 100 % or such an estimate is used – see Bannister 1968.)

All other samples are now located along the axis between end-point samples 8 and 6. Fig. 3 illustrates a simple graphical technique by which sample 7 is located on the axis (Bray & Curtis 1957). The similarity values for sample 7 with samples 8 and 6 are each subtracted from 100 %, to obtain distances (D_1, D_2) of 73.5 and 60.5. A compass is set to these distances by the scale of the axis, and a line is dropped onto the axis from the point at which the arcs for the distances from samples 8 and 6 intersect. The distance of sample 7 along the axis from sample 8 (x) is 57.2; its vertical distance from the axis (e) is 46.1. These values are easily computed from the Pythagorean theorem (Beals 1960, 1965):

$$x = (L^2 + D_1^2 - D_2^2) / 2L, \text{ and } e = \sqrt{(D_1^2 - x^2)},$$

in which L is the distance between the end-points, and D_1 and D_2 are the distances of the sample being located from the first and second end-point respectively. All the samples of our example have been thus located in Fig. 3. In many such ordinations the samples

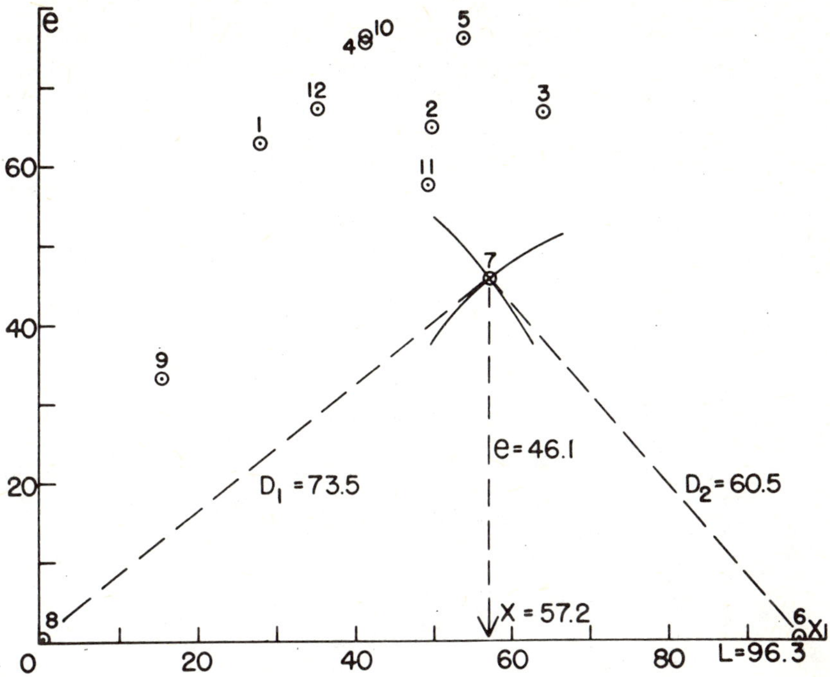

Fig. 3. The Bray & Curtis (1957) graphic technique for ordinating samples along an axis defined by end-point samples. For each sample its distances from the end-points intersect at a point defining relative location along the axis (x) and apparent departure from the axis (e) as shown for sample 7. The 12 samples of Tables I-III have been ordinated along an axis of length 96.3 between samples 8 and 6 by distance values (100 minus the percentage similarities in Table III) from end-point samples 8 and 6. Results from calculation by the formulas in the text are the same, as given in the first columns of Table IV.

form an arch over the axis (article 10). There may be little significance to the values for e, but the values for x should express relative position along a gradient of community composition between the end-point samples chosen.

A second axis of ordination is now desired. A second pair of end-point samples should be sought which are: (i) both among the samples in the middle part of the first axis, and (ii) close to one another in their positions along that axis, but (iii) most distant from one another among such pairs of samples. They should also (iv) have high values for e, but samples with the highest e values may not be the best choice of end-points. In our example samples 1 vs. 12 and 4 vs. 10 are likely choices of second end-point pairs; both of these

Table IV

New Jersey saltmarsh sample ordination, positions on (x) and distances from (e) two axes.

	First axis (8-6)		Second axis (1-12)	
Sample	x	e	x	e
1	28.0	63.0	0	0
2	49.6	64.8	14.6	44.7
3	64.2	66.7	33.5	68.1
4	41.1	74.8	30.0	70.4
5	54.0	76.4	38.4	76.8
6	96.3	0	48.9	79.1
7	57.2	46.1	54.9	66.8
8	0	0	41.4	55.1
9	15.6	33.2	77.1	53.7
10	41.1	75.1	77.3	44.2
11	49.5	57.7	80.5	48.0
12	35.2	67.4	93.7	0

pairs were tried but 1 and 12 both had the lower similarity (6.2) and gave an ordination more interpretable in terms of tide-levels and other relations. All the remaining samples are now located between samples 1 and 12, along an axis of length $L = 100 - 6.2 = 93.8$, by either the graphical technique or the computation given above. Results of ordination along both axes are given in Table IV. The two axes may now be used for a coordinate system, in which each sample is located by its positions on both axes (Fig. 4). The second axis is usually not strictly perpendicular to the first, but may be so treated as in Fig. 4.

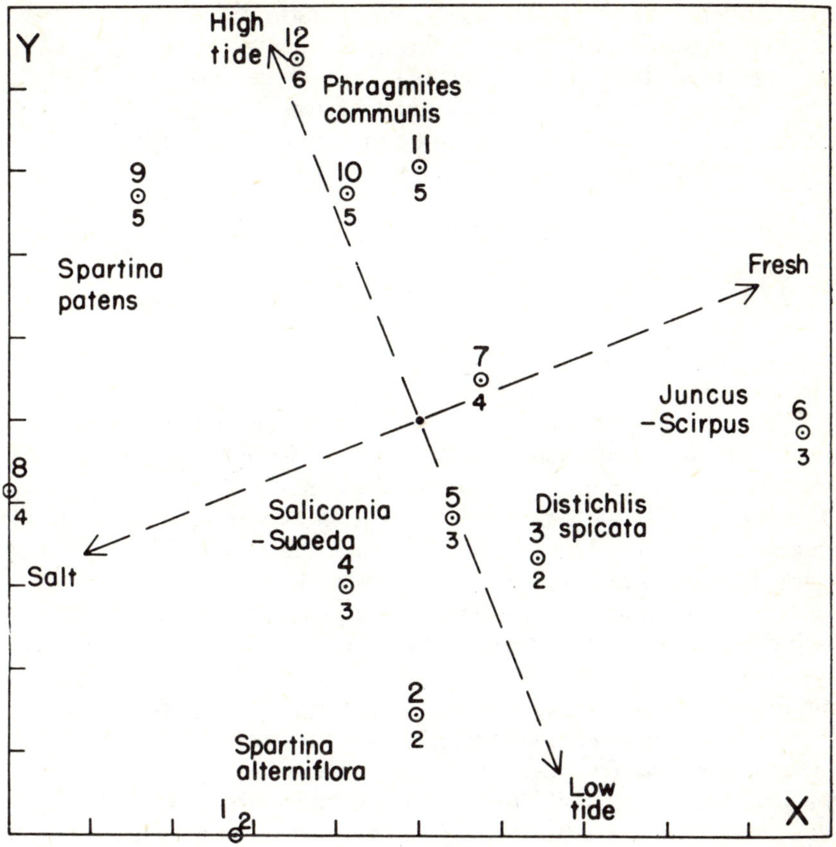

Fig. 4. A Wisconsin comparative ordination of 12 samples from New Jersey salt-marshes (unpublished data of L. F. M. Fresco). Coverage values for species (Table I) were doubly standardized (Table II); percentage similarities of samples were computed (Table III), and distances (100 minus these similarities) were used to locate samples in relation to two pairs of end-points, samples 8 and 6, and 1 and 12 (Table IV). Sample numbers are given above their points, tide levels on a scale from 1 (lowest) to 7 (highest) below the points. The names are the dominant species by which the samples may be classified into types; these types relate to major environmental gradients as indicated by the broken lines.

The relationships of samples in the ordination field should now be inspected. If samples that are located close together are similar to one another in composition, the two-dimensional ordination may be sufficient, and no third axis may be needed. If samples close together are markedly different, this difference suggests the existence of a third direction of community variation and the need for a third ordination axis. The end-points for the third axis may best be chosen from among the samples in the central area of the two-dimensional ordination field, in the same way the second end-points were chosen. Our example will not be carried beyond the second axis. It is sometimes possible, but not usually profitable, to carry a polar ordination to a fourth or even a fifth axis.

7.3.2 Results

The axes and the community pattern are now to be interpreted. Environmental measurements may be plotted at the sample points in the ordination field. In Fig. 4 the numbers below the sample points are tide-levels on a scale from 1 (lowest, including pure *Spartina alterniflora* marshes not represented in our example) to 7 (highest, samples with *Phragmites* and *Iva* and additional species not given in Table I). The second axis is related to the tide-level gradient, but the sample relationships (especially samples 3 vs. 4 and 6 vs. 8) suggest that the gradient is obliquely related to the *Y* axis, as indicated. Field observations, and the presence of *Scirpus olneyi* and *Juncus gerardii* as indicators of fresh-water influence, imply that the first axis is related to salinity. The oblique sequence of samples 1, 2, 3, and 6 in particular appears to represent the gradient of increasingly strong fresh-water dilution. Names of dominance-types, for a preliminary classification of the samples, are entered in Fig. 4 (samples 7 and 8 are transitional to the lower and higher tide-level types).

The ordination thus unifies the samples into an understandable pattern of communities in relation to a range of environmental variation. Further characteristics of the pattern can be shown by plotting other data in the ordination field. In Fig. 5 coverage estimates for four major species are plotted and outlined by contour lines. Each species has a center of maximum coverage in one of the community-types, from which its coverage tapers into other types. Characteristics of the samples as wholes, estimated total coverage and richness in species, are plotted in the bottom panels of Fig. 5. The coverage data are irregular both in the example and the larger sample set; the suggested increase (apart from sample 9) toward lower tide levels is questionable. Species diversities, in contrast, show a clear increase from low to high tide levels. The ordination can thus relate to one another environmental factors, species populations, and community trends, in an intergrading pattern of community-types as recognized by the investigator (cf. Whittaker 1967).

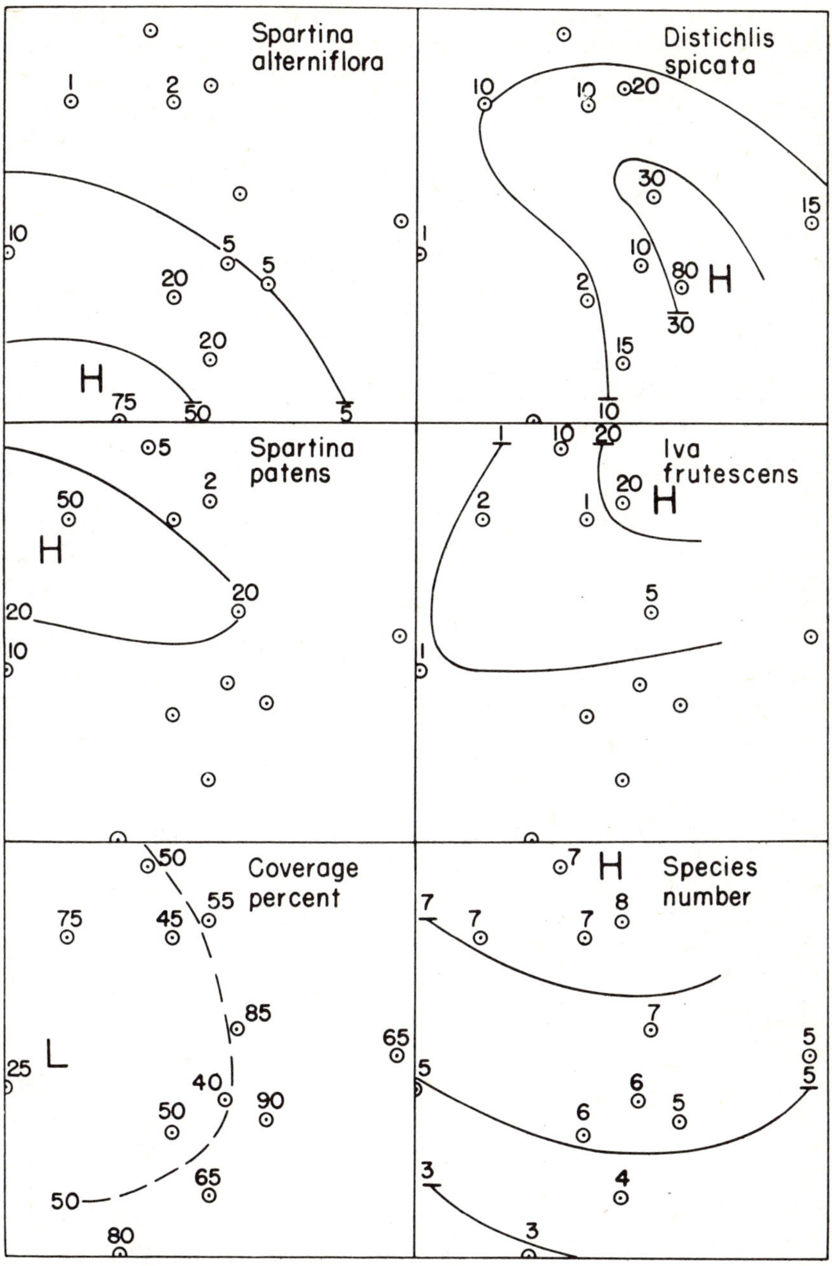

Fig. 5. Coverages of four major species, and sample coverages and species numbers, plotted for New Jersey saltmarsh samples as ordinated in Fig. 4.

Figs. 6–8 illustrate some other results from polar ordination. Fig. 6 plots symbols for community-types at positions for a larger number of samples in an ordination field (REAM 1963). The types occupy different parts of the field but are not sharply segregated or clustered. Fig. 7 indicates by sizes of circles the importance values for two species at sample positions in relation to three axes (AYYAD & DIX 1964). The sample positions may be conceived as a cloud of points in a three-dimensional space. Each species has its center of maximum importance in the space; in all directions away from this center its importance values decrease. It is consequently possible to ordinate species in relation to the same directions of community variation as those used for sample ordination. Fig. 8 represents an ordination of species by their centers in a three-dimensional space (BRAY & CURTIS 1957).

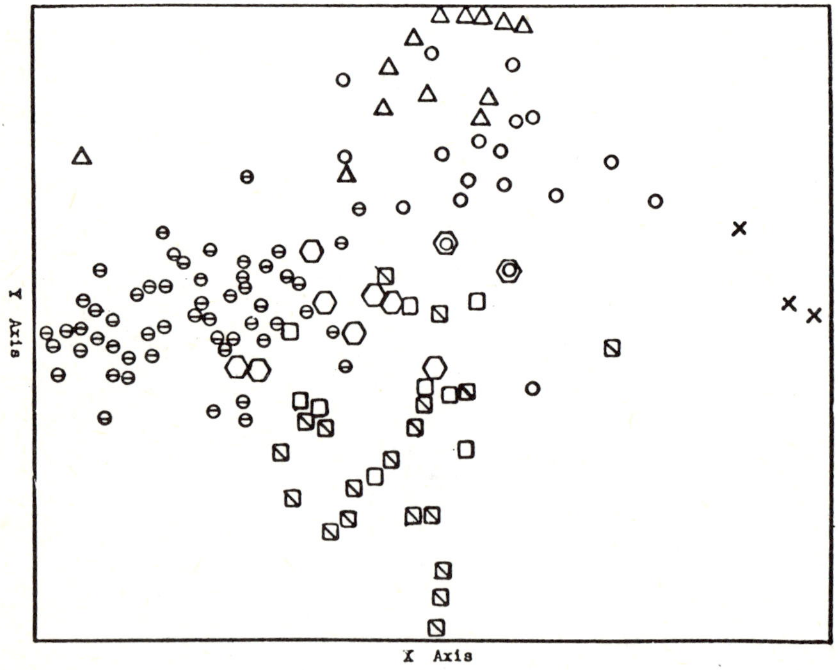

Fig. 6. A Wisconsin polar ordination of forest samples from the Wasatch Mountains, Utah (REAM 1963). Samples have been classified into dominance-types that are represented by different symbols: triangles *Abies concolor*, circles *Pseudotsuga menziesii*, barred circles *Populus tremuloides*, squares *Pinus contorta*, barred squares *Picea engelmanni*, hexagons *Abies lasiocarpa*, x's *Pinus flexilis*.

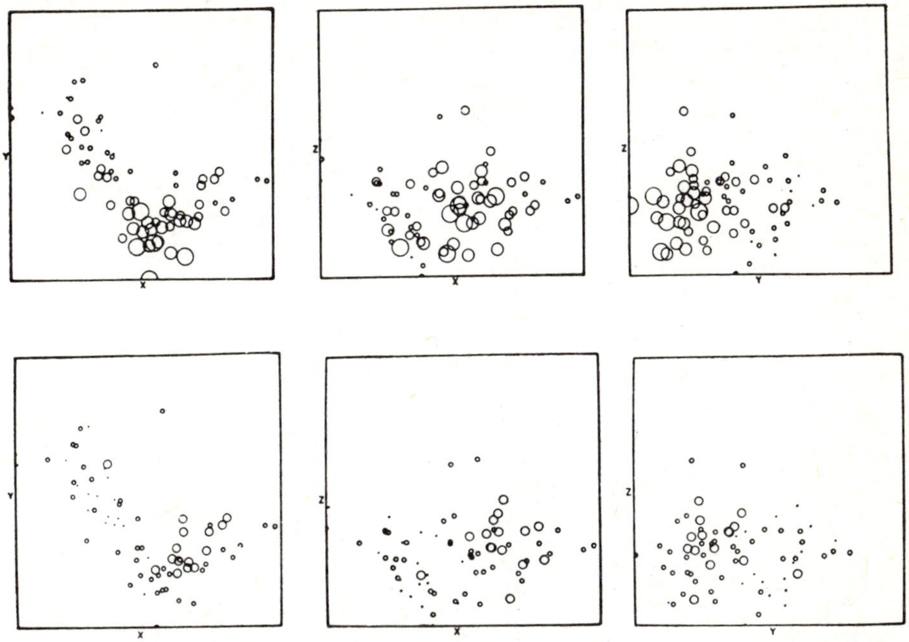

Fig. 7. Sample positions and patterns of species importance values in a three-dimensional ordination space for Saskatchewan prairies (Ayyad & Dix 1964, Fig. 2). Centers of circles locate samples in relation to the *x* and *y* axes on the left, the *x* and *z* axes in the middle, and the *y* and *z* axes on the right. Sizes of circles represent relative density values in these samples for two species — *Agropyron dasystachum* above, and *Koeleria cristata* below.

Ordinations thus provide useful graphical summaries of vegetational variation. A large amount of information can be conveyed in forms representing coherently and at a glance the interrelations of species, communities, and environments. An ordination may be colloquially referred to as a 'coat-rack for ideas' about vegetation, or a conceptual 'skeleton' by which the complex relationships in the field can be held together and given recognizable form. Once the ordination is prepared, the vegetation can be described by referring to segments or areas within the vegetational hyperspace, without the necessity for classification. On the other hand the vegetational hyperspace can be divided to give a classification that is based on the relationships revealed by ordination, or an ordination of samples that have been classified into units can reveal the relations of these units to one another and environment.

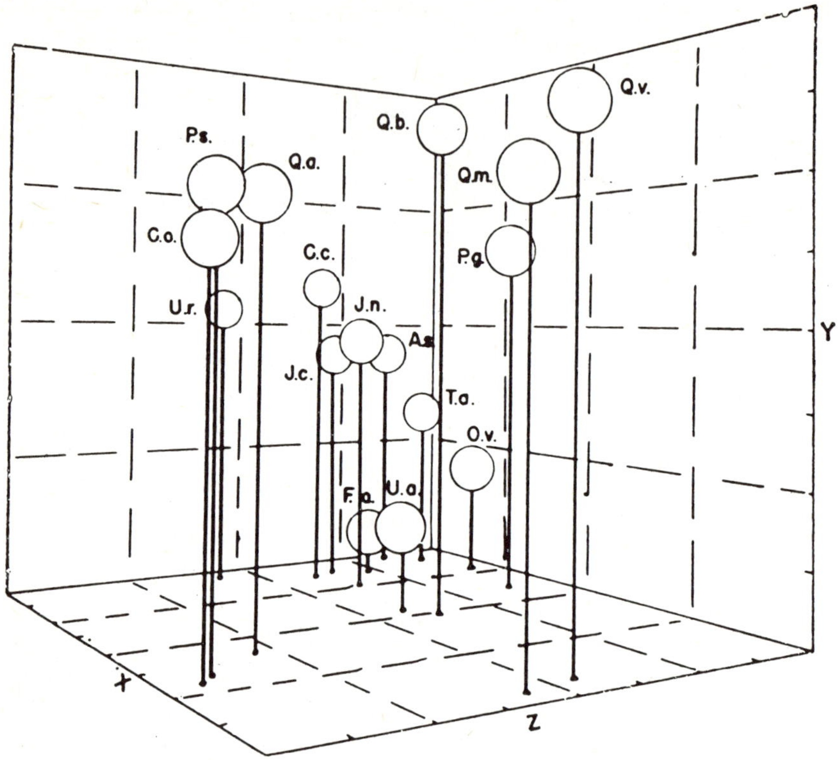

Fig. 8. An ordination of centers of species distributions in a three-dimensional space for Wisconsin upland forests (BRAY & CURTIS 1957, Fig. 8). Tree species are indicated by genus and species initials for: *Acer saccharum*, *Carya cordiformis*, *C. ovata*, *Fraxinus americana*, *Juglans cinerea*, *J. nigra*, *Ostrya virginiana*, *Populus grandidentata*, *Prunus serotina*, *Quercus alba*, *Q. borealis*, *Q. macrocarpa*, *Q. velutina*, *Tilia americana*, *Ulmus americana*, *U. rubra*.

7.3.3 PROBLEMS AND ALTERNATIVES

Ideally, a polar ordination may yield a pattern with one major community-type of intermediate environments in the center, and other communities radiating from it along axes interpretable as complex-gradients. In some cases the central community-type is the prevailing climax for an area (FRYDMAN & WHITTAKER 1968). Most ordination techniques, however, under some circumstances produce distorted representations of community relationships to environment (article 10). Although the BRAY & CURTIS technique is less subject to such distortion than most others, curvilinear ordinations of communities forming a 'straight' ecological series along a gradient in the field are sometimes produced. Thus in the study by BEALS & COTTAM (1960) the central part of the ordination field was occupied by samples with successional species of *Populus* and *Betula*, while other samples representing a moisture gradient were curved around these with the end-points of the gradient, the wettest and driest samples, close together on one side of the field. The arrangement may have some interest for its expression of floristic relationships among samples, but neither the axes nor diagonals to them can be interpreted as environmental gradients. It is not uncommon for some axes of a polar ordination, particularly those extracted after the first two or three, to be difficult to interpret. In general the likehood of curvilinear arrangement and or ininterpretable axes is increased by, (i) increase in the range of communities sampled and degree of difference among them and (ii) increase in the number of environmental factors affecting these communities, particularly (iii) increase in the intensity and variety of disturbance factors.

These effects emphasize first, that the Wisconsin polar ordination is by no means a fool-proof technique and second, the importance of ecological understanding from other sources to interpret its results. These effects may have for some investigators a third implication: the desirability of choosing end-point samples on the basis of known ecological relationships. Choice of end-points by maximum dissimilarities may seem more objective. In favorable cases it results in a first axis that parallels the main direction of change in community composition in the sample set in response to environment. It may also, however, result in a first axis the direction of which is determined by chance inclusion of a sample that is extreme in relation to the rest of the set. When environmental relationships are reasonably understood, there may be advantage in choosing the pairs of end-points, or at least the first one or two pairs, to represent the poles of the environmental gradients judged to be most significant. Such choice may have as a welcome, secondary effect reduction in computation; if the end-points are chosen the whole secondary matrix is not needed, for only the similarities of end-points with other samples are used. When the investigator does not feel the environmental relationships should be thus assumed, end-points may be chosen by maximum dissimilarity and the axes interpreted later as in the example. There may also be advantage in trying both or balancing together dissimilarities and known or suspected environmental relations in the choice of end-points. Research purposes are often best served by experimenting with different end-point pairs and ordinations.

A variety of similarity measurements, values from which these are subtracted to obtain distances, and transformations of distance values have been experimented with (BANNISTER 1968, LOUCKS 1962, BEALS 1960, ORLOCI 1966, GIMINGHAM et al. 1966, FRYDMAN & WHITTAKER 1968, MAAREL 1969). Because of the simple geometry of sample location (Fig. 3), most of these have only modest effects on the resulting ordination; and they will not be discussed in detail (see articles 5.5 and 10.4.3). Probably the most significant choice is that of the similarity measurement itself. Choices include coefficient of community (*CC*), percentage similarity (*PS*) applied to the raw or relativized data, percentage similarity applied to the doubly standardized data (*PS′*), and Euclidean distance (*ED*). (If species rather than samples are to be ordinated, as is also possible, there are comparable choices among measures of species association and correlation, see article 5.5.) Some preliminary suggestions on choice of measurement (see also article 10) are: (i) *CC* may be preferred to the unstandardized *PS* for sample sets of high alpha and beta diversity, and for sets with diverse and unsystematic effects of disturbance on quantitative representation of species, effects that are not themselves to be the subject of ordination axes. (ii) The unstandardized *PS* may be preferred to *CC* for sample sets of low alpha and beta diversity, particularly if species populations are not irregular (differing in samples from similar biotopes in consequence of disturbance or fluctuation). (iii) In some cases use of both *CC* and *PS* should be part of the experimenting to see what ordination best clarifies community and species relationships (article 10). (iv) Experience with Wisconsin ordinations suggests, however, that the doubly standardized *PS′* often gives an optimum combination of the advantages of *CC* and *PS*. *PS′* retains the effect of quantitative difference in species representation between samples, while avoiding the over-weighting of a very few dominant species in the similarity value; and it permits the use for ordination of different kinds of importance values for different species and strata (e.g. density, or basal area, or the Wisconsin synthetic importance value for trees, frequencies or coverages for herbs). (v) It is uncertain that *ED* is ever to be preferred for polar ordination.

There remain problems of excessive numbers of species, numbers of samples, and range of community difference. These problems of sample-set dimensions may overtax either the computer, or the ordination capacity of the BRAY-CURTIS technique, or both. There is in general little or no loss in ordination effectiveness from reducing the number of species by eliminating those occurring in the fewest samples. Too wide a range of beta diversity, or community difference, is a more serious problem and may confront the investigator with a dilemma. If extreme samples are chosen as end-points, many other samples will have very low or zero similarities with the end-points and cannot be effectively ordinated by comparison with them. If less extreme samples, from within the range of community variation in the set, are chosen, there may be risk that more extreme samples will be reflected back to locations within the ordination field, and that ecological relationships may thereby be made obscure.

Three directions for solving problems of excessive number of samples or range in compositional difference are possible: (i) The number of samples may be reduced by random selection, or by a systematic selection of samples representing the range of different biotopes or community-types in the set. Such selection, that may be combined with reduction in number of species, may make a set of excessive size manageable for computer treatment, or may make a hand ordination possible if a computer is not available. It will not generally solve the problem of too wide a range of community difference. (ii) The samples may be classified into subsets representing community-types, composite samples (each averaging or summarizing the characteristics of a subset) may be obtained, and the composite samples may be ordinated. The procedure has the same advantages as (i) and may more effectively utilize the information available. Because the composite samples are richer in species than the individual samples their composition may overlap more, and the composite samples may be more effectively ordinated than individual samples when the range of community difference is wide. (iii) The sample set may be divided into two or more larger subsets, each comprising some range of environmental and community difference. Ordination may now be carried out within these manageable subsets to give results that are clearer and less subject to distortion. If the approach is direct, end-points for the first axes in each subset may be so chosen as to link together the subset ordinations; and end-points of the second (and higher) axes may be chosen to give parallel ordinations in relation to the second (and higher) axes. Such coordination of subsets is not likely to be easy but in principle, in favorable cases, it may permit comparative ordination of a sample set otherwise too diverse to handle by the BRAY & CURTIS or any other technique.

7.3.4 MODIFICATIONS

Polar ordination differs from compositional gradient analysis in that: (i) Stands are ordered directly rather than by species indices, and (ii) multidimensional models are constructed rather than single-dimensional 'gradients.' It is worth noting that the gradient analysis approach can easily be extended to multiple directions. However, this was not understood at the time of the first polar ordination, and the extension from unidimensional to multidimensional models was one of the primary incentives for the new technique. Further examples of polar ordination include Ayyad & Dix (1964), McIntosh & Hurley (1964), Ashton (1964), Gittins (1965a, c), White (1965), Larsen (1965), Swan & Dix (1966), Frydman & Whittaker (1968), Maarel (1969), Rogers (1970), Arno & Habeck (1972), Gemborys (1974), Mueller-Dombois & Ellenberg (1974), and Vitt & Slack (1975). A number of experiments with and modifications of the technique deserve mention.

Maycock & Curtis (1960) used species adaptation values and compositional indices to select the end-point samples for the first (direct) ordination axes, and sample dissimilarities to select the end-points for the second and third (indirect) axes. Bray (1956) and Swindale & Curtis (1957) used comparative ordination to ordinate species instead of samples and to derive adaptation values for compositional ordinations; Gittins (1965a, b) has also used the technique to ordinate species. Over the past 4 or 5 years, students at Wisconsin have applied the polar ordination procedure to several different types of stand similarity matrices, including correlation, covariance, association, root mean square, Chi-square, and average absolute difference matrices. Goff & Cottam (1967) made use of residual similarity matrices for non-primary dimensions and based the selection of axis ends on standard deviation criteria rather than simply proximity on a former dimension. They also outlined an extension of ordination, termed 'index iteration,' which amounts essentially to characteristic vector analysis on an arbitrary numerical scale. This technique arose as an extension of the polar ordination technique.

Recent literature continues discussion of evaluation and modification of the technique. Swan (1970) has considered means of compensating for the effects of zero values in ordination. Gauch (1973a, b, article 3.4.2) shows that the relation of sample similarity values to separation along an environmental gradient fits a non-standardized complement of an error function and analyzes the performance of polar ordination. The inverse of the error function and other transformations of similarity values seem not to produce significant improvement in ordinations of field data (article 10.4.3). LaFrance (1972) found that sampling and analytic procedures of polar ordination do not necessarily bias the results toward continuity as suggested by critics. Orlóci (1966) and Swan (1969) have suggested ordinations employing ED and orthogonal axes; Orlóci (1974, 1975, article 11) has further developed corrections for orthogonality of axes. Studies of ordination performance using different similarity values indicate that ED generally gives less effective ordination than PS′, PS, or CC (Bannister 1968, Gauch & Whittaker 1972, article 10). The requirement of strict orthogonality of axes for the representation of field relationships seems questionable; we doubt that the additional computations necessary to produce strictly orthogonal axes are justified. The use of single endpoints for orthogonal second and higher axes in Orlóci's (1966) 'simple ordination' produces severe distortion (Gauch et al. 1976, article 10, Fig. 9). Scruggs (1976) has suggested replacing the Pythagorean determination of sample position by a simple proportionality, $x' - D_1/(D_1 - D_2)$, which gives smaller displacements in sample positions when axis length is less than 3.7 halfchanges. A series of evaluations of polar ordination against other techniques (Gauch & Whittaker 1972, Kessell & Whittaker 1976, Gauch et al. 1976) have given results generally favorable to polar ordination (article 10). However, more recent indirect techniques, particularly reciprocal averaging, seem likely to have the advantage over polar ordination for some uses (articles 10 and 11).

Polar ordination has been applied in several studies involving other than phytosociological data. For example, Hole & Hironaka (1960) and Monk (1965) have used the technique to order samples by soil characteristics. Beals (1960) applied parallel ordination procedures to birds and forest tree community data and then compared the patterns to infer relationships between these two biological strata. Loucks (1962) used a similar approach to compare the vegetation and environment of New Brunswick forests. His environmental model was constructed by numerically combining several environmental variables into three synthetic 'scalars' which were then plotted orthogonally to form the three-dimensional model. This was compared to a polar ordination of vegetation.

Knight used polar ordination in his analysis of the 'structural-functional' aspects of Wisconsin prairies (1965) and forests (Knight & Loucks 1969). In place of species he used such features as life-form, bark thickness, time of blooming, and drought resistance as the basis for stand comparison and ordination. After constructing the ordination he expressed stand positions in terms of new axes that seemed to account for the greatest amount of variation in the characteristics initially used to construct the model. These characteristics were then scaled according to the positions of their peaks along the new axes. This amounts to an axis rotation not greatly different from the type commonly used in factor analysis, but again it was independently derived and is mathematically less demanding than other possible methods.

In '*The Vegetation of Wisconsin*' CURTIS (1959) employed ordination to order entire community-types, rather than individual stands (article 6). All species that were 'prevalent' in any community were included in the analysis. Each of 28 terrestrial community-types was represented by the percentage presence of these prevalent species in samples of the type. Index of similarity values were calculated between these composite samples for community-types, but the model was constructed by means of direct triangulation, starting with the most central community and building outward, rather than by use of end-points. REAM (1963) likewise constructed a model of community-types, in this case those of the Wasatch mountain range in central Utah. He used polar ordination for both the initial stand model and the subsequent model of types. Polar ordination has also been applied to community-types by FRYDMAN & WHITTAKER (1968) and MAAREL (1969).

7.4 **Conclusion**

In conclusion, the principal contributions of the Wisconsin school appear to fall into three major categories. The first is in contributing to and stimulating research toward analysis of vegetation as a more or less continuous variable rather than as categorized types. The approach used at Wisconsin was based on the philosophy that the plants themselves are their own best indicators of environmental conditions, and that when ordered as described above the relationships of stands or species to the environmental complexes to which the plants were responding should be discernible. Much remains to be done in defining these relationships. It seems clear that the plants are seldom ordered along single-factor gradients, and that modeling of the operating of environment through time upon growth of individuals and populations will be necessary to explain in detail the correlations between species populations and environmental complexes. Yet the attempt to relate phytosociological ordinations and synthetic environmental variables, including stand structural variables containing information related to development of the included populations through time, represents a first-order attempt to define and deal quantitatively with ecological reality at the regional or landscape level.

The second major contribution is in the variety of techniques that were devised or adapted to refine the ordination of the various communities to which the methods were applied. No two of the major papers were based on identical techniques (see also McIntosh 1967, Whittaker 1967). It appeared at one time that principal components analysis or other more formal techniques might replace polar ordination as the preferred technique of vegetation analysis (Orloci 1966, Austin & Orloci 1966). Further experience with principal components analysis has, however, been disappointing because of the technique's vulnerability to distortion (Noy-Meir & Austin 1970, Jeglum et al. 1971, Austin & Noy-Meir 1972, article 10). Wisconsin polar ordination may well be the sturdiest and most versatile of ordination techniques, equally adaptable to direct and indirect applications, as well as the least demanding of computational effort.

A third and most important contribution to date was the application of ordination methods to the entire vegetation of a major geographical unit, the state of Wisconsin. J. T. Curtis's book, '*The Vegetation of Wisconsin*' (1959, 1971), which is a synthesis of the work of many Wisconsin students, stands as a landmark in vegetational studies and provides a comprehensive treatment of the whole vegetation of an area.

7.5 Summary

Early ordination in Wisconsin was based on compositional gradient analysis. Vegetation samples were classified into dominance-types, and samples representing these types were arranged into an ecological series by the slopes of the populations of dominant species ('method of leading dominants'). Species were given numbers by their relative positions in the ecological series, and the numbers were used as weights to obtain compositional indices for samples expressing their positions along the gradient. The samples were then ordinated and compiled into transects by means of the compositional indices .

Later work has been based on the Wisconsin comparative or polar ordination of Bray & Curtis (1957) and variants thereof. Given a set of vegetation samples to be ordinated, similarity measures may be computed for each sample compared with every other sample. Two end-point samples are chosen either by maximum dissimilarity from the rest of the set and one another, or because they represent extremes of an environmental gradient. The similarity measures are used to arrange all other samples in sequence between the end-point pair along the axis. A second pair of end-point samples dissimilar from the first pair and one another are chosen, and all samples are arranged along the axis between the second pair. The two axes become coordinates of an ordination field in which all the samples may be located, and in which values for environmental variables, species populations, and community characteristics may be plotted at the points for samples. Samples may be ordinated in relation to third and higher axes when this is appropriate.

The technique is illustrated with a simple example. A variety of modifications that have been experimented with in the school of Wisconsin or suggested by others are mentioned. Experience and research comparing different ordinations suggest: (i) the continued value of the Wisconsin polar ordination as a most versatile and least vulnerable technique, (ii) its equal effectiveness for direct and indirect ordination, and (iii) its real value in research as a framework for investigating and understanding of vegetational relationships. Principal contributions of Wisconsin ecology include the development of the technique, the exploration of various alternatives, and the achievement by CURTIS (1959) of an ordination and description of the vegetation of the state.

REFERENCES

ANDERSON, D. J., – 1963 – The structure of some upland plant communities in Caernarvonshire. III. The continuum analysis. *J. Ecol.* 51: 403–414.

ANDERSON, D. J., R. C. COOKE, T. T. ELKINGTON, & D. J. READ, – 1966 – Studies on structure in plant communities. II. The structure of some dwarf-heath and birch-copse communities in Skjaldfannardalur, north-west Iceland. *J. Ecol.* 54: 781–793.

ARNO, S. F. & J. R. HABECK, – 1972 – Ecology of alpine larch (*Larix lyallii* Parl.) in the Pacific Northwest. *Ecol. Monogr.* 42: 417–450.

ASHTON, P. S., – 1964 – Ecological studies in the mixed dipterocarp forests of Brunei State. *Oxford For. Mem.* 25: 1–75.

AUSTIN, M. P. & I. NOY-MEIR, – 1972 – The problem of non-linearity in ordination: experiments with two-gradient models. *J. Ecol.* 59: 763–774.

AUSTIN, M. P. & L. ORLOCI, – 1966 – Geometric models in ecology. II. An evaluation of some ordination techniques. *J. Ecol.* 54: 217–227.

AYYAD, M. A. G. & R. L. DIX, – 1964 – An analysis of a vegetation-microenvironmental complex on prairie slopes in Saskatchewan. *Ecol. Monogr.* 34: 421–442.

BANNISTER, P., – 1968 – An evaluation of some procedures used in simple ordinations. *J. Ecol.* 56: 27–34.

BEALS, E. W., – 1960 – Forest bird communities in the Apostle Islands of Wisconsin. *Wilson Bull.* 72: 156–181.

BEALS, E. W., – 1965 – Ordination of some corticolous cryptogamic communities in south-central Wisconsin. (Russ. summ.) *Oikos* 16: 1–8.

BEALS, E. W. & G. COTTAM, – 1960 – The forest vegetation of the Apostle Islands, Wisconsin. *Ecology* 41: 743–751.

BOND, R. R., – 1957 – Ecological distribution of breeding birds in the upland forests of southern Wisconsin. *Ecol. Monogr.* 27: 351–384.

BRAY, J. R., – 1956 – A study of mutual occurrence of plant species. *Ecology* 37: 21–28.

Bray, J. R., – 1960 – The composition of savanna vegetation in Wisconsin. *Ecology* 41: 721–732.

Bray, J. R. & J. T. Curtis, – 1957 – An ordination of the upland forest communities of southern Wisconsin. *Ecol. Monogr.* 27: 325–349.

Brown, R. T. & J. T. Curtis, – 1952 – The upland conifer-hardwood forests of northern Wisconsin. *Ecol. Monogr.* 22: 217–234.

Buell, M. F., A. N. Langford, D. W. Davidson, & L. F. Ohmann, – 1966 – The upland forest continuum in northern New Jersey. *Ecology* 47: 416–432.

Christensen, E. M., J. J. Clausen, & J. T. Curtis, – 1959 – Phytosociology of the lowland forests of northern Wisconsin. *Am. Midl. Nat.* 62: 232–247.

Christensen, M., W. F. Wittingham, & R. O. Novak, – 1962 – The soil microfungi of wet-mesic forests in southern Wisconsin. *Mycologia* 54: 374–388.

Clausen, J. J., – 1957 – A phytosociological ordination of the conifer swamps of Wisconsin. *Ecology* 38: 638–646.

Culberson, W. L., – 1955 – The corticolous communities of lichens and bryophytes in the upland forests of northern Wisconsin. *Ecol. Monogr.* 25: 215–231.

Curtis, J. T., – 1955 – A prairie continuum in Wisconsin. *Ecology* 36: 558–566.

Curtis, J. T. –1959 – The Vegetation of Wisconsin: An Ordination of Plant Communities. Univ. Wisconsin, Madison., Reprinted 1971. 657 pp.

Curtis, J. T. & R. P. McIntosh, – 1951 – An upland forest continuum in the prairie-forest border region of Wisconsin. *Ecology* 32: 476–496.

Frydman, I. & R. H. Whittaker, – 1968 – Forest associations of southeast Lublin Province, Poland. (Germ. summ.) *Ecology* 49: 896–908.

Gauch, H. G., Jr., – 1973a – The relationship between sample similarity and ecological distance. *Ecology* 54: 618–622.

Gauch, H. G., Jr., – 1973b – A quantitative evaluation of the Bray-Curtis ordination. *Ecology* 54: 829–836.

Gauch, H. G., Jr. & R. H. Whittaker, – 1972 – Comparison of ordination techniques. *Ecology* 53: 868–875.

Gauch, H. G., Jr., R. H. Whittaker & T. R. Wentworth, – 1976 – A comparative study of reciprocal averaging and other ordinations. *J. Ecol.* (in press).

Gehlbach, F. R., – 1967 – Vegetation of the Guadalupe Escarpment, New Mexico-Texas. *Ecology* 48: 404–419.

Gemborys, S. R., – 1974 – The structure of hardwood forest ecosystems of Prince Edward County, Virginia. *Ecology* 55: 614–621.

Gemborys, S. R. & E. J. Hodgkins, – 1971 – Forests of small stream bottoms in the Coastal Plain of southwestern Alabama. *Ecology* 52: 70–84.

Gilbert, M. L. & J. T. Curtis, – 1953 – Relation of the understory to the upland forest in the prairie-forest border region of Wisconsin. *Trans. Wis. Acad. Sci. Arts Lett.* 42: 183–195.

Gimingham, C. H., N. M. Pritchard, & R. M. Cormack, – 1966 – Interpretation of a vegetational mosaic on limestone in the island of Gotland. *J. Ecol.* 54: 481–502.

Gittins, R., – 1965a – Multivariate approaches to a limestone grassland community. I. A stand ordination. *J. Ecol.* 53: 385–401.

Gittins, R., – 1965b – Multivariate approaches to a limestone grassland community. II. A direct species ordination. *J. Ecol.* 53: 403–409.

Gittins, R., – 1965c – Multivariate approaches to a limestone grassland community. III. A comparative study of ordination and association-analysis. *J. Ecol.* 53: 411–425.

Goff, F. G. & G. Cottam, – 1967 – Gradient analysis: the use of species and synthetic indices. *Ecology* 48: 793–806.

GREIG-SMITH, P., – 1964 – Quantitative Plant Ecology. 2nd ed. Butterworths, London. 256 pp.

HALE, M. E., Jr., – 1955 – Phytosociology of corticolous cryptogams in the upland forests of southern Wisconsin. *Ecology* 36: 45–63.

HOLE, F. D. & M. HIRONAKA, – 1960 – An experiment in ordination of some soil profiles. *Proc. Soil Sci. Soc. Am.* 24: 309–312.

JEGLUM, J. K., C. F. WEHRHAHN, & J. M. A. SWAN, – 1971 – Comparisons of environmental ordinations with principal component vegetational ordinations for sets of data having different degrees of complexity. (French summ.) *Can. J. For. Res.* 1: 99–112.

KESSELL, S. R. & R. H. WHITTAKER, – 1976 – Comparisons of three ordination techniques. *Vegetatio* 32: 21–29.

KENDALL, M. G., – 1965 – A Course in Multivariate Analysis. Griffins, London.

KNIGHT, D. H., – 1965 – A gradient analysis of Wisconsin prairie vegetation on the basis of plant structure and function. *Ecology* 46: 744–747.

KNIGHT, D. H. & O. L. LOUCKS, – 1969 – A quantitative analysis of Wisconsin forest vegetation on the basis of plant function and gross morphology. *Ecology* 50: 219–234.

LAFRANCE, C. R., – 1972 – Sampling and ordination characteristics of computer-simulated individualistic communities. *Ecology* 53: 387–397.

LARSEN, J. A., – 1965 – The vegetation of the Ennadai Lake Area, N. W. T.: studies in subarctic and arctic bioclimatology. *Ecol. Monogr.* 35: 37–59.

LINDSEY, A. A., R. O. PETTY, D. K. STERLING, & W. VAN ASDALL, – 1961 – Vegetation and environment along the Wabash and Tippecanoe Rivers. *Ecol. Monogr.* 31: 105–156.

LOUCKS, O. L., – 1962 – Ordinating forest communities by means of environmental scalars and phytosociological indices. *Ecol. Monogr.* 32: 137–166.

MAAREL, E. VAN DER, – 1969 – On the use of ordination models in phytosociology. (Germ. summ.) *Vegetatio* 19: 21–46.

MAYCOCK, P. F. & J. T. CURTIS, – 1960 – The phytosociology of boreal conifer-hardwood forests of the Great Lakes region. *Ecol. Monogr.* 30: 1–35.

MCINTOSH, R. P., – 1958 – Plant communities. *Science, N.Y.* 128: 115–120.

MCINTOSH, R. P., – 1967 – The continuum concept of vegetation. *Bot. Rev.* 33: 130–187.

MCINTOSH, R. P. & R. T. HURLEY, – 1964 – The spruce-fir forests of the Catskill Mountains. *Ecology* 45: 314–326.

MONK, C. D., – 1965 – Southern mixed hardwood forest of northcentral Florida. *Ecol. Monogr.* 35: 335–354.

MUELLER-DOMBOIS, D. & H. ELLENBERG, – 1974 – Aims and Methods of Vegetation Ecology. Wiley, New York. 547 pp.

NOY-MEIR, I. & M. P. AUSTIN, – 1970 – Principal component ordination and simulated vegetational data. *Ecology* 51: 551–552.

ORLOCI, L., – 1966 – Geometric models in ecology. I. The theory and application of some ordination methods. *J. Ecol.* 54: 193–215.

ORLÓCI, L., – 1974 – Revisions for the Bray and Curtis ordination. (French summ.) *Can. J. Bot.* 52: 1773–1776.

ORLÓCI, L., – 1975 – Multivariate Analysis in Vegetation Research. Junk, The Hague. 276 pp.

REAM, R. R., – 1963 – The vegetation of the Wasatch Mountains, Utah and Idaho. *Thesis, Univ. of Wisconsin.*

ROGERS, D. J., – 1970 – A preliminary ordination study of forest vegetation in the

Kirchleerau area of the Swiss Midland. (Germ. summ.) *Ber. geobot. Inst. ETH, Stiftg. Rübel*, Zürich, 1969, 40: 28–78.
ROWE, J. S., – 1956 – Uses of undergrowth plant species in forestry. *Ecology* 37: 461–473.
SCRUGGS, W. M., – 1976 – Revised coordinate determination for the Bray and Curtis ordination (manuscript).
SEAL, H. L., – 1964 – Multivariate Statistical Analysis for Biologists. John Wiley and Sons, Inc., New York.
SWAN, J. M. A., – 1970 – An examination of some ordination problems by use of simulated vegetational data. *Ecology* 51: 89–102.
SWAN, J. M. A. & R. L. DIX, – 1966 – The phytosociological structure of upland forest at Candle Lake, Saskatchewan. *J. Ecol.* 54: 13–40.
SWAN, J. M. A., R. L. DIX, & C. F. WEHRHAHN, – 1969 – An ordination technique based on the best possible stand-defined axes and its application to vegetational analyss. *Ecology* 50: 206–212.
SWINDALE, D. N. & J. T. CURTIS, – 1957 – Phytosociology of the larger submerged plants in Wisconsin lakes. *Ecology* 38: 397–407.
TRESNER, H. D., M. P. BACKUS, & J. T. CURTIS, – 1954 – Soil microfungi in relation to the hardwood forest continuum in southern Wisconsin. *Mycologia* 46: 314–333.
VITT, D. H. & N. G. SLACK, – 1975 – An analysis of the vegetation of Sphagnum-dominated kettle-hole bogs in relation to environmental gradients (French summ.) *Canad. J. Bot.* 53: 332–359.
WARING, R. H. & J. MAJOR, – 1964 – Some vegetation of the California coastal redwood region in relation to gradients of moisture, nutrients, light, and temperature. *Ecol. Monogr.* 34: 167–215.
WHITE, K. L., – 1965 – Shrub-carrs of southeastern Wisconsin. *Ecology* 46: 286–304.
WHITTAKER, R. H., – 1951 – A criticism of the plant association and climatic climax concepts. *NW. Sci.* 25: 17–31.
WHITTAKER, R. H., – 1956 – Vegetation of the Great Smoky Mountains. *Ecol. Monogr.* 26: 1–80.
WHITTAKER, R. T., – 1967 – Gradient analysis of vegetation. *Biol. Rev.* 42: 207–264.

FACTOR ANALYSIS

Pierre Dagnelie

Contents

8 FACTOR ANALYSIS

8.1 Definition

At the beginning, factor analysis encompassed methods taken from experimental psychology and mainly used in the statistical analysis of intelligence or behavioral tests (Burt 1949, Harman 1967, Vincent 1953). Progressively, however, these methods were integrated within the frame of classical statistical methods (Kendall 1950, Kendall & Lawley 1956, Lawley 1955, Lawley & Maxwell 1963), and at present, factor analysis stands among the methods of multivariate statistical analysis, along with component analysis, discriminant analysis and analysis of variance.

8.2 A Simple Example

The best way to understand the meaning and the main elements in factor analysis is to take a simple example from the field of psychology.

Let us consider three tests to which several persons have been subjected and let us call z_1, z_2, and z_3, the observed variables corresponding to the results of the three tests. We shall assume that the scores z_j ($j = 1,2,3$) depend linearly upon two fundamental variables x_1 and x_2 common to two or more tests and, in each case, upon one fundamental variable v_j peculiar to the test. Thus, the mathematical model will be:

$$z_j = a_{j1}x_1 + a_{j2}x_2 + u_j v_j \qquad (j = 1,2,3),$$

the coefficients a_{j1}, a_{j2} and u_j, called loadings, indicating to what extent the scores z_j respectively depend on the variables x_1, x_2 and v_j.

We shall also assume that all the variables considered are reduced, having zero means and unit standard deviations, and that the fundamental variables x_1, x_2, and v_j ($j = 1,2,3$) are independent of each other. Under these conditions, the loadings are actually the correlation coefficients of the fundamental variables with the observed variables.

For instance, the three tests under consideration could be composed of reasoning problems (z_1), arithmetic problems (z_2) and addition computations (z_3). The fundamental aptitudes which could influence the results obtained by the different persons could be, in this case, a reasoning aptitude (x_1) and a computation aptitude (x_2). The variable z_1 will then be very closely related to x_1; in the same way, z^1 would depend mainly on x_2, whereas z_2, would be greatly influenced by both x_1 and x_2. We could for example observe the following relations:

$$z_1 = 0{,}70x_1 + 0{,}10x_2 + 0{,}71v_1$$
$$z_2 = 0{,}40x_1 + 0{,}70x_2 + 0{,}59v_2$$
$$z_3 = 0{,}10x_1 + 0{,}70x_2 + 0{,}71v_3,$$

where v_1, v_2 and v_3 are the fundamental variables peculiar to each one of the tests.

The loadings of the common factors x_1 and x_2 can be put together in a two-way table (Table I), to which we usually add the common variances:

$$h_j^2 = a_{j1}^2 + a_{j2}^2 \qquad (j = 1,2,3),$$

and the variances resulting from each of the factors:

$$\lambda_1 = a_{1i}^2 + a_{2i}^2 + a_{3i}^2 \qquad (i = 1,2).$$

The common variances, often called communalities, measure the importance of all the common factors in determining the results of each of the tests, whereas the factor variances measure the relative importance of the different common factors in determining the results of all the tests.

TABLE I. Loadings, common variances and variances of common factors x_1 and x_2, for three psychological tests.

j	a_{j1}	a_{j2}	h_j^2
1	0,70	0,10	0,50
2	0,40	0,70	0,65
3	0,10	0,70	0,50
λ_i	0,66	0,99	1,65

The loadings in Table I can also be presented as a correlation matrix, gathering all correlation coefficients between the results from various tests taken two at a time. Indeed, it will be easily proved that for each pair of tests (j, k):

$$r_{jk} = a_{j1}a_{k1} + a_{j2}a_{k2},$$

which simply means that two tests are more closely correlated when they depend on the same fundamental aptitudes. The coefficients obtained this way for the example under consideration are indicated in Table II.

TABLE II. Correlation matrix between three psychological tests.

j \ k	1	2	3
1	1	0,35	0,14
2	0,35	1	0,53
3	0,14	0,53	1

But actually, the problem of factor analysis is opposite to what we have considered thus far: once we have subjected some persons to various tests, we find the observed correlation coefficients and we try to get a loadings table. In these conditions, the aim of factor analysis is, on one hand, to explain the correlations observed between the tests by spotting some fundamental underlying aptitudes, and, on the other hand, to identify so far as possible these fundamental aptitudes.

Moreover, as soon as we are considering observed values instead of theoretical values, we must add an uncertainty or error term to the mathematical mode. This term is often encompassed in the specific factors, which could then be divided for each test into a true specific component (systematic component) and an error component (random component). In the case of our example, which is not all artificial, the correlations actually observed for a group of 145 persons are given by HARMAN (1967), and scarcely differ from those of Table II.

However, when looked at in this way, the problem of factor analysis can generally have an infinite number of solutions. The observed correlations can not only be explained by supposing the presence of the common factors x_1 and x_2, which would come respectively in the proportions a_{j1} and a_{j2} (Table I), but also by supposing for example the presence of the common factors x_1 and x_2 and of the corresponding loadings shown in Table III. From this last table, we can in fact recalculate the correlation coefficients, as was done previously, by the following relation:

$$r_{jk} = a'_{j1}a'_{k1}+a'_{j2}a'_{k2}.$$

TABLE III. Loadings, common variances and variances of common factors x'_1 and x'_2, for three psychological tests.

j	a'_{j1}	a'_{j2}	h^2_j
1	0,50	−0,50	0,50
2	0,80	0,10	0,65
3	0,62	0,34	0,50
λ'_1	1,27	0,38	1,65

It is therefore necessary to choose the most interesting solution(s) out of the innumerable possible ones. Various methods of loading computations help us make this choice.

The existence of an infinite number of solutions is not consequential, as we can see on the graphic representation of the tests. By assigning to each factor a coordinate axis in a rectangular cartesian system, it is possible to make each test correspond to a point for which the coordinates equal the loadings. For example, Fig. 1 represents the three tests studied so far (1, 2 and 3) in two systems of coordinates, corresponding to two possible solutions (Tables I and III).

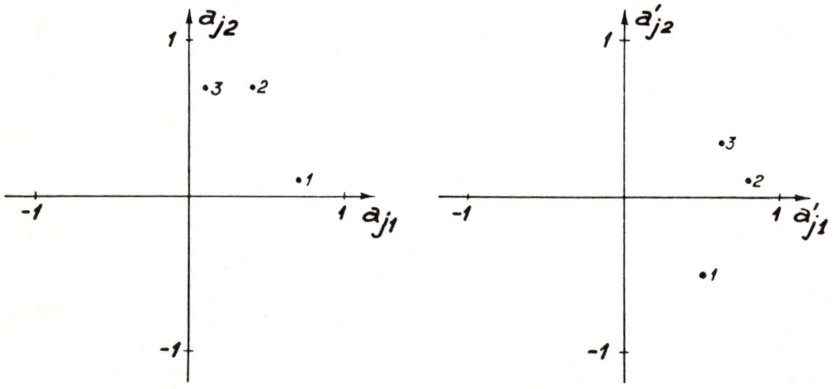

Fig. 1. Graphic representation of three psychological tests in two systems of coordinates.

The comparison of these two graphs shows that the relative positions of the origin and the three points representing the tests are identical: in order to switch from one system of factors to the other, we simply need to rotate the axes around the origin. Conversely, the loadings obtained by projecting the test-points on any orthogonal axis system with the same origin enable us to reconstruct the correlation matrix, as shown above.

Finally, since we know the initial scores for a given number of persons, and when the loadings are computed, we may also want to estimate, for the persons under consideration, what the values of the fundamental variables would be, that is the scores of these persons in the various fundamental aptitudes. This procedure, called estimation or measurement of factors, allows us to draw a graphic representation of the persons by having each correspond to one point, the coordinates of which are the estimated fundamental scores.

It is interesting to compare these two types of graphic representation: the graph of tests in what we will call the loadings space and the graph of persons in the space of common factors or factorial space. The first is such that two tests indicating similar aptitudes are represented by different points located close to one another. Moreover, we can easily show that the distances separating the test-points from the origin are equal to the square roots of the common variances, and therefore reflect the precision involved in the tests indicating the common fundamental aptitudes. In the second graph, two persons with similar aptitudes are also represented by neighbouring points, but the distances between these points and the origin do not have the same particular meaning.

To make the interpretation easier, it is also possible to replace, in both graphs, orthogonal axes with oblique axes.

8.3 General Principles

Everything given above can be generalized to the case of n tests depending on r factors ($n \geqq r$). For each test ($j = 1, \ldots, n$) we have:

$$z_j = a_{j1}x_1 + a_{j2}x_2 + \ldots + a_{jr}x_r + u_j v_j = \sum_{i=1}^{r} a_{ji}x_i + u_j v_j$$

and
$$h_j^2 = a_{j1}^2 + a_{j2}^2 + \ldots + a_{jr}^2 = \sum_{i=1}^{r} a_{ji}^2;$$

for each couple of tests (j and $k = 1, \ldots, n$):

$$r_{jk} = a_{j1}a_{k1} + a_{j2}a_{k2} + \ldots + a_{jr}a_{kr} = \sum_{i=1}^{r} a_{ji}a_{ki};$$

and for each factor ($i = 1, \ldots, r$):

$$\lambda_i = a_{1i}^2 + a_{2i}^2 + \ldots + a_{ni}^2 = \sum_{j=1}^{n} a_{ji}^2.$$

In practice, the factor analysis of the scores z_{jp}, obtained by subjecting a number of persons ($p = 1, \ldots, N$) to a number of tests ($j = 1, \ldots, n$), is then realized through the following process.

Starting from the scores z_{jp}, one computes the table or matrix of the observed correlations r_{jk}, as well as a first estimate of the common variances h_j^2. This first estimate can be computed with the help of multiple correlation coefficients, especially for small correlation matrices, or with the help of the highest correlation coefficient values, especially for large matrices.

From these observed correlation coefficients and from these common variances, one can then deduce the loadings, for instance by using the principal factor or the centroid method. In the first method, one must use the smallest number of factors by bringing to light the factors which have the greatest variances. In matrix terms, the factor variances λ_i are in fact the eigenvalues or latent roots of the correlation matrix in which the diagonal elements are the common variances and the loadings a_{ji} are the elements of the corresponding vectors[1]. The centroid method does not have the optimal properties of the principal factor analysis; however, it does give very similar results even though it requires much less work.

As the work on loadings proceeds, one must test as much as possible the meaning of the factors thus in evidence, in order to know at what point the computation process should be interrupted. It must be noted, however, that this test of significance problem has not yet been satisfactorily solved in the frame of principal factor and centroid methods. In the absence of completely valid tests, attention must particularly be paid to whether the various factors can or cannot find a logical interpretation.

Once the number of factors to be considered is decided upon, and once the loadings have been estimated in a first approximation, we can eventually get a better estimate of the common variances and deduce from it, as has been said previously, a better estimate of the loadings. However, the use of this repeated process is not always indispensable, especially for large correlation matrices.

Finally, it is possible to compute the estimate of factors, through multiple regression, and to step to the interpretation of results with the help of various types of graphic representation.

Additional information relative to these different problems can be found in a previously published paper (DAGNELIE 1960) or in some specialized works, among which the most interesting one today, for non-psychologists, seems to be the book of HARMAN (1967). Other works, older or more oriented towards psychological

1) The computation of eigenvalues and eigenvectors can also be done by keeping the values 1 in the main diagonal: this will give the principal components or principal axes in the frame of what is called component analysis. The aim of this method is to sum up the essential content of some interdependent variables into a more restricted number of independent variables (DAGNELIE 1966, KENDALL 1957, KENDALL & LAWLEY 1956). Although fundamentally different factor analysis and component analysis are very often mistaken; in both cases, the results obtained for large correlation matrices are very similar (DAGNELIE 1960).

problems, were written by BURT (1940), CATTELL (1952), FAVERGE (1954), HOLZINGER & HARMAN (1941), HORST (1956), THOMSON (1956), THURSTONE (1947) and VERNON (1952).

8.4 A Phytosociological Example

As an example, we shall summarize here the study that was previously made from 80 relevés of beech forests (DAGNELIE 1956, 1960), carried out according to the method of BRAUN-BLANQUET (1951) in a well-known type of vegetation (NOIRFALISE 1956).

8.4.1 CORRELATION BETWEEN SPECIES, LOADINGS AND COMMON VARIANCES

Aiming only to reduce as much as possible the amount of computation, we have discarded from this analysis the rarest species, which cannot provide us with really interesting information. For the 38 species which appeared in at least 10 % of the releves, we have determined a point correlation coefficient matrix, by taking into consideration either the presence or absence of each species, or, for the most common of them, their more or less high occurrence. Table IV lists all considered species and Table V shows a portion of the correlation matrix.

The loadings of the first three factors were determined through the centroid method by using the highest correlation coefficients as first estimates of the common variances; table IV gives the results thus obtained. Moreover, the χ^2 signifiance test of BURT (1952) shows that the first two factors are highly significant, the third factor being just significant. Later on, in order to simplify graphic representations, we shall essentially consider the interpretation of the first two factors.

The first remark bears on the existence of a close relation between the common variances of species and their differential or indicator power, as it is recognized by phytosociologists. On the basis of the three factors, the species are classified as follows, in the decreasing order of common variances:

0,40 and above: *Festuca silvatica*, *Acer pseudoplatanus* (A), *Oxalis acetosella* and *Deschampsia caespitosa*;

from 0,39 to 0,30; *Deschampsia flexuosa*, *Anemone nemorosa*, *Fagus silvatica* (h), *Acer pseudoplatanus* (h), *Milium effusum*, *Juncus* sp., *Agrostis vulgaris*, *Dicranum scoparium* and *Vaccinium myrtillus*;

from 0,29 to 0,20: *Quercus sessilis* (A), *Luzula nemorosa*, *Carex pilulifera*, *Picea excelsa* (h), *Polygonatum verticillatum*, *Carex remota*, *Leucobryum glaucum*, *Rubus idaeus*, *Epilobium spicatum*, *Fagus silvatica* (A), *Dryopteris spinulosa*, *Carex leporina*, *Dicranella heteromalla* and *Polytrichum formosum*;

from 0,19 to 0,10: *Quercus sessilis* (h), *Athyrium filix femina*, *Rubus* sp., *Hypnum loreum*, *Fagus silvatica* (a), *Sorbus aucuparia* (h), *Thuidium tamariscinum*, *Luzula silvatica* and *Eurynchium myosuroides*;

less than 0,10: *Mnium hornum* and *Hypnum cupressiforme*.

The species generally considered as companions are gathered at the bottom of the scale of common variances. Joining them in the sections from 0,19 to 0,10 and below 0,10 are only *Athyrium filix femina*, *Luzula silvatica* and some woody species from lower layers (*Quercus sessilis*, *Fagus silvatica* and *Sorbus aucuparia*). However, we must note that common variances are to be considered as indications of the differential value of species within the framework of the releves under consideration, and not indications of the value of species as character-species or, in a more general way, as habitat indicators: indeed, a character-species, randomly distributed in the community under consideration, is not in theory related to any other species inside such a community. Theoretically, its common variance should be zero and it would then be wrong to assume that weak common variances are typical of species with weak indicator potentialities.

8.4.2 Ecological and Sociological Groups

Figure 2, drawn from the loadings in Table IV, shows how the selected species are distributed in the factorial space relative to the first two factors. From this diagram, we could then define four groups of species.

A first group (A) includes the differential species from fresh and fertile soils: *Anemone nemorosa*, *Festuca silvatica*, *Milium effusum*, *Polygonatum verticillatum*, *Athyrium filix femina*, *Oxalis acetosella* and *Deschampsia caespitosa*.

The second group (B) essentially includes the differential species from fresh to moist, but not necessarily fertile soils: *Carex leporina*, *Epilobium spicatum*, *Luzula nemorosa*, *Dryopteris spinulosa*, *Rubus idaeus*, *Rubus* sp., *Fagus silvatica* (h), *Juncus* sp. and *Carex remota*.

As for group C, it includes the species from poor and frequently dry soils; *Dicranum scoparium*, *Deschampsia flexuosa*, *Vaccinium myrtillus*, *Quercus sessilis* (A) and *Leucobryum glaucum*.

TABLE IV. List of species, loadings and common variances for 80 relevés of beech forests.

j	Species (§)	Loadings			Common variances (h_j^2)
		a_{j1}	a_{j2}	a_{j3}	
1	*Fagus silvatica* (A)	—0,06	—0.35	—0,34	0,24
2	*Acer pseudoplatanus* (A)	0,39	—0,50	0,33	0,51
3	*Quercus sessilis* (A)	—0,48	0,18	0,17	0,29
4	*Fagus silvatica* (a)	—0,15	0,17	0,30	0,14
5	*Fagus silvatica* (h)	0,54	0,29	0,09	0,38
6	*Acer pseudoplatanus* (h)	0,23	—0,16	0,54	0,38
7	*Quercus sessilis* (h)	—0,09	0,41	0,13	0,19
8	*Sorbus aucuparia* (h)	—0,31	0,10	—0,12	0,12
9	*Milium effusum*	0,55	—0,21	0,11	0,36
10	*Festuca silvatica*	0,65	—0,26	0,21	0,54
11	*Luzula nemorosa*	0,50	0,15	—0,11	0,29
12	*Oxalis acetosella*	0,59	—0,11	0,36	0,48
13	*Deschampsia caespitosa*	0,65	—0,08	0,17	0,46
14	*Juncus* sp.	0,51	0,30	0,08	0,36
15	*Carex remota*	0,34	0,29	—0,25	0,26
16	*Carex leporina*	0,44	0,10	0,15	0,23
17	*Vaccinium myrtillus*	—0,47	0,27	0,18	0,32
18	*Deschampsia flexuosa*	—0,53	0,30	—0,11	0,39
19	*Carex pilulifera*	0,14	0,24	—0,46	0,29
20	*Polytrichum formosum*	0,09	0,35	0,25	0,20
21	*Dicranum scoparium*	—0,33	0,33	0,35	0,34
22	*Dicranella heteromalla*	0,17	—0,31	—0,29	0,21
23	*Agrostis vulgaris*	0,36	0,42	—0,23	0,35
24	*Rubus idaeus*	0,41	0,26	—0,26	0,25
25	*Rubus* sp.	0,31	0,25	—0,03	0,16
26	*Dryopteris spinulosa*	0,33	0,16	0,32	0,24
27	*Mnium hornum*	—0,15	—0,22	0,13	0,09
28	*Thuidium tamariscinum*	—0,13	0,13	0,29	0,12
29	*Eurynchium myosuroides*	—0,26	—0,18	—0,11	0,11
30	*Hypnum cupressiforme*	—0,13	0,08	0,19	0,06
31	*Hypnum loreum*	—0,22	0,16	0,29	0,16
32	*Picea excelsa* (h)	0,16	0,43	—0,24	0,27
33	*Anemone nemorosa*	0,51	—0,30	0,19	0,39
34	*Polygonatum verticillatum*	0,47	—0,16	0,16	0,27
35	*Athyrium filix femina*	0,38	—0,08	0,21	0,19
36	*Luzula silvatica*	0,18	0,27	0,11	0,12
37	*Leucobryum glaucum*	—0,38	0,11	0,32	0,26
38	*Epilobium spicatum*	0,47	0,13	0,13	0,25
	Variances (λ_i)	5,56	2,48	2,23	10,27

(§) A = tree layer; a = shrublayer; h = herb layer.

Finally, a last group (D), much less well defined, includes the companions, which are close to the origin of the axes: the distance from the origin to the point that represents a species is indeed equal

TABLE V: Part of the correlation matrix of 38 species studied in 80 relevés of beech forests (only the first ten species are indicated).

Species (j)	1	2	3	4	5	6	7	8	9	10
1	1									
2	0,059	1								
3	—0,285	—0,261	1							
4	—0,234	—0,040	0,175	1						
5	—0,183	0,156	—0,193	0,096	1					
6	—0,150	—0,515	—0,020	0,033	0,227	1				
7	—0,058	—0,088	0,082	0,282	0,275	0,255	1			
8	0,063	—0,109	0,152	0,089	—0,128	—0,194	0,181	1		
9	0,008	0,389	—0,319	0,015	0,200	0,119	—0,126	—0,148	1	
10	0,029	0,427	—0,302	—0,041	0,358	0,338	—0,164	—0,334	0,444	1
⋮	⋮	⋮	⋮	⋮	⋮	⋮	⋮	⋮	⋮	⋮

to the square root of the common variance of this species. The circumference which was drawn with the origin as a center corresponds to a common variance of 0,10; its radius is equal to $\sqrt{0,10}$. Within this circle and just outside it, we can find the following species: *Carex pilulifera*, *Luzula silvatica*, *Polytrichum formosum*, *Fagus silvatica* (a), *Thuidium tamariscinum*, *Hypnum cupressiforme*, *Hypnum loreum*, *Sorbus aucuparia* (h), *Eurynchium myosuroides*, *Mnium hornum*, *Fagus silvatica* (A), *Dicranella heteromalla* and *Acer pseudoplatanus* (h).

Each of the first three groups includes species which are closely related to one another and most likely also linked to particular ecological conditions; they can be considered as ecological or sociological groups. On the other hand, the species from the last group probably have no particular ecological meaning within the type of vegetation under consideration, with the possible exception of those which are the furthest from the origin, especially *Luzula silvatica* and *Sorbus aucuparia*. Finally, some species have not yet been mentioned: *Agrostis vulgaris*, *Picea excelsa* (h), *Quercus sessilis* (h) and *Acer pseudoplatanus* (A). Some of these can eventually be added to the sociological groups which were just defined, especially *Acer pseudoplatanus* (group A) and *Agrostis vulgaris* (group B).

The ecological meaning given to species is only valid for the set of relevés under study. This is the reason why *Epilobium spicatum*, which is usually considered as a species of logged areas, is grouped with fresh but less fertile differentials in the beech forests under study.

The orthogonal factors shown through factor analysis cannot, however, be interpreted as such: in this particular case, they can very well be replaced with two other factors which are no longer

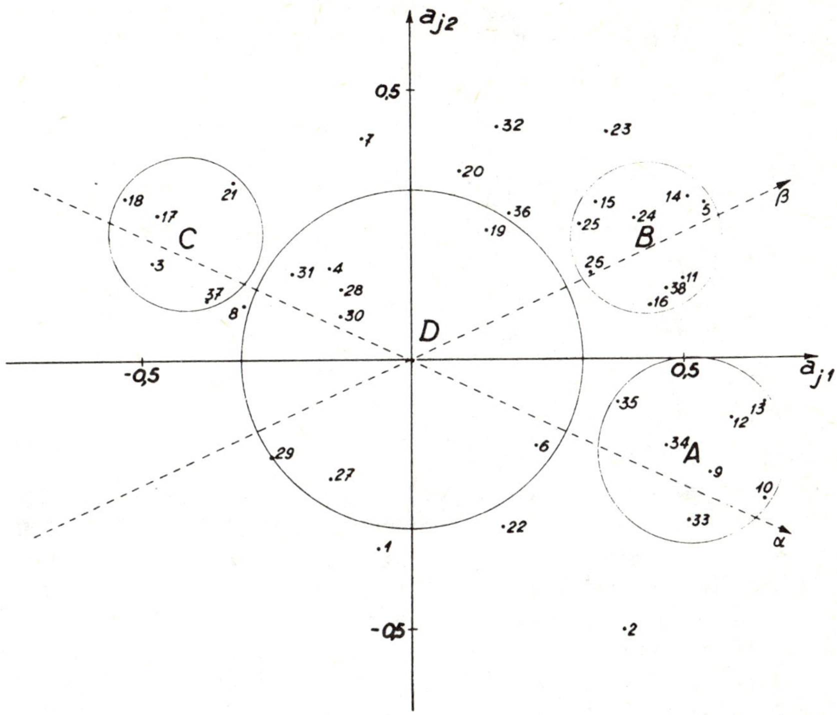

Fig. 2. Distribution of the 38 species according to their loadings on the first two factors.

independent of one another. Those two oblique factors are identified on figure 2 by the axes α and β, each of them giving a 25° angle with the axis of the first initial factor (a_{j1}). The first one (α) enables us to compare in direct opposition the species from poor and generally dry soils (group C) with those from fresh and fertile soils (group A); the other one (β), is linked with the group of differentials for fresh to moist soils (group B).

8.4.3 Classification of Relevés

Once the sociological groups are defined, we can set up a classification of relevés according to the relative importance of the various groups. We can therefore identify four types of relevés; the first ones will include species from both rich and fresh to moist soils (groups A and B); the second ones include mainly differential species from fresh to moist soils (group B); the third ones include

TABLE VI. Average frequency (in %) and average presence of the species of four sociological groups (A to D) in four types of relevés (1 to 4)[1].

	1	2	3	4
A	53 III	9 I	5 I	1 I
B	43 III	45 III	31 II	1 I
C	10 I	14 I	55 III	70 IV
D	30 II	30 II	35 II	30 II

these last species to which we add the species from poor to usually dry soils (groups B and C); whereas the last relevés essentially include species from the last group (C). Moreover, the companions (D) are included in all relevés.

Table VI sums up the essential characteristics of this classification, by giving the average frequency (in %) and the average occurrence of each group in each type of releves. This succession of four types of beech forests is in fact very similar to the one given by NOIRFALISE (1956): *Luzulo-Fagetum festucetosum*, *Luzulo-Fagetum caricetosum*, *Luzulo-Fagetum typicum* and *Luzulo-Fagetum vaccinietosum*.

Moreover, the estimation of factors can be carried out, and the results then enable us to draw a scatter diagıam in which each relevé is represented by a point (Fig. 3). The study of this diagram will first show that the distribution of the relevés is very irregular, and more particularly, that the relevés do not appear in clusters or groups clearly separated from one another.

The estimation of factors also enables us to test the value of a previously established classification and to compare two or more classifications. Fig. 4 was drawn with this in mind; it represents, with various symbols, the relevés which, on one hand, belong to the various phytosociological types previously identified (DAGNELIE 1956) and, on the other hand, the different types of vegetation which were must presented.

In the first classification, we note that each type is more or less localized into a given section of the floristic factor space, but this localization is far from being perfect. As for the new classification

1) The Roman numerals refer to presence classes, as used by BRAUN-BLANQUET (1951).

which was just introduced, the connection between the types of vegetation and the positioning of relevés in the factor space is slightly better, which is not surprising since these types have been precisely defined with the help of the sociological groups brought to light through factor analysis. The diagram referring to this last classification also shows in which sectors these various groups of species have a preponderant importance: group A is located in the lower right part (relevés bearing symbol o); group C in the left part (symbols + and *I*); whereas group B is everywhere except in the far left part (symbols o, — and +). Thus, for each group of species we define a zone of dominance in the floristic factor space, and it is possible to see the relative extents of the various zones.

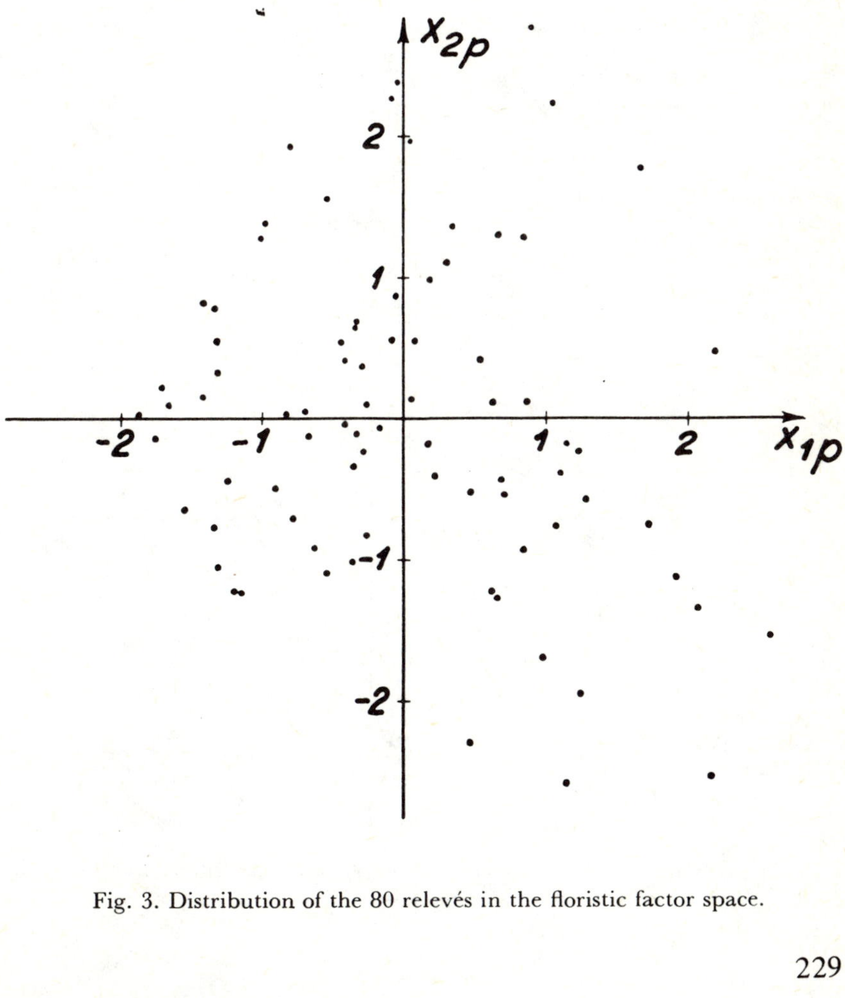

Fig. 3. Distribution of the 80 relevés in the floristic factor space.

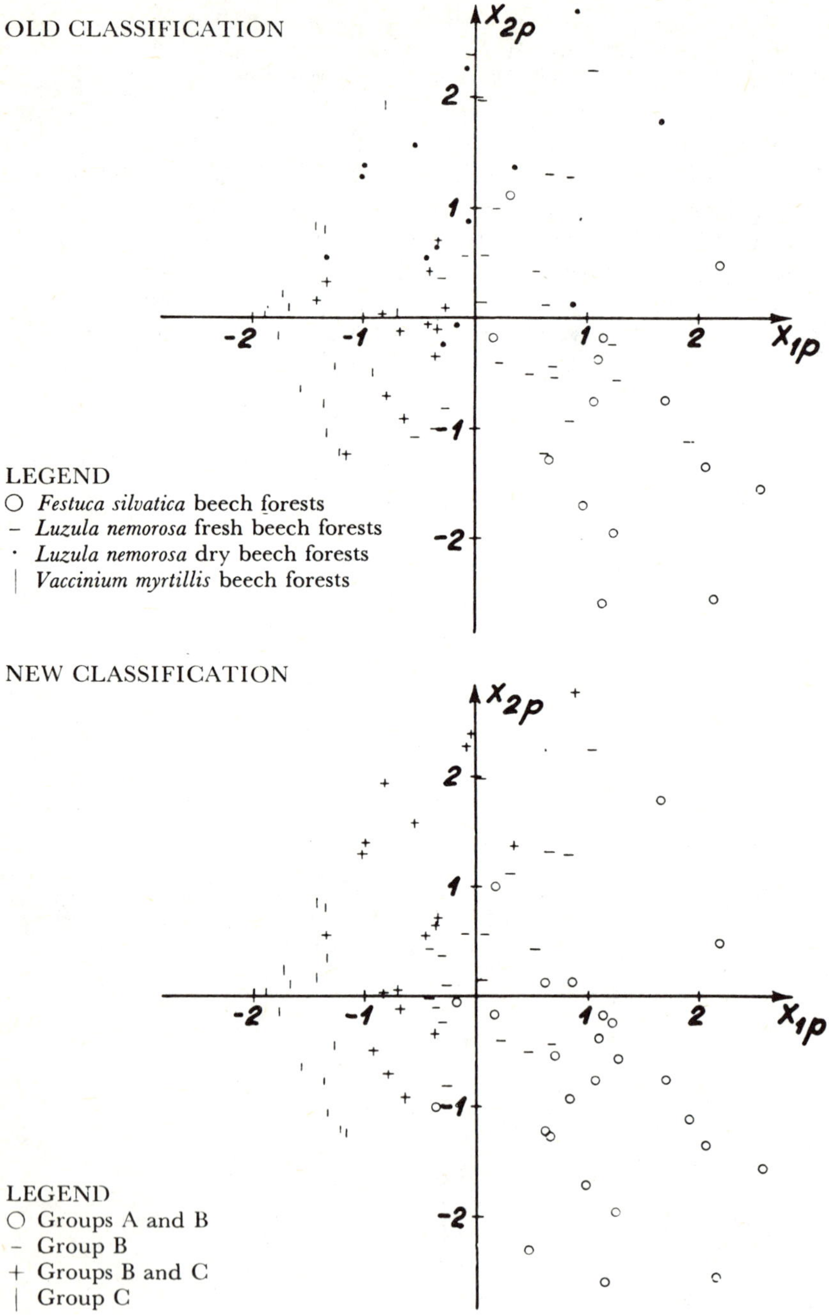

Fig. 4. Distribution of the main types of relevés in the floristic factor space, for two different classifications.

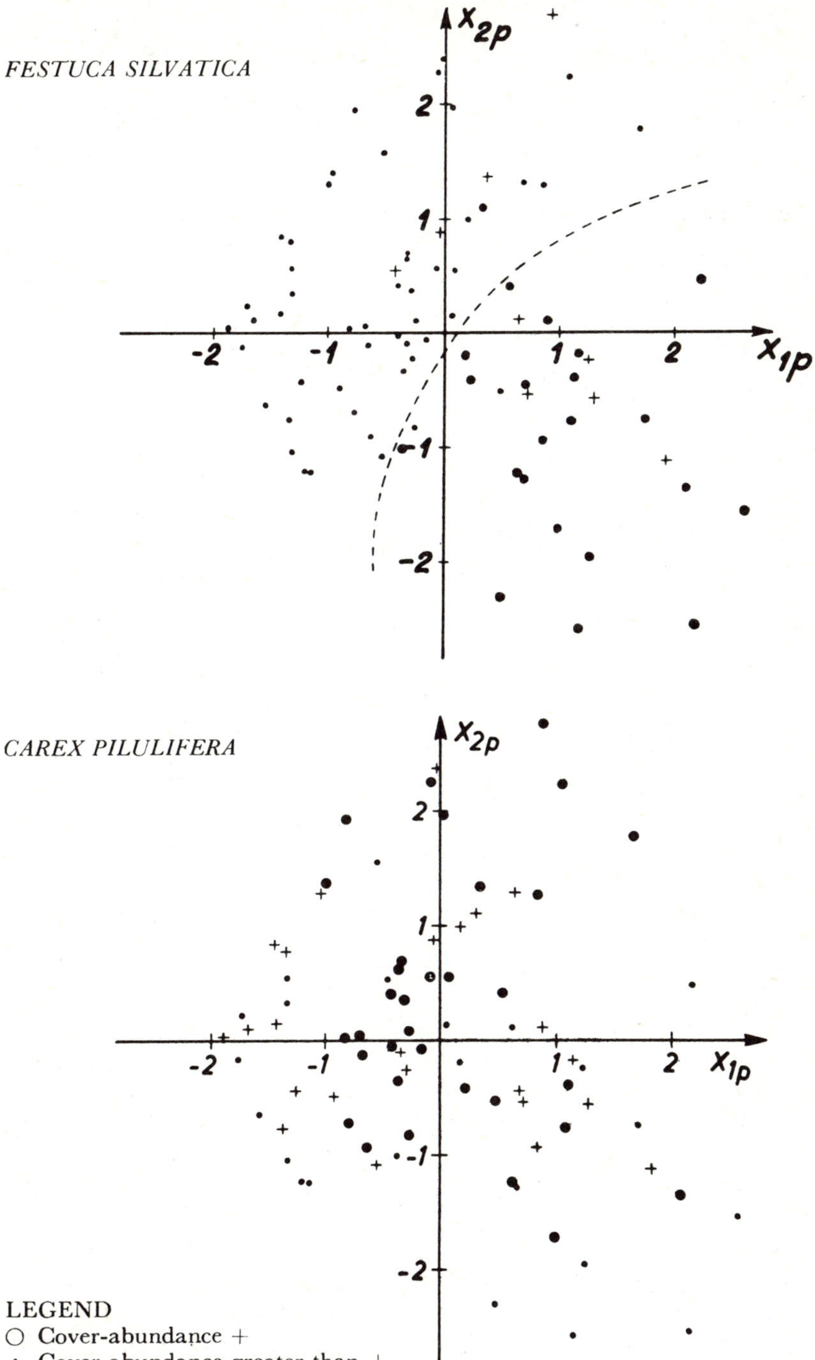

Fig. 5. Distribution of *Festuca silvatica* and *Carex pilulifera* in the floristic factor space.

8.4.4 DISTRIBUTION OF THE SPECIES IN THE FACTOR SPACE

Finally, this last notion can also be applied to species by defining for each of them a distirubtional zone in the factor space. As was expected, we can note that for the better differential species, or the species with the highest common variances, we get well-defined zones. As examples, we give here the distributions of two species (Fig. 5). The diagram for *Festuca silvatica* shows the existence of a relatively tight zone, which in fact approximately corresponds with the dominance zone for group A (Fig. 4). Within this area, the fescue appears in 93 % of the relevés (26 out of 28); outside of the area, in only 10 % of the relevés (5 out of 52). On the other hand, a species such as *Carex pilulifera*, which is only slightly related to the first two observed factors ($h_j^2 = 0{,}08$ for the first two factors), appears almost in the entire factor space. Nevertheless, this species is rather abundant in a certain part of the releves.

8.5 **Discussion and Conclusions**

In the example given above, factor analysis allowed us to solve the following problems:
— the study of the floristic homogeneity in a group of relevés;
— the estimation of the differential value of species within a group of relevés;
— the definition of sociological groups of species closely related to one another;
— the setting up of a floristic classification of relevés and the comparison between two or more classifications;
— the study of the distribution of species and groups of species within a given set of relevés.

Factor analysis can also be of use in the interpretation of ecological observations; in this case, the various observed ecological characteristics replace the species (DAGNELIE 1960, 1965b). Here the problems that arise are mainly;
— the study of ecological homogeneity in a group of relevés;
— the identification of preponderant ecological factors for a group of relevés;
— the estimation of the indicator power and the habitat amplitude of the species, and, more generally, the study of relations existing between the ecological factors and the distribution of species.

On a smaller scale, factor analysis can also be helpful in the detailed analysis of a particular community (DAGNELIE 1962b), especially for:

— the analysis of floristic or ecological homogeneity and of the structure of a vegetal community;
— the comparison of various types of relevés.

Finally, factor analysis can also be used in the study of similarity between relevés or groups of relevés; therefore, the matrix of correlations between species is replaced by a matrix of similarities between relevés or groups of relevés (Dagnelie 1960)[1].

Of all multivariate statistical methods, factor analysis could then have the most important place in ecological and phytosociological studies. Indeed, it ranks high among methods for analyzing relations between species and ecological variables (Dagnelie 1962a, 1965a, Gounot 1961, Greig-Smith 1964), and it fits perfectly in the frame of gradient analysis methods as defined by Whittaker (1967). Although it is mainly useful in this last field, factor or component analysis can also well take part in some classification processes (Groenewoud 1965).

However, for various reasons, we must limit the use of factor analysis in phytosociology and in ecology. On one hand it is a group of methods often ill-known and rather complex for many biologists. On the other hand, any use of factor analysis raises numerous problems, related to the coding of data and the choice of different methods of computation: choice of a correlation or association coefficient, computation of common variances, computation of loading, tests of significance, etc. (Dagnelie 1960).

These various problems can be solved without too great difficulty through teamwork or through collaboration between research workers in different fields (botanists and statisticians for example). In the same way, the difficulties encountered in the practical realization of computations become every day less and less important, thanks to the ever more frequent use of modern data processing equipment (punch cards equipment and computers). However, the necessity of teamwork or collaboration and the frequent use of powerful means of computation may imply that only a few large-scale problems can justify the use of factor analysis.

Finally, very serious limitations, which appeared with the first studies (Goodall 1954), are due to the linear characteristic of the models used for both factor and component analysis. This is a very important point, because the distribution of species is rarely related in a linear manner, even approximately, to the

1) The opposition between the study of correlations between species and the study of similarities between relevés is sometimes shown by using the letters R and Q ("R-techniques" and "Q-techniques"). However, this convention can also be confusing (Dagnelie 1968).

ecological factors that influence them. Another limitation results from the fact that factor analysis appears to be successful only with sets of data that do not have a high degree of heterogeneity.

Other comments about factor analysis and other multivariate techniques are to be found also in chapters 9 (Ordination by resemblance matrices), 10 (Evaluation of ordination techniques) and 11 (Recent developments) of this book.

8.6 **Addendum**

Since the writing of this text, in 1967, years have passed and the present situation is no more like 10 years ago.

New books have been published on general multivariate analysis (COOLEY et LOHNES 1971, DAGNELIE 1975, MORRISON 1967, PRESS 1972), as well as on factor analysis (LAWLEY and MAXWELL 1971, TORRENS-IBERN 1972, UBERLA 1968), on quantitative ecology (GOUNOT 1969, PIELOU 1969, POOLE 1974), or even on multivariate analysis applied to plant ecology (ORLÓCI 1975).

New computers — much more powerful — and new methods are now available. Among these, special attention must be paid to maximum likelihood factor analysis and to factor analysis of correspondences (= reciprocal averaging, articles 10 and 11).

Owing to the work of JÖRESKOG (1963, 1967), maximum likelihood factor analysis is no more a nice statistical method with very limited use. The centroid method seems to be more and more superseded and even the principal factor method has now to reckon with the competition of maximum likelihood.

The "*analyse factorielle des correspondances*" has been developed by a team of French statisticians (BENZECRI *et al.* 1973, LEBART and FENELON 1971) and, among others, has the advantage of giving a single scatter diagram for both species and relevés (or tests and persons), due to the symmetric way species and relevés (or stands and persons) are considered. Examples of this method, also called reciprocal averaging in the field of plant ecology, are given by HILL (1973), LACOSTE and ROUX (1971—1972) and RITTER (1972). A brief treatment of our beech forests example has to be found in DAGNELIE (1975).

The applications of classical factor analysis are very numerous and could hardly be enumerated here. Most of them have been published in journals such as *Ecology*, the *Journal of Ecology* and the *Canadian Journal of Botany*. Some information is also given in the reports of the International Symposium on Statistical Ecology in New Haven (PATIL *et al.* 1971).

8.7 Summary

Factor analysis consists of a set of methods developed in experimental psychology and used especially in the interpretation of results of psychological tests. Together with principal components analysis, factor analysis is one of the most effective methods of multivariate analysis (paragraph 8.1).

The principal concepts of factor analysis are introduced with a simple example taken from psychology: results of three tests (problem reasoning, arithmetic problems, and addition), that can be applied to various persons, are related to two fundamental abilities of these persons (reasoning ability and calculation ability). In such a situation the aim is an explanation of the observed correlations between the test results, making evident various underlying fundamental abilities and identifying those abilities as far as possible (paragraph 8.2).

The general model of factor analysis is then presented, and a list of references is given (paragraph 8.3).

Factor analysis can be applied in phytosociology, by considering that the plant species take the place of the tests, the relevés of vegetation replace the persons, and the ecological factors correspond to the fundamental abilities. The objective is then, on the basis of observed correlations between species, to make evident and to identify as far as possible the underlying ecological factors. In addition, other problems can be approached such as the study of the indicator value of species, the definition of sociological groups of closely associated species, the erection of a floristic classification of the relevés, the comparison of two or more classifications of the relevés, and the study of the distribution of species and species groups in the set of relevés studied. An example involving 38 species and 80 relevés of beech forests is presented in detail (paragraph 8.4).

In conclusion, other possible uses of factor analysis are outlined, notably those concerning the interpretation of ecological observations and the study of the structure of vegetation. Finally the main factors limiting the utilization of factor analysis in phytosociology and ecology are listed and discussed (paragraph 8.5).

An addendum gives some bibliographic references that appeared after the preparation of the main text (paragraph 8.6).

ANDERSON, T. W., – 1958 – An Introduction to Multivariate Statistical Analysis. Wiley, New York, 374 p.

AUSTIN, M. P., – 1968 – An ordination study of a chalk grassland community. *J. Ecol.* 56: 739—757.

BENZECRI, J. P. *et al.*, – 1973 – *L'analyse des donnees* (vol. 2: l'analyse des correspondances). Dunod, Paris, 619 pp.

BRAUN-BLANQUET, J., – 1951 – Pflanzensoziologie. Grundzüge der Vegetationskunde. Springer, Wien, 631 p.

BURT, C., – 1940 – The Factors of the Mind. Univ. Press, London, 509 p.

BURT, C., – 1949 – Alternative methods of factor analysis and their relations to Pearson's method of principal axes. *Br. J. Psychol., Statist. Sect.* 2: 98—121.

BURT, C., – 1952 – Tests of significance in factor analysis. *Br. J. Psychol., Statist. Sect.* 5: 109—133.

CATTELL, R. B., – 1952 – Factor Analysis. Harper, New York, 462 p.

COOLEY, W. W. and LOHNES, P. R., – 1971 – *Multivariate Data Analysis.* Wiley, New York, 364 pp.

DAGNELIE, P., – 1956 – Recherches sur la productivité des hêtraies d'Ardenne en relation avec les types phytosociologiques et les facteurs écologiques. *Bull. Inst. agron. Stat. Rech. Gembloux* 24: 369—410.

DAGNELIE, P., – 1960 – Contribution à l'étude des communautés végétales par l'analyse factorielle. *Bull. Serv. Carte phytogéogr., Sér. B,* 5: 7—71 et 93—195.

DAGNELIE, P., – 1962a – L'application de l'analyse multi-variable à l'étude des communautés végétales. *Bull. Inst. int. Statist.* 39(2): 265—275.

DAGNELIE, P., – 1962b – Etude statistique d'une pelouse à *Brachypodium ramosum*: les liaisons interspécifiques. *Bull. Serv. Carte phytogéogr., Sér. B,* 7: 85—97 et 149—160.

DAGNELIE, P., – 1965a – L'étude des communautés végétales par l'analyse statistique des liaisons entre les espèces et les variables écologiques: principes fondamentaux. *Biometrics* 21: 345—361.

DAGNELIE, P., – 1965b – L'étude des communautés végétales par l'analyse statistique des liaisons entre les espèces et les variables écologiques: un exemple. *Biometrics* 21: 890—907.

DAGNELIE, P., – 1966 – Introduction à l'analyse statistique à plusieurs variables. *Biom.-Praxim.* 7: 43—66.

DAGNELIE, P., – 1968 – Introduction aux problèmes et aux méthodes de classification numérique. *Biom.-Praxim.* 9: 87—111.

DAGNELIE, P., – 1975 – *Analyse statisque a plusieurs variables.* Presses Agronomiques, Gembloux, 362 pp.

FAVERGE, J. M., – 1954 – Méthodes statistiques en psychologie (2 vol.). Presses Univ. de France, Paris.

GOODALL, D. W., – 1954 – Objective methods for the classification of vegetation. III. An essay in the use of factor analysis. *Austral. J. Bot.* 2: 304—324.

GOUNOT, M., – 1961 – Les méthodes d'inventaire de la végétation. *Bull. Serv. Carte phytogéogr., Sér. B,* 6: 7—73.

GOUNOT, M., – 1969 – Méthodes d'étude quantitative de la végétation. Masson, Paris, 314 p.

GREIG-SMITH, P., – 1964 – Quantitative Plant Ecology. Butterworths, London, 2nd ed. 256 p.

GROENEWOUD, H. VAN, – 1965 – Ordination and classification of Swiss and Canadian coniferous forests by various biometric and other methods. (Germ. summ.) *Ber. geobot. Inst. ETH, Stiftg. Rübel,* Zürich, 36: 28—102.

HARMAN, H. H., – 1967 – Modern Factor Analysis. Chicago Univ. Press, 474 p.

HILL, M. O., – 1973 – Reciprocal averaging: an eigenvector method of ordination. *J. Ecol.* 61, 237–249.

HOLZINGER, K. J. & HARMAN, H. H., – 1941 – Factor Analysis. A Synthesis of Factorial Methods. Chicago Univ. Press, 417 p.

HORST, P., – 1965 – Factor Analysis of Data Matrices. Holt, Rinehart et Winston, New York, 730 p.

JORESKOG, K. G., – 1963 – *Statistical Estimation in Factor Analysis: a New Technique and its Foundation.* Almqvist and Wicksell, Stockholm, 145 pp.

JORESKOG, K. G., – 1967 – Some contributions to maximum likelihood factor analysis. *Psychometrika* 32, 443–482.

KENDALL, M. G., – 1950 – Factor analysis as a statistical technique. *Jl. R. statist. Soc., Ser. B,* 12: 60—73.

KENDALL, M. G., – 1957 – A Course in Multivariate Analysis. Griffin, London, 185 p.

KENDALL, M. G. & LAWLEY, D. N., – 1956 – The principles of factor analysis. *Jl. R. statist. Soc., Ser. A,* 119: 83—84.

LACOSTE, A. and ROUX, M., – 1971–1972 – L'analyse multidimensionnelle en phytosociologie et en ecologie: application a des donnees de l'etage subalpin des Alpes maritimes. *Oecol. Plant.* 6, 353–369 and 7, 125–146.

LAWLEY, D. N., – 1955 – A statistical examination of the centroid method. *Proc. R. Soc. Edinb., Sect. A,* 64: 175—189.

LAWLEY, D. N. & MAXWELL, A. E., – 1963 – Factor Analysis as a Statistical Method. Butterworths, London, 117 p.

LAWLEY, D. N. and MAXWELL, A. E., – 1971 – *Factor Analysis as a Statistical Method* (2nd ed.). Butterworths, London, 153 pp.

LEBART, L. and FENELON, J. P., – 1971 – *Statistique et informatique appliquees.* Dunod, Paris, 426 pp.

MORRISON, D. F., – 1967 – Multivariate Statistical Methods. Mc Graw Hill, New York, 338 p.

NOIRFALISE, A., – 1956 – La hêtraie ardennaise. *Bull. Inst. agron. Stat. Rech. Gembloux* 24: 208—239.

ORLOCI, L., – 1968 – A model for the analysis of structure in taxonomic collections. *Can. J. Bot.* 46: 1093—1097.

ORLÓCI, L., – 1975 – *Multivariate Analysis in Vegetation Research.* Junk, The Hague, 276 pp.

PATIL, G. P., PIELOU, E. C. & WATERS, W. E., – 1971 – Statistical Ecology, Vol. 3: Many Species Populations, Ecosystems, and Systems Analysis. Pennsylvania State Univ. Press, 462 p.

PIELOU, E. C., – 1969 – *An Introduction to Mathematical Ecology.* Wiley, New York, 000 pp.

POOLE, R. W., – 1974 – *An Introduction to Quantitative Ecology.* McGraw Hill, New York, 480 pp.

PRESS, S. J., – 1972 – *Applied multivariate analysis.* Holt, Rinehart and Winston, New York, 521 pp.

RITTER, J., – 1972 – Les groupements vegetaux des etages subalpin et alpin du Vercors meridional. Essai d'interpretation statistique. *Vegetatio* 24, 313–403.

SEAL, H. L., – 1964 – Multivariate Statistical Analysis for Biologists. Methuen, London, 207 p.

THOMSON, G. H., – 1956 – The Factorial Analysis of Human Ability. London Univ. Press, 383 p.

Torrens-Ibern, J., – 1972 – *Modeles et mathodes del'analyse factorielle.* Dunod, Paris, 202 pp.

Thurstone, L. L., – 1947 – Multiple-Factor Analysis. Chicago Univ. Press, 535 p.

Überla, K., – 1968 – Faktorenanalyse. Springer, Berlin, 399 p.

Vernon, P. E., – 1952 – La structure des aptitudes humaines (Trad.: M. Reuchlin). Presses Univ. de France, Paris, 195 p.

Vincent, D. F., – 1953 – The origin and development of factor analysis. *Appl. Statist.* 2: 107—117.

Whittaker, R. H., – 1967 – Gradient analysis of vegetation. *Biol. Rev.* 42: 207—264.

9 ORDINATION BY RESEMBLANCE MATRICES

László Orlóci

Contents

9 ORDINATION BY RESEMBLANCE MATRICES

9.1 Introduction

The term *ordination* is used in vegetation studies to describe techniques in which given ecological entities are ordered according to one or several of their properties in such a manner that their arrangement will reveal some useful information about their relationships (cf. GREIG-SMITH 1964). The relationships in question are those of composition between communities or with environmental variables. Ordination methods are as varied as continuum analysis, component analysis, factor analysis, and many more, which differ greatly both in underlying theory and also in practical effectiveness (cf. DAGNELIE 1960, WHITTAKER 1967, MCINTOSH 1967, PRÉCSÉNYI & SCÖCS 1969, ORLÓCI 1975 and references therein).

Although ordination is sometimes treated as a part of gradient analysis (article 1, WHITTAKER 1967), it is appropriate to discuss the methodology in the broader context of multivariate techniques (ORLÓCI 1975). In this, ordination can be construed as a structuring procedure (cf. LAMBERT & DALE 1964) in which the total variability in an ecological collection is apportioned efficiently among new axes. Efficient structuring is, of course, a concept closely related to the notion of parsimonious summarization (SEAL 1964).

The ordination techniques, with which we are concerned in the present paper, proceed by analysis of a matrix of resemblances. These are measurements of similarity or dissimilarity between the entities to be ordinated. Our discussions will be concerned with summarization techniques, such as for instance component analysis, and also techniques of other kinds, such as factor analysis and multidemensional scaling. This broad class of methods includes the open-model of factor analysis, component analysis, position vectors ordination, polar ordination (several methods), principal axes analysis, and techniques commonly referred to as multidimensional scaling. The common property of these techniques is their ability to extract metric information from a given resemblance matrix. This is new information on the basis of which specific phytosociological relationships, or trends in these relationships, can more easily be recognized.

We shall consider some general concepts first, before turning to a discussion of the admissible resemblance functions and methods for ordination.

9.2 Purposes of Ordination

The actual objectives of ordinations may differ considerably. Factor analysis, for instance, is aimed at the isolation of the common variation from the specifics in a population (Lawley & Maxwell 1963, Seal 1964, Cattell 1965), while another commonly used technique, component analysis, is intended to summarize the total variation (Seal 1964). Whereas these are often recognized as the direct goals in ordinations, the primary results of an ordination, such as the component scores for instance, may serve as input data in classifications (e.g. Goodall 1954, Ivimey-Cook 1968), or in further ordinations (e.g. Quadling 1967). Ordinations may also provide a basis for other specialized analyses such as the quantification of successional trends (Maarel 1969), group identification (Dagnelie 1960, 1965), cluster recognition (e.g. Gittins 1965a,b, Greig-Smith et al. 1967), or graphical discriminant analysis (Orlóci 1968). Most often, however, ordinations are concerned with the identification of trends and patterns of variation, the explanation of these, or their generalization as hypotheses in terms of correlations with external variables (see Whittaker 1967, Gittins 1969, and references therein). Other examples include Bray & Curtis (1957), Curtis (1959), Maycock & Curtis (1960), Maycock (1963), Mc Intosh & Hurley (1964), Ashton (1964), Gittins (1965b), van Groenewoud (1965), Ivimey-Cook & Proctor (1967), Yarranton (1967a,b,c), Kershaw (1968), Austin & Greig-Smith (1968), Chandapillai (1970), Grigal & Goldstein (1971), Moore et al. (1970), Phipps (1970), and Austin et al. (1972), among others. The secondary analyses may themselves rely on complex analytical techniques such as trend surface analysis (Sneath 1967, Gittins 1968), canonical correlation analysis (e.g. Austin 1968), curve seeking (Sneath 1966), or axis rotation (e. g. Sokal 1958, Sokal & Rohlf 1962).

9.3 Stochastic vs. Descriptive Ordinations

It is most desirable to consider ordinations as stochastic (probabilistic) techniques. If, for instance, the assumption of an underlying multivariate normal distribution is acceptable then it is possible in component analysis to test the hypothesis that the last k ordination axes, corresponding to the k smallest eigenvalues of the covariance matrix, are statistically indistinguishable. Factor analysis, under the same assumptions regarding the nature of the underlying population, will provide a test for the hypothesis that a set of m common-factors completely describes the covariance structure of a given population.

A stochastic use of ordinations, while most desired for its hypothesis-testing aspects, may however be inappropriate in many practical situations when the specific assumptions that are implicit in the model are unwarranted .The ordination, in such situations, reduces to a deterministic use of the mathematical model, aimed at precise descriptions and hypothesis generation.

9.4 Hypothesis Testing in Ordinations

Different methods may be used by phytosociologists for testing given hypotheses .These may be based on (i) consistency of the observed ordination results with known facts, (ii) consistency of ordination results in repeated field surveys or experiments, and (iii) the use of standard statistical indices with known probability distributions. The first of these methods may not seem too rigorous. It can, nevertheless, prevent the phytosociologist from attributing much significance to observed results which, according to past scientific experience, could have arisen with a relatively high probability as a consequence of sampling or experimental error. Repeated ordinations of materials derived from successive surveys in the same area may provide a most reliable basis of significance tests. This method, however, can be very uneconomical in most types of practical applications. The most common methods of hypothesis testing are based on the use of standard statistical indices, which presumably possess specific distributional properties. The use of these indices, however, can be very hazardous in the hands of the practicing phytosociologist who is required to accept specific assumptions purely on an intuitive basis without adequate facilities required for a reliable theoretical or experimental verification.

9.5 Ordination Efficiency

Different criteria are suggested by the various authors to evaluate the relative efficiency of an ordination. Van Groenewoud

(1965) mentions (i) the need to account for the largest possible portion of variation in the sample by the simplest possible ordering of individuals, and (ii) meaningfulness in ecological terms (see also AUSTIN & NOY-MEIR 1971). He further considers (iii) the nature of the resemblance function used, and (iv) the existence of facilities whereby new individuals can be placed in a completed ordination.

We may add at this point that component analysis is expected to be the most efficient method for the summarisation of linear variation, using the smallest possible number of dimensions. The question of ecological meaningfulness can be answered on the basis of the potentials of the model itself, whether or not it is capable of detecting the kind of variability which would interest the phytosociologist, and on the basis of the correlation of the ordination results with given ecological factors. It is obvious that the nature of the resemblance function is also very important for the simple reason that it must qualify as a valid spatial parameter in the assumed sample space of operations of a given technique. And further, the results of the ordination may greatly depend on the resemblance function. The placement of newly-acquired individuals in a completed ordination is usually accomplished by the transformation $\mathbf{Y} = \boldsymbol{\alpha}'\mathbf{A}$ which produces coordinates for an individual (vector) $\mathbf{A}$ given $\boldsymbol{\alpha}$, the matrix of transformation coefficients. Such a transformation is readily available in component analysis, and can also be derived in other ordination methods.

There are further criteria that should be considered. Among these we may mention (v) the stochastic properties of the model and (vi) the type of variation, whether linear or curvilinear, that can be detected by the model. The conventional ordination methods are deterministic in nature with the exception of factor analysis and perhaps component analysis, which can be used in association with standard statistical tests based on the postulated normal distribution of joint frequencies in the underlying population. The conventional methods, however, appear to have one really serious defect regarding criterion (vi): they treat all variation as if it were linear. This implies that ecological properties with curvilinear variation cannot be handled by these ordination methods without distortion (see GOODALL 1954, VAN GROENEWOUD 1965, SWAN 1970, AUSTIN 1968, NOY-MEIR & AUSTIN 1970, AUSTIN & NOY-MEIR 1971, GAUCH & WHITTAKER 1971, 1972, ORLÓCI 1975). What is meant by the inability to account for non-linear variation can be illustrated by the ordination axes of component analysis. These axes are linearly orthogonal (in terms of zero covariances) but they may be curvilinearly related. The position of these axes in no way is affected by the existence of higher order variation in the experimental data. There seems to be no satisfactory solution to the problem of non-linear ordinations, although many attempts have been made to respond to this problem. AUSTIN & NOY-MEIR (1971) gives a summary of several possibilities. They suggest to limit the scope of ordinations to situations of narrow vegetational and environmental variation where the assumption of linear variation is more acceptable, and to always take into account the linear property of the model when interpreting the ordination results in ecological terms. Another approach may include manipulations of the ordination results, aimed at the extraction of the non-linear component of variation, by techniques such as trend surface analysis (e. g. GITTINS 1968) and other regression methods concerned with curve-seeking (e.g. SNEATH 1966). A further possibility is the standardization, or transformation of the data (e. g. SWAN 1970), or the use of some specialized multivariate technique such as for instance MC DONALD's (1962) non-linear factor analysis and KRUSKAL's (1964a,b) method of multidimensional scaling. The method of SHEPARD & CARROLL (1965) has also been mentioned as another possibility (see AUSTIN & NOY-MEIR 1971, 1974, DALE 1975, ORLÓCI 1975).

It seems that, in need of a better method, the practicing phytosociologist may continue using the linear models of ordination, and when it appears necessary, he may further scrutinize the ordination results for non-linear variation on the basis of the use of some other method. We must, of course, assume that he will use ordinations, or as a matter of fact any other method, after careful consideration of their relevance to the problems on hand. The ordination methods described in the present paper should not be regarded as general strategies. They should rather be considered as highly specialized techniques intended to answer specific questions. The failure of the phytosociologist to recognize this will inevitably result in a wasteful analysis.

9.6 A Model for Ordination

We shall define ordination *entities* in a general sense, as *individuals* or *characters* . An individual, the basic distinguishable unit for observation, will be identified with a site (quadrat, sample plot), and the characters will be designated as entities on the basis of which the individuals are described. Among the characters we shall consider only those which vary among the individuals. The *variates* may represent the quantity or presence of different species within the sites, or any other properties, vegetational or environmental, on the basis of which the characteristics of the sites can be effectively described. It should be noted that although we designate the sites as individuals whose ordination is contemplated, there should be no conceptual difficulty in identifying entities of some other kind as individuals and using them accordingly in an ordination.

We may picture an individual as a fixed point in space whose position is determined by the size or state of the characters which it possesses. It is implicit in the foregoing analogy that the characters are regarded as axes of a reference system. This system with all of its properties constitutes the sample space for ordination. It may be noted that the *sample space*, as we define it, is an abstraction.

Our ordination model consists of (i) a collection (sample, population) **X** of c individuals described on the basis of r variates, (ii) a resemblance function f which generates the resemblance matrix **R** from **X**, and (iii) statements in a formal algebraic language specifying the steps in the analysis. The c individuals (column vectors of **X**) identify the entities to be ordinated, and the r variates (row vectors of **X**) are those properties on the basis of which the pairwise resemblance of individuals is determined. We operate the model on the assumption that the variates are commensurable (see PIELOU 1969), and with purpose to extract metric information from **R**. In some instances our model will have stochastic properties. More often, however, we shall use the model in a deterministic context.

9.7 Ordination by Euclidean Matrices

9.7.1 The Sample Space

9.7.1.1 *Definition of a Metric Space*

Firstly, we shall describe the properties of a general metric space. The metric properties are quite relevant in developing our idea of ordinations in Euclidean sample space which completely possesses these properties. A metric space is defined as a collection of objects **X** and a metric (distance) function d. We note in this connection that (i) a metric function cannot be negative, (ii) it must be symmetric so that it will not be affected by the sequence in which the individual objects are compared, and (iii) it must possess the triangle inequality property, implying that given three individuals as points of a sample space the sum length of any two sides of the triangle which they define must not be less than the length of the third side. In other words, any resemblance function which can produce negative values is automatically disqualified as a metric. And further, a resemblance function which gives one result when the comparison is from A to B, and another when the comparison is from B to A is useless as a metric function. A function which does not conform with the triangle inequality axiom will also be rejected as a metric measure. Although we must reject any function as a metric if it fails on any one of the three metric space axioms, we must also note that the metric use of a non-metric function does not necessarily result in serious distortion in some ordinations. This problem will be considered later when we discuss the concept of distortion which results from the Euclidean use of given non-Euclidean measures of resemblance in phytosociological ordinations.

9.7.1.2 *Some Notes on Resemblance Functions*

Several measures of resemblance which can be used in ordinations are discussed in article 6 by GOODALL (also ORLÓCI 1972, 1975). With this point in mind we shall limit our discussion of the resemblance measures to those properties which must be considered by the phytosociologist when selecting an appropriate measure for ordination. The resemblance functions of the present chapter, with few exceptions, are meaningful in Euclidean sample space. This space is distinguished from other metric spaces by the fact that it is only the Euclidean space where a scalar product of vectors exists.

A commonly used resemblance function is the Euclidean distance, or common distance, which we know from every day experience. The simplest definition of such a distance between two quadrats of vegetation is given as the square root of the sum of squared differences between their corresponding species. It, however, should be noted at this point that (i) when we decide to define a distance in this way we also implicitly assume that the individual species, whose squared differences are summed, are statistically independent, and (ii) when the original data are not adjusted or standardized in some appropriate manner (see ORLÓCI 1967a) then this distance will not have a fixed upper limit — that is to say that it will not be a relative measure on the basis of which we can make direct comparisons between ordinations.

Our initial definition of Euclidean distance is seriously weakened by its total reliance on absolute quantity, without any regard to the relative importance or weight of the different species. In this connection we should mention that different weighting procedures were suggested which used the total species correlations (see MACNAUGHTON-SMITH et al. 1964) or the total chi square (see WILLIAMS et al. 1964). We may note that (i) these weighting techniques will both have the same effect when the data represents species presence, and (ii) the Euclidean distance so weighted will not evenly retain its metric properties in sets of quadrats in which the weight of a species does not remain constant. In more specific terms of our ordination problem, such a 'distance' measure will not be allowed to serve as a definition of resemblance if the goal is a simultaneous analysis of several sets of quadrats which associate with greatly different weights of the species.

It is a peculiarity of the Euclidean distance that it may show a low degree of resemblance between two quadrats in which all species are identical, if the species values are greatly different, and conversely, it may indicate a high degree of resemblance between two quadrats which have no species in common, if the species values are generally low. A procedure which will help overcoming this weakness of the Euclidean distance involves the standardization of the data in such a manner that each observation is divided by the square root of the sum of squared observations in the quadrat (column vector of **X**) of its occurrence. This kind of standardization has a definite geometric meaning (see ORLÓCI 1967a, PIELOU 1969), and it also has the advantage of being independent from the set in which a quadrat occurred. For these reasons when the distance is computed by this method it will remain a valid Euclidean spatial parameter, and it will continue to be admissible as a resemblance function in ordinations.

In some statistical applications distance is defined in the manner of HOTELLING's T^2 statistic, or as a fraction of this statistic (see MORRISON 1967). This definition is equivalent to our first definition of Euclidean distance if the species are statistically independent and have unit variance. Due to standardization and transformation implicit in the T^2 statistic, its application as a resemblance function in Euclidean ordinations is restricted to sets of quadrats in which the pattern of covariance relationships of species remains constant. This statistic is further limited to sets of quadrats for which the assumption of multivariate normality holds.

Throughout the preceding discussions, regarding the Euclidean distance, we implicitly assumed that the species can be represented by perpendicular axes. This assumption, however, is unnecessary, and in some instances, it may not even be desired. As an alternative assumption we may stipulate that the species are correlated, and as such, they can be represented by oblique axes. The obliqueness can be measured by the product-moment correlation coefficient. A definition of Euclidean distance on the basis of the assumption that the reference axes are oblique has been given by GENGERELLI (1963) and its use has also been demonstrated by CARLSON (1970). The principal advantage of their definition of distance is that it removes the effect of linear species correlations from the value of absolute distance. In a sense this property makes this definition quite similar, though not identical in general, to a fraction of the T^2 statistic, also known as generalized distance. Since in this distance we use the correlation coefficient as an element, it will have similar locally metric properties as T^2. This property should not be ignored when the application of these measures is considered in a simultaneous ordination of sets of quadrats with unequal correlation matrices.

The Euclidean distance will simplify to $\sqrt{(b+c)}$ when the data represent presence scores of species. In this and following expressions for comparison of two quadrats, the symbol a is the number of species present in both quadrats, b the number in the first quadrat only, and c the number in the second quadrat only.

The $\sqrt{(s)}$ fraction of this distance, where s is the total number of species in the sample, represents the one-complement of the simple matching coefficient (SOKAL AND MICHENER 1958). The sampling distribution of this coefficient was derived by GOODALL (1967). While division by s will not interfere with the admissibility of this distance as a resemblance function in Euclidean ordination, division by $(a+b+c)$ definitely will. Such a division is implicit in the JACCARD (1901) coefficient (39). (Numbers in parentheses are those of formulas in article 5.) For some interesting properties of this coefficient the reader should consult a paper by HOLGATE (1971). The one complement of JACCARD's coefficient is admissible as a resemblance function in Euclidean ordinations in sets of quadrats in which no species is missing in more than one quadrat (see ORLÓCI 1971).

A variant of the JACCARD coefficient which has been used in ordination is the SØRENSEN (1948) index (40), $CC=2a/(2a+b+c)$. It is interesting to point out that the one complement of this index, given by $CD=1-CC=(b+c)/(2a+b+c)$, can be regarded as an Euclidean distance in sets of quadrats for which $(2a+b+c)$ is a constant quantity. This function and a related function $2CD$, which was treated by VAN DER MAAREL (1969) as a metric essentially in the same context as discussed earlier by GOWER (1966), have been used in ordinations.

A function, $PD=1-PS$, has been used for quantitative data (e.g. BRAY & CURTIS 1957) where $PS=2\sum\min(x_i,y_i)/\sum(x_i+y_i)$. The 100 multiple of PS is a measure of percentage similarity (50); x_i and y_i represent the quantities of the ith species in the two quadrats (x and y), and the summations are taken over all the species from $i=1$ to r. It can be shown that when the data represent presence scores of species PD is the same as CC, related to the Euclidean distance. When the data represent quantities other than presence scores the function PD will be a metric function, equivalent to the absolute value function and not to the Euclidean distance, in sets of quadrats for which $\sum(x_i+y_i)$ is a constant quantity for all pairs of quadrats. To show this relationship we rewrite PD in the form $PD = \sum|x_i-y_i|/\sum(x_i+y_i)$, where the quantity in the numerator is the familiar absolute value function. We conclude that PD, when computed from quantitative data, will be a fraction of the absolute value function in sets where the sum of species values is constant for any pair of quadrats. A variant of PD (51) has been suggested by WHITTAKER (1952) in which the sum of relativized species values is a constant at 1.0, so that for any pair of quadrats $PD = \sum|x_i-y_i|/2. = 1-\sum\min(x_i,y_i)$.

In sets where $(2a+b+c)$, or $\sum(x_i+y_i)$, is not a constant, the use of CD, or PD, as a metric ordination parameter is not strictly justified. It seems though that CD, and PD, will retain traces of the metric properties, and even when the sums of species values differ considerably between the quadrat pairs, the distortion in the ordination resulting from the metric use of CD, or PD, may not be too extensive. A frequently used variant of these indices is a quantity defined by $D = \max L - L$ in which L is either CC or PS, and the highest value observed (ideally a value for replicate samples) becomes max L. This definition has been used as a substitute for Euclidean distance (e.g. BEALS 1960). Such a use of D, however, is not quite appropriate (see AUSTIN & ORLÓCI 1966) for the following reasons: If we put $D = 1-L+e$, where $e = \max L - 1$, we can clearly see that an adjustment is implicit in D which takes the form of adding a constant quantity e to the function $1-L$. A consequence of such an adjustment in sets of quadrats where $1-L$ is a metric function, but e is not a zero value, is that D may fail as a metric measure of resemblance. This failure will distort the ordination in those circumstances in which the metric nature of $1-L$ is assumed. In sets where e is zero and $2a+b+c$, or $\sum(x_i+y_i)$, is a constant, the quantity $\max L - L = 1-L$ can be used as an Euclidean distance in the case of presence data, or as an absolute value function in the case of quantitative data (See also ORLÓCI 1973, 1974a, b, 1975).

Ordinations often are based on resemblance measures which can be derived from Euclidean distance. An especially important group of such measures is represented by the scalar products. A scalar product of two vectors (species or quadrats) is defined as the product of their lengths and the cosine of their subtending angle. The relevant expressions are given by ORLÓCI (1967b), including a detailed discussion of their effect upon principal component analysis.

Regarding the general topic of scalar products we may comment further on the similarity coefficients which can be represented in terms of the general function $\rho = a/m$. The similarity coefficient of RUSSELL & RAO (1940) is of this type in which $m = a+b+c+d$ (symbols as above, with $d=$ number of species in neither quadrat). We can similarily express the JACCARD coefficient (39) in which $m = a+b+c$, and the coefficient (43) of OCHIAI (1957) in which $m = \sqrt{((a+b)(a+c))}$. The coefficient of RUSSELL & RAO represents an average scalar product of vectors of binary elements. The coefficient of JACCARD, however, qualifies as a scalar product only in sets of quadrats in which m is a constant quantity. The coefficient of OCHIAI, on the other hand, is a true scalar product between vectors of unit length. Note that these coefficients can be converted into a standard (Euclidean distance) measure of dissimilarity in the manner of $ED = \sqrt{(2(1-\rho))}$ with values occurring between zero and $\sqrt{(2)}$.

None of the a/m coefficients so far mentioned could be rightly labelled as a product-moment correlation coefficient since none of them achieves a translation of the co-ordinate origin of the reference system to the common centroid (center of gravity). A peculiarity of the product-moment correlation coefficient is that it uses the centroid of the system about which the scalar product of vectors is computed. This property will qualify the correlation coefficient as an admissible function in, for instance, component analysis or position vectors ordination. The Euclidean distance corresponding to the product-moment correlation coefficient ρ is given by $\sqrt{(2(1-\rho))}$. This distance has values between 0 and 2.

9.7.1.3 *A Classification of Resemblance Functions*

The resemblance functions which we have so far discussed can be associated with different ordination methods. We summarize some of their important properties in Table I. Shown in this table is the classification of the different coefficients according to their distance or scalar product properties, and according to admissibility as resemblance functions in the different ordination methods .The idea of local admissibility, indicated by an L between parentheses, is introduced in the table to indicate that in some sets of quadrats these functions may not be acceptable as valid definitions of resemblance for Euclidean ordination. Because the metric properties of some coefficients change depending on the data, we used the symbol P

TABLE I. Distance and scalar product properties of coefficients, and their admissibility in ordinations.

Coefficient	Classification				Ordination method of admissibility
	Euclidean distance	Scalar product centered	Scalar product not centered	Absolute value function	
Simple matching			X		
One-complement of simple matching*	X				C, Po
JACCARD			X (L)		
One-complement of JACCARD*	X (L)				C, Po
SØRENSEN			X (L,P)		
One-complement of SØRENSEN*	X (L,P)			X (L)	C,Po
RUSSELL and RAO			X		
One-complement of RUSSELL and RAO*	X				C,Po
OCHIAI			X		
One-complement of OCHIAI*	X				C,Po
Simple product moment correlation		X			C,P
One-complement of simple product moment correlation*	X				C,Po
Variance/Covariance		X			C,P,F

Legends to symbols: (L) - locally, (P) - presence data, C - component analysis, P - position vectors ordination, Po - polar ordination, F - factor analysis ,* — $\sqrt{2(1-\rho)}$,

between parentheses to signify that the use of presence data is assumed.

9.7.1.4 *Resemblance Functions and Sample Structure*

The phytosociologist often thinks of sample structure as a covariance structure which he sees in the raw data. He may not realize that such a structure can completely be destroyed by the different adjustments or standardizations usually implicit in the resemblance function. For example, when he uses the product-moment correlation coefficient as a measure of species resemblance, he is in fact introducing a standardization which makes the variance of every species equal to unity. Thus after applying the correlation coefficient to the data, it is no longer valid to think of structure as it may have existed in the original data, since then that type of structure will have been completely destroyed by the correlation coefficient. This effect of the resemblance function on the sample structure may be visualized as a scrambling of the original spatial configuration of individuals, shifting their position relative to one another, in a manner depending on the nature of the resemblance function.

9.7.2 Factor Analysis

Factor analysis is discussed by Dagnelie in article 8. It, however, is necessary to briefly describe some aspects of this method on the basis of which we will be able to make a comparison between factor analysis and component analysis. We characterize factor analysis as a statistical method aimed at the estimation of the underlying common covariance structure of a multivariate normal population. In this respect factor analysis differs from component analysis which is a variance-oriented method of summarization. Both factor analysis and component analysis, when used as stochastic techniques, require the assumption that the underlying joint probability distribution is multivariate normal, and that the variates are commensurable.

The basic model in factor analysis can be given by the expression $x_i = \sum a_{ij} Y_j + e_i$ which implies that each observed variate x_i is considered to be a composite of the effect of m common factors Y_j, $j=1$ ..., m, and a specific factor e_i. The symbol a_{ij} represents factor loadings. It follows from this definition that the basic computational problem actually amounts to finding the factor loadings and the variances of the specific factors. In contrast with the extraction of component scores in component analysis, the factor scores (coordinates) cannot be derived by a simple linear transformation in factor analysis, but they must be estimated by one of several possible methods (see Lawley & Maxwell 1963). The factor analysis model, like the component analysis model, is linear.

There is another important difference between factor analysis and component analysis. regarding the covariance structure. This structure is completely destroyed in the course of factor analysis. The common factor space thus cannot be regarded as an orthogonal projection of the original r dimensional configuration (as depicted by the covariance matrix) within the ordination space of the m common factors.

The assumption that the underlying multivariate distribution is normal, and the fact that the phytosociologist must specify the number of common factors before commencing with the analysis, will impose limitations upon the applicability of factors analysis in phytosociology. The assumption of normality may require data transformations which may or may not be particularly helpful. The choice of an initial value for m, the number of postulated common factors, is a matter of personal judgment. Several trials may be required before the postulated m-factor model will give a sufficiently close fit to the data. The application of factor analysis to problems in phytosociology has apparently been limited to the use of centroid (principal) factor analysis (DAGNELIE 1960, 1965, SZÖCS 1971). The phytosociological value of the maximum-likelihood solution in association with factor rotation, however, has yet to be clearly demonstrated.

9.7.3 COMPONENT ANALYSIS

9.7.3.1 *General Characteristics of the Method*

While factor analysis attempts to elicit the underlying common covariance structure of a given population in terms of m common factors, component analysis is aimed at the efficient summarization of the total variation in the sample. The two techniques are further distinguished by other properties in addition to those which we already mentioned: (i) The principal components are variant under changes in the scale of the variates. This implies, for example, that the ordination axes (components) extracted from a correlation matrix are expected to differ from those which are extracted from a covariance matrix. A peculiar effect of the scale used is that after standardization of each variate by a factor $1/S_i$ the component scores are expected to differ, by a factor other than $1/S_i$, from the component scores extracted from the unstandardized variates. There is no direct transference between the two sets of component scores in the manner of a simple scaling of the results. This is not so in factor analysis where such a transference is readily available. (ii) There is a simple linear relationship between the components and the original variates in component analysis. This is not so with the factor scores in factor analysis. (iii) The concepts of common and specific factors do not apply to component analysis. (iv) The variates are assumed to be simple transforms $x_i = \sum a_{ij} Y_j$ of r components, $j=1,...,r$. Apparently no criteria are available which would specify the number of components needed to account for a significant portion of the total variance.

9.7.3.2 *The Basic (R-type) Algorithm*

The technique of component analysis is illustrated by an example at the end of this chapter. Assuming that the sample consists of r species and c quadrats we can outline an algorithm for component analysis in the following manner: (i) Adjust or standardize the original data to derive an rxc matrix **A** with elements $a_{ij}=(x_{ij}-\bar{x}_i)/S_i$ where x_{ij} is the original value of species i in quadrat j, $\bar{x}_i$ is the mean value of species i in the sample set, and S_i is a factor of standardization chosen by the phytosociologist (see ORLÓCI 1967b). (ii) Generate an rxr matrix **R** of scalar products in accordance with $\mathbf{R}=\mathbf{AA}'$; the elements of **R** relate the species in pairs. (iii) Find the eigenvalues (λ_1, ..., λ_r) and the corresponding eigenvectors (α_1,...,α_r) of **R**. It is assumed that the eigenvectors are so standardized that their lengths will be equal to unity, i.e. $\alpha_i' \alpha_i=1$. Note that any element a_{hi} is regarded as a direction cosine of species h and component i. (iv) The component scores are computed for quadrats by the transformation $\mathbf{Y}=\alpha'\mathbf{A}$. A characteristic element y_{ij} of **Y** represents the component score (co-ordinate) of the jth quadrat on the ith component. The components are the new ordination axes subject to the condition of mutual linear independence. The dispersion on the ith component is defined by the ith eigenvalue λ_i, and the efficiency of the ith component in accounting for variation in the sample is given by $\lambda_i/\sum\lambda_h$, the summation taken from $h=1$ to r. The correlation of species h and component i is defined by $a_{hi}\sqrt{(\lambda_i/\rho(h,h))}$ where $\rho(h,h)$ is the hth element in the principal diagonal of **R**. The basic computational problem in component analysis is concerned with the extraction of the eigenvalues and eigenvectors of the scalar product matrix. For computer use the JACOBI (NEUMANN) iterative method for which an ALGOL 60 algorithm is described by GRENCH & THACHER (1965), or IBM's EIGEN subroutine in Fortran IV (IBM 1968) is available. The underlying concepts are discussed by GREENSTADT (1960).

9.7.3.3 *A Simplified (Q-type) Algorithm Based on Scalar Products*

The basic algorithm, which has just been described, is not a particularly practical one for deriving component scores when the number of species is relatively large as compared to the number of quadrats. When the number of quadrats is relatively small we can obtain the component scores more simply by analysis of a scalar product matrix of quadrats, $\mathbf{Q}=\mathbf{A}'\mathbf{A}$, in which **A** is identical to the **A** matrix from which the **R** matrix is computed. The analysis may be called in this case as the *Q-type component analysis*, signifying that, in contrast with the *R-type component analysis*, here we are analysing a **Q** matrix .This analysis has been described in detail by Orlóci (1966, 1967b), Gower (1966), and Pielou (1969). These authors point out that the two analyses can produce identical component scores. This duality is a consequence of the relationship $\mathbf{Y}=\beta=\alpha'\mathbf{A}$, where the column vectors of β are the eigenvectors of **Q** so adjusted that their squared lengths are equal to the corresponding eigenvalues, i.e. $\beta_j'\beta_j=\lambda_j$.

9.7.3.4 *A Simplified (D-type) Algorithm Based on Distance*

In both the R-type and Q-type component analysis we have used the scalar product function as the basic definition of phytosociological resemblance. It, however, has been shown by Gower (1966) that the same results can be obtained from the analysis of Euclidean distance. The distance matrix which we need to define is $\mathbf{D}^1$ with elements given by $-0.5d_{jk}^2$ where the d_{jk}^2 is the squared Euclidean distance computed between quadrat j and quadrat k. The matrix $\mathbf{D}^1$ has cxc elements, its non-negative eigenvalues and the associated eigenvectors are identical to those of the **Q** matrix. It directly follows from these identities that the analyses of $\mathbf{D}^1$, **Q**, or **R** will produce identical components scores for the quadtats. The user may find that in some ordinations $\mathbf{D}^1$ is much simpler for computation than either **R** or **Q** matrix.

9.7.3.5 *Mapping new Individuals in Scatter Diagrams*

It is possible to map new individuals in a scatter diagram on the basis of the transformation $y_{jk}=\alpha_j'\mathbf{A}_k$ where y_{jk} is the co-ordinate of the individual $\mathbf{A}_k$ on the jth ordination axis, and the elements of α_j represent direction cosines. We note that the species values in $\mathbf{A}_k$ must be adjusted similarily as in the original data which formed the basis of the ordination. The direction cosines can be obtained directly in a component analysis of **R**, or computed by $a_{hi}=(\sum a_{hj}y_{ij})/\lambda_i$ where a_{hi} is the direction cosine of the hth species relative to the ith ordination axis, a_{hj} is the value of species h in the jth quadrat, y_{ij} is the co-ordinate of the jth quadrat on the ith ordination axis, and λ_i is a measure of dispersion on the ith ordination axis. The summation is taken from $j=1$ to c.

9.7.3.6 *Number of Component Axes Required*

The phytosociologist may want to limit the analysis only to those components which correspond to the first *m* largest eigenvalues. Alternatively he may discard all components from the end of the list that are statistically indistinguishable (see SEAL 1964). He may also consider the ecological meaningfulness of the components and reject those to which such a meaning cannot be attached (see GOODALL 1954, GREIG-SMITH 1964, 1971).

9.7.3.7 *Remarks on Scatter Diagrams*

The joint distribution of the principal components is plotted in scatter diagrams with the purpose of graphical illustration of some ordination properties or the relationship of ordination results to external factors. It must, however, be noted that a scatter diagram represents an orthogonal projection in two dimensions of a potentially very complex multidimensional configuration of points. A degree of distortion may thus occur in the projection depending on the original dimensions of the configuration. The distortion will occur in the sense that points, separated by great distances in the multidimensional configuration, may occupy a common position in the projection. It follows, therefore, that gaps between points in a scatter diagram will always signify separations at least as large as the real separations in the multidimensional configuration, whereas the lack of such gaps will not necessarily mean a similar lack of real separations.

9.7.3.8 *An Example*

The extraction of component scores is illustrated on an example so devised that the computational steps can be easily reconstructed. Consider a data matrix of 5 species (rows) and 4 quadrats (columns):

$$\mathbf{X} = \begin{bmatrix} x_{11} & x_{12} & x_{13} & x_{14} \\ x_{21} & x_{22} & x_{23} & x_{24} \\ x_{31} & x_{32} & x_{33} & x_{34} \\ x_{41} & x_{42} & x_{43} & x_{44} \\ x_{51} & x_{52} & x_{53} & x_{54} \end{bmatrix} = \begin{bmatrix} 1 & 3 & 4 & 4 \\ 3 & 3 & 1 & 1 \\ 5 & 3 & 5 & 3 \\ 1 & 1 & 3 & 3 \\ 4 & 2 & 1 & 1 \end{bmatrix}.$$

Each entry x_{ij} in this matrix represents the quantity of a species in a given quadrat. The sample means for species are given by

$$\bar{\mathbf{X}} = \begin{bmatrix} \bar{x}_1 \\ \bar{x}_2 \\ \bar{x}_3 \\ \bar{x}_4 \\ \bar{x}_5 \end{bmatrix} = \begin{bmatrix} 3 \\ 2 \\ 4 \\ 2 \\ 2 \end{bmatrix}.$$

From **X** we derive **A** according to $a_{ij} = (x_{ij} - \bar{x}_i)/S_i$ in which for reasons of simplicity we set S_i equal to 1. The **A** matrix is given by

$$\mathbf{A} = \begin{bmatrix} a_{11}\, a_{12}\, a_{13}\, a_{14} \\ a_{21}\, a_{22}\, a_{23}\, a_{24} \\ a_{31}\, a_{32}\, a_{33}\, a_{34} \\ a_{41}\, a_{42}\, a_{43}\, a_{44} \\ a_{51}\, a_{52}\, a_{53}\, a_{54} \end{bmatrix} = \begin{bmatrix} -2 & 0 & 1 & 1 \\ 1 & 1 & -1 & -1 \\ 1 & -1 & 1 & -1 \\ -1 & -1 & 1 & -1 \\ 2 & 0 & -1 & -1 \end{bmatrix}.$$

The **R** matrix of scalar products for species is computed according to

$$\mathbf{R} = \mathbf{AA}' = \begin{bmatrix} 6 & -4 & -2 & 4 & -6 \\ -4 & 4 & 0 & -4 & 4 \\ -2 & 0 & 4 & 0 & 2 \\ 4 & -4 & 0 & 4 & -4 \\ -6 & 4 & 2 & -4 & 6 \end{bmatrix},$$

and the **Q** matrix of scalar products for quadrats according to

$$\mathbf{Q} = \mathbf{A'A} = \begin{bmatrix} 11 & 1 & -5 & -7 \\ 1 & 3 & -3 & -1 \\ -5 & -3 & 5 & 3 \\ -7 & -1 & 3 & 5 \end{bmatrix}.$$

Note the equality of the sum of elements in the diagonal cells of **R** and **Q**. The distance matrix of quadrats is given by

$$\mathbf{D}^1 = \begin{bmatrix} 0 & -6 & -13 & -15 \\ -6 & 0 & -7 & -5 \\ -13 & -7 & 0 & -2 \\ -15 & -5 & -2 & 0 \end{bmatrix}$$

where the elements represent $-0.5d_{jk}^2$ values.

The eigenvalues of **R**, **Q**, and $\mathbf{D}^1$ are identical (within rounding errors) with the exception of the fourth eigenvalue of $\mathbf{D}^1$ which is a negative value and should be ignored. The non-zero eigenvalues include $\lambda_1 = 18.591$, $\lambda_2 = 4.672$, and $\lambda_3 = 0.737$. Two more eigenvalues of **R** and one more of **Q** are zeros. These indicate that the four-point configuration is embedded in a three dimensional space. A relatively large λ_1 clearly is an indication that regarding the linear variation in **X** the summarization is quite successful. The eigenvectors (containing the transformation coefficients) of **R**, adjusted to unit length, are given by

$$\alpha' = \begin{bmatrix} 0.558 & -0.421 & -0.153 & 0.421 & -0.558 \\ -0.143 & -0.343 & 0.850 & 0.343 & 0.143 \\ 0.411 & 0.452 & 0.503 & -0.452 & -0.411 \end{bmatrix}.$$

The elements in each row of this matrix represent the direction cosines of species relative to a given principal component axis. The component scores are obtained according to

$$\mathbf{Y} = \beta = \alpha'\mathbf{A} = \begin{bmatrix} -3.226 & 0.735 & -0.235 \\ -0.690 & -1.537 & 0.401 \\ 1.805 & 1.252 & 0.420 \\ 2.110 & -0.449 & -0.587 \end{bmatrix}$$

The elements in the columns of **Y** are the component scores (coordinates) of quadrats on the new ordination axes. These scores are identical for the **R**, **Q**, and $\mathbf{D}^1$ matrices. In the present example the component scores can be produced least tediously on the basis of the $\mathbf{D}^1$ matrix.

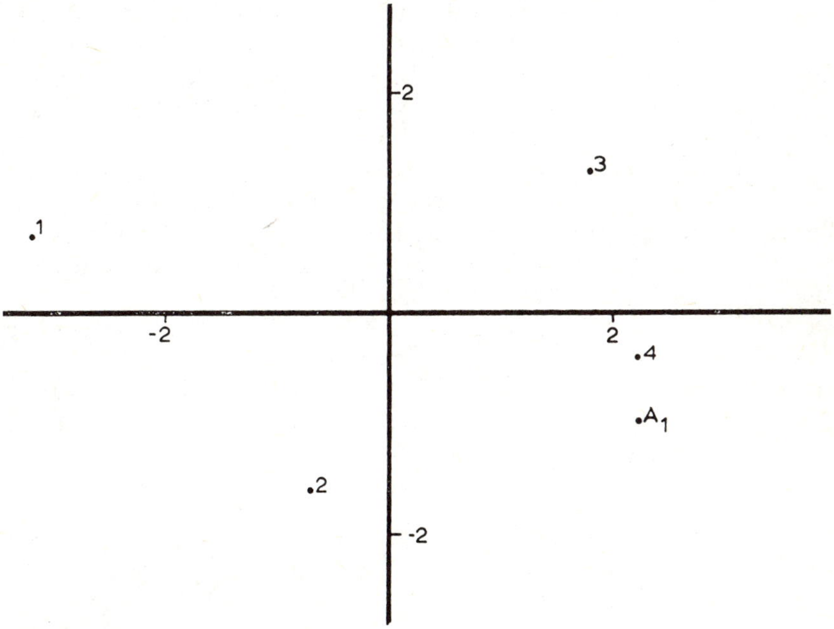

Fig. 1. Scatter diagram produced by component analysis from data given in the main text.

The scatter diagram of the first and second ordination axis is given in Fig. 1. Point $\mathbf{A}_1$ in this diagram represents an external quadrat in which the species values are the elements in $\mathbf{X}_1' = (5\ 2\ 3\ 3\ 1)$, so adjusted that $\mathbf{A}_1' = \mathbf{X}_1' - \overline{\mathbf{X}}' = (2\ 0\ -1\ 1\ -1)$. The component scores for individual $\mathbf{A}_1$ are given by $y_{11} = \alpha_1' \mathbf{A}_1 = 2(0.558) + 0(-0.421) + (-1)(-0.153) + 1(0.421) + (-1)(-0.558) = 2.248$ and $y_{21} = \alpha_2' \mathbf{A}_1 = 2(-0.143) + 0(-0.343) + (-1)(0.850) + 1(0.343) + (-1)(0.143) = -0.936$.

9.7.3.9 *Summary of Component Analysis*

Component analysis as a mathematical method of data analysis is aimed at the extraction of a new set of co-ordinates (component scores) from a resemblance matrix whose elements qualify as Euclidean spatial parameters. These components are linearly independent, and as such, they provide the most efficient means for a parsimonious summarization of linear vegetational or environmental variation in a given sample. The field of phytosociological applications of component analysis includes trend seeking, search for clusters, discriminant-type assignments, and production of new data for input in further analyses.

We have outlined different algorithms on the basis of which phytosociological entities can be ordinated. While these algorithms are associated with different resemblance functions, which generate the coefficient matrices (**R**, **Q**, or $\mathbf{D}^1$) for analysis, they nevertheless represent alternative avenues of approach leading to the same set of component scores in their final results.

It may be pointed out that the application of component analysis as a general strategy is seriously weakened by its linear model Because the assumption of linearity can be expected to be a relatively reasonable assumption in the description of species and environmental correlations within relatively short segments of their gradients, it is reasonable to suggest that the application of component analysis should be limited to sample sets of relatively high homogeneity. Alternatively, component analysis may be followed by secondary analyses aimed at the extraction of higher-order variation, not detected by linear ordination.

9.7.4 Miscellaneous Methods

9.7.4.1 *Polar Ordinations*

Component analysis, or factor analysis, is totally unsuited for longhand calculations, even in the case of relatively small sets of data.

It is the computational difficulty in these methods which probably stands out as one of the main reasons for the introduction of simplified methods, such as for instance the BRAY & CURTIS (1957) ordination and its many variants, also including those introduced to improve upon these simplified methods (see BEALS 1960, AUSTIN & ORLÓCI 1966, SWAN et al. 1969, VAN DER MAAREL 1969, and ORLÓCI 1974 a, b, 1975). The improvements were concerned with (i) the nature of the resemblance function, (ii) the perpendicularity and common origin of the ordination axes, and (iii) the choise of end stands, i.e. ordination poles (hence the term *polar* ordination) to mark out the direction of the ordination axes. In the method of AUSTIN and ORLOCI the resemblance functions $1-L$, or max $L-L$, have been replaced by Euclidean distance, and the axes were selected in perpendicular directions. SWAN and his collaborators defined a first axis as that line, connecting two strategically chosen stands (quadrats) in sample space, on which the sum of squared projections of the inter-stand distances is a possible maximum. They extracted the subsequent ordination axes on the basis of the same principle from the residual distances. A computationally less tedious method was proposed by VAN DER MAAREL to achieve similar goals. In his method the first ordination axis is comparable to the first axis in the SWAN ordination, but the subsequent axes are produced from the original distances rather than the residuals.

Polar ordinations, similarly to component analysis, may represent linear models, although when used with certain types of resemblance functions they may be non-linear (see GAUCH & WHITTAKER 1972). However, they are less precise mathematically, and also, they are less efficient as methods of summarization than component analysis. These ordinations, nevertheless, are computationally simple, or relatively so, and can be handled with relative ease by the less mathematically minded phytosociologist. These properties, and the fact that they have produced meaningful ecological results, seem to justify the continued use of polar ordinations in phytosociology.

9.7.4.2 *Position Vectors Ordination*

The SWAN ordination as well as the method of VAN DER MAAREL are potentially less efficient than the position vectors ordination (ORLÓCI 1966, FIELD & ROBB 1970, MARTIN 1969) in accounting for linear variability in a given sample. This is a consequence of the fact that the axes in position vectors ordination, similarily to the component axes, originate at the centroid of the sample space, and that each axis extracted by position vectors ordination approximates a component axis as closely as possible, under the condition that each ordination axis must go through an existing point (quadrat in our case) in sample space.

We shall illustrate position vectors ordination on the data given by **X** in paragraph 10.7.3.8. The analysis begins with computation of a **Q** matrix:

$$\mathbf{Q} = \mathbf{A'A} = \begin{bmatrix} q_{11} q_{12} q_{13} q_{14} \\ q_{21} q_{22} q_{23} q_{24} \\ q_{31} q_{32} q_{33} q_{34} \\ q_{41} q_{42} q_{43} q_{44} \end{bmatrix} = \begin{bmatrix} 11 & 1 & -5 & -7 \\ 1 & 3 & -3 & -1 \\ -5 & -3 & 5 & 3 \\ -7 & -1 & 3 & 5 \end{bmatrix}.$$

A characteristic element q_{jk} in this matrix defines the Q-type scalar product of quadrat j and k in a manner described in detail in paragraph 10.7.3.8.

Position vectors ordination derives its name from the use of position vectors as ordination axes. The position vector of a quadrat is defined as the line which connects the centroid of the sample space with the quadrat as a spatial point The initial goal is to identify the position vector which lies nearest to the first principal component axis. The criterion to be used is the sum of squared projections of all position vectors on the line of a given position vector. This sum for the jth position vector (quadrat) is given by $S_j = \sum q_{jk}^2/q_{jj}$ where the summation is from $k=1$ to c. The numerical values for the first and subsequent quadrats are given by $S_1 = 11^2/11 + 1^2/11 + (-5)^2/11 + (-7)^2/11 = 17.818, S_2 = 6.667, S_3 = 13.600$, and $S_4 = 16.8$ respectively. The maximum occurs on the line of the position vector of the first quadrat. We thus chose the first position vector as the first ordination axis. The co-ordinates on the first axis are computed according to $y_{1k} = q_{1k}/\sqrt{(q_{11})}$ for which the numerical values are given by $y_{11} = 11/\sqrt{(11)} = 3.317$, $y_{12} = 1/\sqrt{(11)} = 0.302$, $y_{13} = -5/\sqrt{(11)} = -1.508$, and $y_{14} = -7/\sqrt{(11)} = -2.111$. Note the closeness of these co-ordinate values to the component scores in the first column of **Y**, given in paragraph 10.7.3.8. The dispersion on the first ordination axis is 17.818, nearly equal to the first eigenvalue (18.591) of **Q**, **R**, or $\mathbf{D}^1$ matrix After computation of the first axis co-ordinates follows the computation of the first residual scalar product matrix **Q′** with elements defined by $q'_{jk} = q_{jk} - y_{1j} y_{1k}$. The value for instance in the second cell of the third row of **Q′** is $q'_{32} = q_{32} - y_{13} y_{12} = -3 - (-1.508)(0.302) = -2.545$. The complete residual matrix is

$$\mathbf{Q}' = \begin{bmatrix} 0.000 & 0.000 & 0.000 & 0.000 \\ 0.000 & 2.909 & -2.545 & -0.364 \\ 0.000 & -2.545 & 2.727 & -0.183 \\ 0.000 & 0.364 & -0.183 & 0.545 \end{bmatrix}$$

We now repeat the procedure on $\mathbf{Q}'$. The sum of squared projections is a maximum on the second position vectoer ($S_2=5.182$), and the co-ordinates are $y_{21}=0, y_{22}=1.706, y_{23}=-1.492$, and $y_{24}=-0.213$. These results are directly comparable to the second largest eigenvalue (4.672) and corresponding component scores given in paragraph 10.7.3.8. The discrepancy between the two sets of results is explainable by the fact that position vectors ordination extracted a first axis which slightly deviates from the first component axis. This has a cumulative effect resulting in an increasing departure of the subsequent ordination axes from the corresponding components.

The second residual matrix $\mathbf{Q}''$ is derived from the first residual $\mathbf{Q}'$ according to the expression $q''_{jk}=q'_{jk}-y_{2j}y_{2k}$ for which the computed values are given by

$$\mathbf{Q}'' = \begin{bmatrix} 0.000 & 0.000 & 0.000 & 0.000 \\ 0.000 & 0.000 & 0.000 & 0.000 \\ 0.000 & 0.000 & 0.500 & -0.500 \\ 0.000 & 0.000 & -0.500 & 0.500 \end{bmatrix}$$

The sum of squared projections is equal on both non-zero position vectors, $S_3=S_4=1.000$. The corresponding co-ordinates then are $y_{31}=0, y_{32}=0, y_{33}=-0.707$, and $y_{34}=0.707$. These results are directly comparable to the third largest eigenvalue (0.737) and associated component scores. It may be useful at this point if we noted that the above example is rather peculiar in that respect, that the sequence of position vectors happens to be identical with the sequence in which they are extracted as ordination axes. This is not generally the rule, and it would not have happened if the quadrats were listed in a different sequence.

Note that the third residual matrix $\mathbf{Q}'''$ is a zero matrix, indicating that all variation in $\mathbf{Q}$ has been accounted for by the first three ordination axes. The total variation, i.e. the sum of all diagonal elements in $\mathbf{Q}$ is equal to 24. The total dispersion on the three ordination axes ($17.818+5.182+1.000$) is also equal to 24.

Position vectors ordination will be a reliable approximation to component analysis when two specific circumstances happen to occur simultaneously: (i) a relatively uniform phytosociological sample, and (ii) a large number of points (quadrats in our case). We can anticipate position vectors ordination to produce an accurate approximation to component analysis under these circumstances, because when there are many points, and if these points are more-or-less evenly dispersed in the relevant part of sample space, then it is quite probable that there will be at least one point at or very near each of the principal component axes. These points will be found by position vectors ordination. The ordination axes will go through these points and the centroid.

9.8 Ordination by non-Euclidean matrices

9.8.1 The Sample Space

The sample space, in this case, can be derived by negation of the properties of an Euclidean space. It will be sufficient here if we simply stated that the space we have in mind is one in which the spatial parameters are functions other than an Euclidean distance or a scalar product. Examples of such functions have already been discussed in chapter 9.7.1. The present chapter will further elaborate on the use of non-Euclidean resemblance functions in phytosociological ordinations. Two methods are discussed in this connection, including Gower's principal axis analysis and Kruskal's method of multidimensional scaling. It should be mentioned, however, that the use of these methods is not really limited to non-Euclidean functions.

9.8.2 Principal Axes Analysis

The underlying mathematical model in principal axes analysis is the same as in component analysis (see Gower 1966, Pielou 1969). In both of these methods ordination axes are obtained on the basis of the use of an eigenvalue and vector procedure. Principal axes analysis, however, differs from component analysis in one important respect — it is not limited to the use of Euclidean distance or scalar products.

We shall define principal axes analysis in the broadest sense as a method aimed at the extraction of Cartesian co-ordinates from any given resemblance matrix. This does not need to be a metric matrix. The use of a non-metric matrix, however, will lead to distortion related to the inefficiency of ordination distances **M** to account for the observed values in the resemblance matrix **D**. Ordination distances are defined by $m_{jk}^2 = \sum(y_{ij} - y_{ik})^2$ where the y_{ij} and y_{ik} values are the co-ordinates of quadrat j and quadrat k on the ith ordination axis. The distortion so-defined can be measured by the function $\sigma_{MD} = 100(1 - \rho_{MD}^2)$, where ρ_{MD}^2 is the squared product-moment correlation coefficient, computed between the elements of **M** and the corresponding elements of **D**.

We can suggest that degree of distortion in the ordination will be dependent on the metric properties of the resemblance function which generates the **D** matrix. To illustrate the point in question, we shall consider the following functions:

(1) $d_{jk} = 2(1 - 2\sum \min(x_{ij}, x_{ik}) / \sum (x_{ij} + x_{ik})) = \text{PD}$

(2) $d_{jk} = 2(\max L - 2\sum \min(x_{ij}, x_{ik}) / \sum (x_{ij} + x_{ik})) = \text{PD}^*$

(3) $d_{jk} = 2(\sum (x_{ij} + x_{ik}) - 2\sum \min(x_{ij}, x_{ik})) = 2\sum |x_{ij} - x_{ik}|$, where x_{ij} and x_{ik} represent species values in the quadrats, $\sum \min(x_{ij}, x_{ik})$ is the sum of lesser values in quadrats j and k, and max$L$ is a maximum value of $2\sum \min(x_{ij}, x_{jk}) / \sum (x_{ij} + x_{ik})$.

Similar expressions have already been discussed in some respect in chapter 9.7.1. Function (3) is the absolute value function. This metric function is degenerated into a dissimilarity function in (1) and (2) with properties that fail on at least one, but possibly more, of the metric space axioms.

We have generated successive **D** matrices on the basis of functions (1), (2), and (3). The eigenvalues and corresponding eigenvectors were extracted from each **D** matrix after multiplication of its elements by the scalar −0.5. The co-ordinates of quadrats then were obtained by adjustment of the eigenvalues in the same manner as in component analysis of a **Q** or $\mathbf{D}^1$ matrix (see chapter 9.7.3). The data used in the examples are given in Table II.

LE II. Tree counts in a Wisconsin maple forest [1]).

drat number ies	1	2	3	4	5	6	7	8	9	10	11	12	13	14	15	16	17	18	19	20	21	22	23	24	25[1])
ar Maple	8	4	3	7	0	0	4	5	8	1	1	5	8	9	9	10	10	5	6	8	3	2	2	6	4
wood	0	7	4	1	1	6	2	0	0	3	0	4	1	0	0	3	0	0	0	0	4	6	0	0	0
pery Elm	0	1	0	0	5	4	7	0	0	6	5	3	1	0	0	0	0	0	0	0	2	5	5	2	0
wood	0	0	1	0	4	0	0	0	0	1	0	0	0	0	0	0	0	0	0	0	0	0	0	0	0
Oak	0	0	0	0	0	0	0	0	0	0	0	0	3	1	0	0	0	2	1	0	0	0	0	0	1
owbud	0	1	1	0	0	0	0	0	0	0	0	0	0	0	0	0	0	0	0	0	0	0	0	0	0
ernut	1	0	0	0	0	0	0	0	0	0	0	0	0	0	0	0	0	0	0	0	0	0	0	0	0
k Cherry	1	0	0	0	0	0	0	0	0	0	1	0	0	0	0	0	0	0	0	0	0	0	0	0	0
kberry	0	0	0	1	0	0	0	0	0	0	0	0	0	0	0	0	0	0	0	0	0	0	0	0	0
te Oak	0	0	0	0	0	0	0	0	0	0	0	0	0	0	0	0	0	0	0	0	0	0	0	0	1

ta obtained by sampling on a vegetation map given by Curtis (1956). Sampling map units are equivalent) feet square quadrats in natural size.

No distortion occurred in the ordination (in terms of σ_{MD}) when **D** was generated by function (3). For these results the scatter diagram of the first two ordination axes is plotted in Fig. 2. The lack of

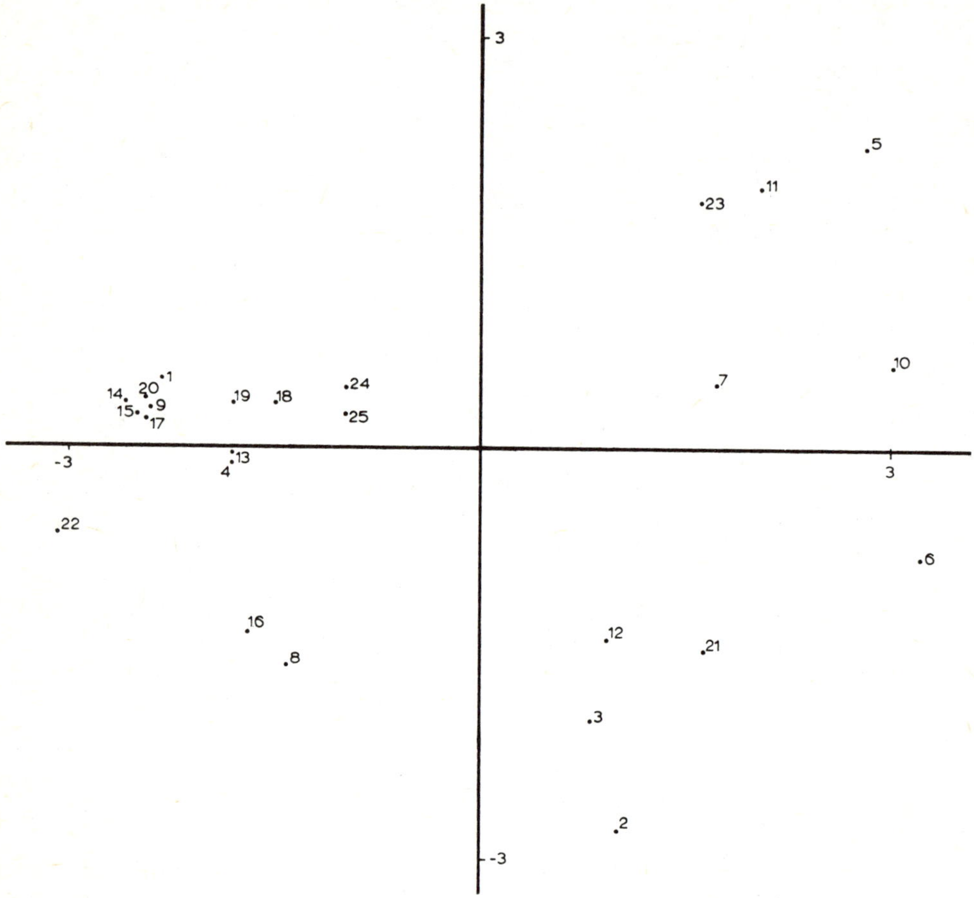

Fig. 2. Scatter diagram produced by principal axes analysis of the absolute value function based on data in Table II. See the main text for explanations.

distortion indicates that the information in **D** can be completely accounted for by the Cartesian co-ordinates of principal axes analysis. This implies an identity of the original configuration, as depicted by **D** (or -0.5**D**), and the derived configuration, as defined by the ordination distances **M**. The analysis of the same data on the basis of function (1), however, resulted in about 17 % distortion. An even larger amount of distortion (about 18 %) occurred when we used function (2) to generate the **D** matrix. The value of max$L$ was set at 0.5 in this example. The scatter diagrams for the ordination of functions (1) and (2) are given in Figs. 3—4. The generality of these results, however, remains to be seen.

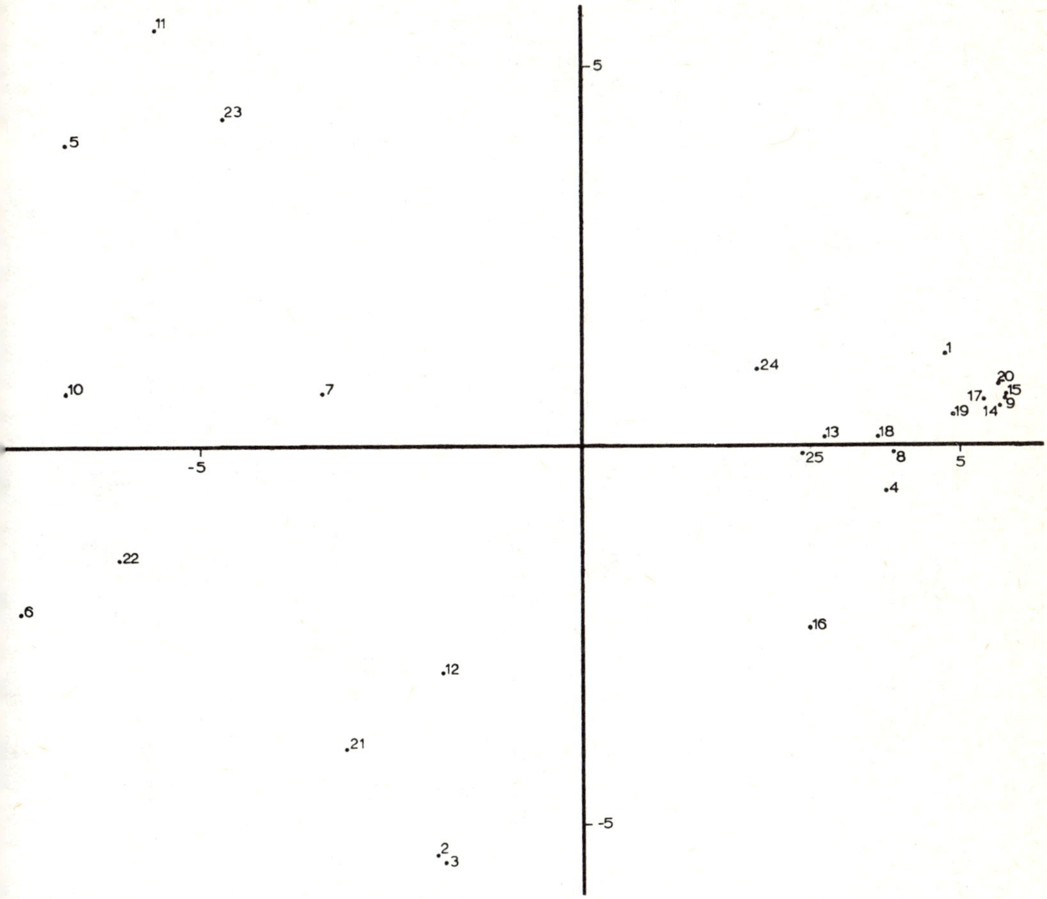

. 3. Scatter diagram produced by principal axes analysis of variant $CD = 1 - CC$ type of the SØRENSEN index based on data in Table II. See the main text for explanations.

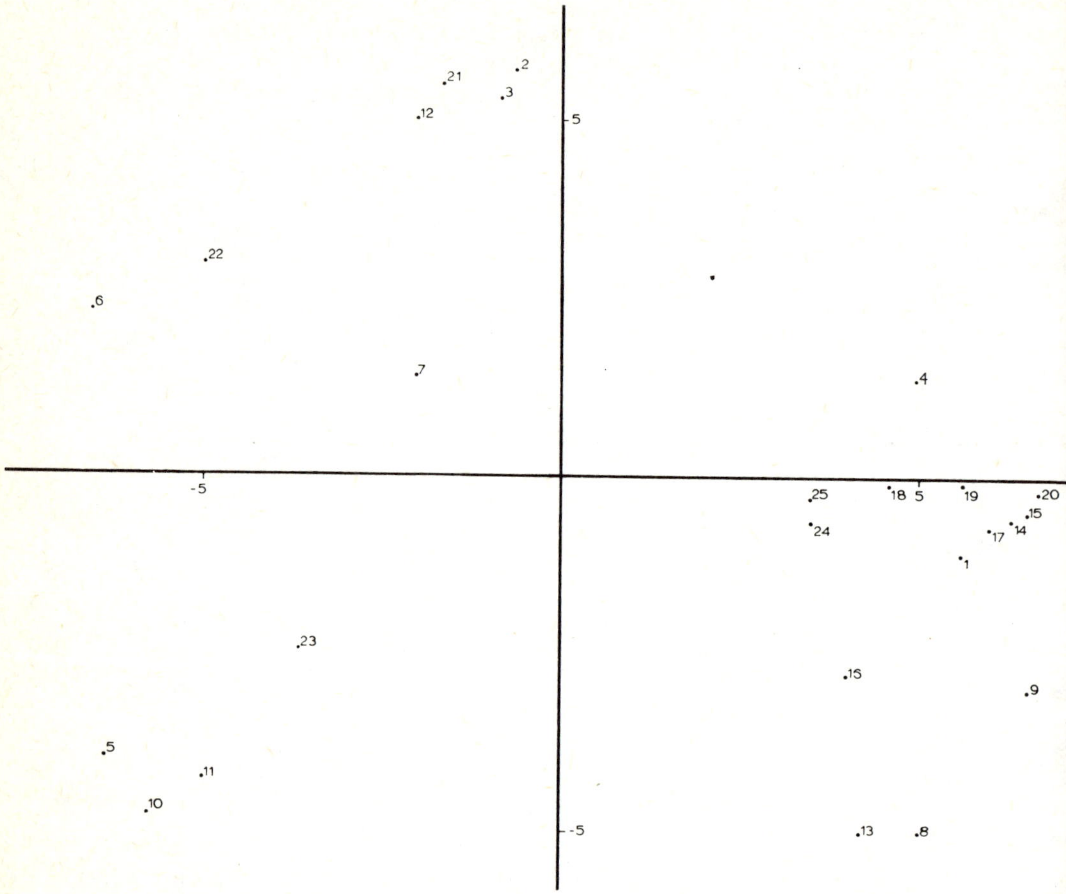

Fig. 4. Scatter diagram produced by principal axes analysis of the $CD' = \max CC - CC$ variant of the SØREN index based on data in Table II. See the main text for explanations.

9.8.3 KRUSKAL'S MULTIDIMENSIONAL SCALING

While the metric configuration **M** is derived by an eigenvalue and vector procedure in principal axes analysis, this type of configuration, or the co-ordinates which describe it, can also be derived by other methods from a given set of experimental (observed) resemblance values. A method particularly well-suited to perform such a task has been proposed by KRUSKAL (1964a,b).

The KRUSKAL method begins by specifying t the number of ordination axes to be extracted. Then an arbitrary set of c points is placed in the t-dimensional ordination space, and the initial ordination resemblance values, often as distances or dissimilarities, are computed. We shall designate these dissimilarities by (d_{jk}). This is followed by fitting the regression line $\hat{d}_{jk}=f(\delta_{jk})$ to the points in the scatter diagram of the ordination dissimilarities (d_{jk}) and the experimental resemblance values (δ_{jk}). We note that the KRUSKAL method is little restricted as far as the definition of d_{jk} or δ_{jk} is concerned, and that this method can handle different types of regression models whose choice is largely a matter of personal preference (see KRUSKAL & CARMONE 1971). Different types of regression models, however, yield different regression estimates $\hat{d}_{jk}$, and consequently, different sets of ordination co-ordinates.

The stress, or distortion, in the ordination configuration, which we seek to minimize in the KRUSKAL method, is measured by the residual sum of squares in (d_{jk}) with respect to the regression estimates $(\hat{d}_{jk})$, and not by the fit of the (d_{jk}) to the experimental resemblance values (δ_{jk}). KRUSKAL & CARMONE have used as a measure of stress $S_{d\hat{d}}=\surd(\sum\sum(d_{jk}-\hat{d}_{jk})^2/\sum\sum(d_{jk}-\bar{d})^2)$. In this expression d_{jk} is an ordination disssimilarity, $\bar{d}$ is the mean of all d_{jk}, $\hat{d}_{jk}$ is a regression estimate of d_{jk} at δ_{jk}, and the summations are taken from $j=1$ to $c-1$ and $k=j+1$ to c. Following the calculation of stress, all points in the initial ordination configuration are moved a bit, i.e. their co-ordinates are changed a little to decrease the value of stress. The procedure is iterated, starting with a new regression analysis, until the value of stress drops below a specified threshold limit. The output includes t sets of c co-ordinates which can be plotted jointly in two dimensional scatter diagrams. The accuracy of the scatter diagrams in representing the ordination configuration (d_{jk}) can be measured by $\sigma_{d^*d}=\surd(1-\rho^2_{d^*d})$ where $\rho^2_{d^*d}$ is the squared correlation of the ordination dissimilarities (d_{jk}) and the distances in the scatter diagram (d^*_{jk}).

There seems to be no example in phytosociology which used the KRUSKAL method for ordination. This method, however, appears to have many attractive properties for the phytosociologist. The method's extreme flexibility to select from different resemblance functions on the basis of which the δ_{jk} and d_{jk} values are computed, and the reliance on regression analysis, the model of which to be chosen by the user himself, can be mentioned in this connection. By using the KRUSKAL method it is possible to derive an ordination configuration (d_{jk}) from a matrix of observed resemblances (δ_{jk}) in such a manner that the former will be some specified non-linear function of the latter. This and the other properties which we already mentioned definitely indicate a good potential of the KRUSKAL method in phytosociological ordinations.

9.9 Comparison of Methods

The ordination methods which we discussed differ greatly on the basis of conceptual grounds and also in terms of the goals which they may attempt to achieve. Among these methods factor analysis stands out as an ordination method principally concerned with estimation of population properties. This method is covariance-oriented and it is burdened by several restrictive assumptions. These, and probably also the lack of sufficient examples, may be reasons why factor analysis, despite its effective use by DAGNELIE (1960, article 8) has not aroused wider interest among phytosociologists.

Component analysis has mainly been used as a descriptive (deterministic) method without much regard of statistical considerations. It has been proved to be a useful method in handling continuous linear variability in phytosociological data in connection with different practical problems which range from trend seeking to classifications. The classification function of component analysis, similarly as in all the ordination techniques, has however been reduced in importance by the availability of more efficient classification techniques.

The principal justification for the use of polar ordinations and position vectors ordination is that they are simple conceptually and easily executed computationally, or relatively so, as compared to component analysis. This advantage, nevertheless, is being fast eroded away as the phytosociologist gains more ready access to the use of digital computers.

The method of principal axes analysis is computionally similar to component analysis. It, nevertheless, permits the use of almost any resemblance function as an ordination parameter. This property is in sharp contrast with component analysis where the admissible resemblance functions must qualify as Euclidean spatial parameters. KRUSKAL's method also can handle the ordination of different non-Euclidean resemblance matrices without much restrictions on the function which generates them. This method has the added advantage of a regression device whereby the ordination configuration can be made to be a non-linear function of the observed resemblances. Such a non-linear derivation of an ordination may very much interest the phytosociologist in his search for an ecologically more meaningful ordination method.

9.10 Summary

Ordination is used as a collective term in the present paper to describe the different methods which can extract metric information, such as for instance a new set of co-ordinates, from a given resemblance matrix. While the elements in such a matrix define the pairwise similarity, or dissimilarity, between phytosociological entities, the entities themselves may represent vegetation units (quadrats, sample plots), individual species, or other kinds of objects, vegetational or environmental, that can be recognized without ambiguity. The present paper is aimed at a comparative review of the different ordination methods which have been used, or have a good potential, as a basis for the mathematical analysis of phytosociological data.

Two general groups of ordination methods are discussed. Those of the first group are restricted to the analysis of resemblance matrices in which every element will qualify as either a common (Euclidean) distance or a centered scalar product. The methods of this group include CATTELL's open model factor analysis, component analysis, several variants of polar ordination, and position vectors ordination.

The ordination methods of the second group, in contrast with those of the first group, are not limited to the use of distances or scalar products. As a matter of fact they can handle resemblance matrices of almost any kind. The different definitions of resemblance, however, will potentially lead to different ordination results. Described in this group are GOWER's method of principal axes analysis, which we broadened in our interpretations of the admissible definitions of resemblance, and KRUSKAL's method of multidimensional scaling.

The first parts of the paper are devoted to a discussion of the general concepts about ordinations, regarding their purpose, the underlying common model, hypothesis testing, and the important efficiency criteria. The discussion of the general concepts is followed by a systematic description of ordination methods with a view at their instrinsic properties and generic relationships as they relate to phytosociological applications. The different methods are illustrated by several numerical examples.

RERERENCES

ASHTON, P. S., – 1964 – Ecological studies in the mixed dipterocarp forests of Brunei State. *Oxf. For. Mem.* 25: 1—75.

AUSTIN, M. P., – 1968 – An ordination study of a chalk grassland community. *J. Ecol.* 56: 739—757.

AUSTIN, M. P., P. S. ASHTON, & P. GREIG-SMITH, – 1972 – The application of quantitative methods to vegetation survey. III. A re-examination of rain forest data from Brunei. *J. Ecol.* 60: 305–324.

AUSTIN, M. P. & P. GREIG-SMITH, – 1968 – The application of quantitative methods to vegetation survey. II. Some methodological problems of data from rain forests. *J. Ecol.* 56: 827—844.
AUSTIN, M. P. & I. NOY-MEIR, – 1971 – The problem of non-linearity in ordination: experiments with two gradient models. *J. Ecol.* 59: 763—773.
AUSTIN, M. P. & L. ORLOCI, – 1966 – Geometric models in ecology. II. An evaluation of some ordination techniques. *J. Ecol.* 54: 217—227.
BEALS, E., – 1960 – Forest bird communities in the Apostle Islands of Wisconsin. *Wilson Bull.* 72: 156—181.
BRAY, R. J. & J. T. CURTIS, – 1957 – An ordination of the upland forest communities of southern Wisconsin. *Ecol. Monog.* 27: 325—349.
CARLSON, K. A., – 1970 – A multivariate classification of reformatory inmates. *Ph. D. Thesis.* University of Western Ontario, London, Canada.
CATTELL, R. B., – 1965 – Factor analysis: an introduction to essentials. *Biometrics* 21: 190—215, 405, 435.
CHANDAPILLAI, M. M., – 1970 – Variation in fixed dune vegetation at Newborough Warren, Anglesey, *J. Ecol.* 58: 193—201.
CURTIS, J. T., – 1956 – Plant Ecology Workbook. Burgess, Minneapolis.
CURTIS, J. T., – 1959 – The vegetation of Wisconsin: An Ordination of Plant Communities. University of Wisconsin, Madison. 657 pp.
DAGNELIE, P., – 1960 – Contribution à l'étude des communautés végétales par l'analyse factorielle. *Bull. Serv. Carte phytogéogr.*, Paris, Sér. B. 5: 7—71, 93—195.
DAGNELIE, P., – 1965 – L'étude des communautés végétales par l'analyse statistique des liaisons entre les espèces et les variables écologiques: un exemple. *Biometrics* 21: 890—907.
DALE, M. B. – 1975 – On objectives of ordination. *Vegetatio* 30: 15–32.
FIELD, J. G. & F. T. ROBB, – 1970 – Numerical methods in marine ecology. 2. Gradient analysis of rock shore samples from False Bay. *Zoologica africana.* 5: 191—210.
GAUCH, H. G. JR. & R. H. WHITTAKER, – 1971 – Coenocline simulation. *Ecology* 53: 446—451.
GAUCH, H. G. JR. & R. H. WHITTAKER, – 1972 – Comparison of ordination techniques. *Ecology.* 53: 868–875.
GENGERELLI, J. A., – 1963 – A method for detecting subgroups in a population and specifying their membership. *J. Psychol.* 55: 457—468.
GITTINS, R., – 1965a – Multivariate approaches to a limestone grassland community. I. A stand ordination. *J. Ecol.* 53: 385—401.
GITTINS, R., – 1965b – Multivariate approaches to a limestone grassland community. III. A comparative study of ordination and association analysis. *J. Ecol.* 53: 411—425.
GITTINS, R., – 1968 – Trend-surface analysis of ecological data. *J. Ecol.* 56: 845—869.
GITTINS, R., – 1969 – The application of ordination techniques. In 'Ecological Aspects of the Mineral Nutrition of Plants,' R. H. RORISON, ed. pp. 37—66. *Symp. Brit. Ecol. Soc.* 1968, Blackwell, London.
GOODALL, D. W., – 1954 – Objective methods for the classification of vegetation. III. An essay in the use of factor analysis. *Aust. J. Bot.* 2: 304—324.
GOODALL, D. W., – 1967 – The distribution of the matching coefficient. *Biometrics* 23: 647—656.
GOWER, J. C., – 1966 – Some distance properties of latent root and vector methods used in multivariate analysis. *Biometrika* 53: 325—338.

GREENSTADT, J., – 1960 – The determination of the characteristic roots of a matrix by the Jacobi method. In 'Mathematical Methods for Digital Computers,' A. RALSTON & H. S. WILF, eds., pp. 84—91. Wiley, New York.

GREIG-SMITH, P., – 1964 – Quantitative Plant Ecology. 2nd ed. Butterworths, London. 256 pp.

GREIG-SMITH, P., – 1971 – Analysis of vegetation data: the user viewpoint. In 'Statistical Ecology,' C. P. Patil ed., 3: 149—166. Pennsylvania State University Press, London.

GREIG-SMITH, P., M. P. AUSTIN, & T. C. WHITMORE, – 1967 – The application of quantitative methods to vegetation survey. I. Association analysis and principal component ordination of rain forest. *J. Ecol.* 55: 483—503.

GRENCH, R. E. & H. C. THACHER, – 1965 – Collected Algorithms 1960—1963 from the Communications of the Association for Computing Machinery. Clearing House for Federal Scientific and Technical Information, U. S. Department of Commerce, Springfield, Virginia.

GRIGAL, D. F. & R. A. GOLDSTEIN, – 1971 – An integrated ordination-classification analysis of an intensively sampled oak-hickory forest. *J. Ecol.* 59: 481–492.

GROENEWOUD, H. VAN, – 1965 – Ordination and classification of Swiss and Canadian coniferous forests by various biometric and other methods. (Germ. summ.) *Ber. geobot. Inst. ETH, Stiftg Rübel*, Zürich, 1964, 36: 28—102.

HOLGATE, P. – 1971 – Notes on the Marcyewski-Steinhaus coefficient of similarity In 'Statistical Ecology,' G. P. Patil, ed., 3: 181—193. Pennsylvania State University Press, London.

IBM, 1968 — System/360 Scientific Subroutine Package (360A-CM-03X) Version III Programmer's Manual Docoment H20-0205-3, IBM, White Plains, N.Y. 454 pp.

IVIMEY-COOK, R. B., – 1968 – Investigations into the phenetic relationships between species of Ononis *L. Watsonia* 7: 1—23.

IVIMEY-COOK, R. B. & M. C. F. PROCTOR, – 1967 – Factor analysis of data from an east Devon heath: comparison of principal component and rotated solutions. *J. Ecol.* 55: 405—413.

JACCARD, P., – 1901 – Distribution de la flore alpine dans le Bassin des Dranses et dans quelques régions voisines. *Bull. Soc. vaud. Sci. nat.* 37: 241—272.

KERSHAW, K. A., – 1968 – Classification and ordination of Nigerian Savanna vegetation. *J. Ecol.* 56: 467—482.

KRUSKAL, J. B., – 1964a – Multidimensional scaling by optimizing goodness of fit to a nonmetric hypothesis. *Psychometrika* 29: 1—27.

KRUSKAL, J. B., – 1964b – Nonmetric multidimensional scaling: a numerical method. *Psychometrika* 29: 115—129.

KRUSKAL, J. B. & F. CARMONE, – 1971 – How to use M-D-SCAL (Version 5M) and other useful information. Bell Telephone Laboratories, Murray Hill, New Jersey, U. S. A., and University of Waterloo, Waterloo, Ontario, Canada. (Mimeographed.)

LAMBERT, J. M. & M. B. DALE, – 1964 – The use of statistics in phytosociology. *Adv. ecol. Res.* 2: 59—99.

LAWLEY, D. N. & A. E. MAXWELL, – 1963 – Factor Analysis as a Statistical Method. Butterworths, London. 117 pp.

MAAREL, E. VAN DER, – 1969 – On the use of ordination models in phytosociology. (Germ. summ.) *Vegetatio* 19: 21—46.

MACNAUGHTON-SMITH, P., W. T. WILLIAMS, M. B. DALE, & L. G. MOCKETT, –

1964 – Dissimilarity analysis: a new technique of hierarchical sub-division. *Nature, Lond.* 202: 1034—1035.

Martin, J., – 1969 – Ordination methods in lichenocoenology. In: 'Quantitative Methods of Vegetation Study,' p. 86. Collected Conference Reports, Institute of Zoology and Botany, Academy of Sciences, Tartu State University, Estonian S. S. R.

Maycock, P. F., – 1963 – The phytosociology of the deciduous forests of extreme southern Ontario. *Can. J. Bot.* 41: 379—438.

Maycock, P. F. & J. T. Curtis, – 1960 – The phytosociology of Boreal conifer-hardwood forests of the Great Lakes region. *Ecol. Monogr.* 30: 1—35.

Mc Donald, R. P., – 1962 – A general approach to non-linear factor analysis. *Psychometrika* 27: 397—415.

Mc Intosh, R. P., – 1967 – The continuum concept of vegetation. *Bot. Rev.* 33: 130—187.

Mc Intosh, R. P. & R. T. Hurley, – 1964 – The spruce-fir forests of the Catskill Mountains. *Ecology* 45: 314—326.

Moore, J. H., P. Fitzsimons, E. Lambe, & J. White, –1970– A comparison and evaluation of some phytosociological techniques. (Germ. summ.) *Vegetatio* 20:1—20.

Morrison, D. F., – 1967 – Multivariate Statistical Methods. Mc Graw-Hill, New York. 338 pp.

Noy-Meir, I. & M. P. Austin, – 1970 – Principal component ordination and simulated vegetational data. *Ecology* 51: 551—552.

Noy-Meir, I. – 1974 – Catenation: quantitative methods for the definition of coenoclines. *Vegetatio* 29: 89–99.

Ochiai, A., – 1957 – Zoogeographic studies on the soleoid fishes found in Japan and its neighbouring regions. *Bull. Jap. Soc. scient. Fish.* 22: 526—530.

Orlóci, L., – 1966 – Geometric models in ecology. I. The theory and application of some ordination methods. *J. Ecol.* 54: 193—215.

Orlóci, L., – 1967a – An agglomerative method for classification of plant communities. *J. Ecol.* 55: 193—205.

Orlóci, L., – 1967b – Data centering: a review and evaluation with reference to component analysis. *Syst. Zool.* 16: 208—212.

Orlóci, L., – 1968 – A model for the analysis of structure in taxonomic collections. *Can. J. Bot.* 46: 1093—1097.

Orlóci, L., – 1971 – Notes on the 'Notes on the Marcyewski-Steinhaus coefficient of similarity' - a paper by P. Holgate. In 'Statistical Ecology,' G. P. Patil, ed., 3: 192—193. Pennsylvania State University Press, London.

Orlóci, L. – 1974a – Revisions for the Bray and Curtis ordination. *Can. J. Bot.* 52: 1773–1776.

Orlóci, L. – 1974b – On information flow in ordination. *Vegetatio* 29: 11–16.

Orlóci, L. – 1975 – Multivariate Analysis in Vegetation Research. Junk, The Hague, 286 pp.

Phipps, J. B., – 1970 – Studies in the Arundinelleae (Gramineae). X Preliminary taximetrics. *Can. J. Bot.* 48: 2333—2356.

Pielou, E. C., – 1969 – An Introduction to Mathematical Ecology. Wiley-Interscience, New York. 286 pp.

Précsényi, I. & Z. Szöcs, – 1969 – Javaslat a kontinuum-vizsgálat egy módszerének módosítására. (Magyar with Engl.summ.: Modification of a method of continuum studies). *Bot. Közl.* 56: 189—196.

Quadling, C., – 1967 – Evaluation of tests and grouping of cultures by a two-stage principal component method. *Can. J. Microbiol.* 13: 1379—1400.

Russell, P. F. & T. R. Rao, – 1940 – On habitat and association of species of anopheline larvae in south-eastern Madras. *J. Malar. Inst. India* 3: 153—178.

SEAL, H. L., – 1964 – Multivariate Statistical Analysis for Biologists. Methuen, London. 207 pp.

SHEPARD, R. N. & J. D. CARROLL, – 1965 – Parametric representation of non-linear data structures. In 'Multivariate Analysis,' P. R. Krishnaiah, ed. pp. 561—592. Academic Press, London.

SNEATH, P. H. A., – 1966 – A method for curve seeking from scattered points. *Comput. J.* 8: 383—391.

SNEATH, P. H. A., – 1967 – Trend-surface analysis of transformation grids. *J. Zool.* 151: 65—122.

SOKAL, R. R., – 1958 – Quantification of systematic relationships and of phylogenetic trends. *Proc. Tenth Int. Congr. Ent.*, Montreal 1956, 1: 409—415.

SOKAL, R. R. & C. D. MICHENER, – 1958 – A statistical method for evaluating systematic relationships. *Kans. Univ. Sci. Bull.* 38: 1409—1438.

SOKAL, R. R. & F. J. ROHLF, – 1962 – The comparison of dendrograms by objective methods. *Taxon* 11: 33—39.

SØRENSEN, T., – 1948 – A method of establishing groups of equal amplitude in plant sociology based on similarity of species content, and its application to analyses of the vegetation on Danish Commons. *Biol. Skr. K. danske Vidensk. Selsk.* 5 (4): 1—34.

SWAN, J. M. A., – 1970 – An examination of some ordination problems by use of simulated vegetation data. *Ecology* 51: 89—102.

SWAN, J. M. A., R. L. DIX, & C. F. WEHRHAN, – 1969 – An ordination technique based on the best possible stand-defined axes and its application to vegetation analysis. *Ecology* 50: 206—212.

SZÖCS, Z., – 1971 – A Vértes-Hegység bükkösei: II. Interspecifikus korrelációvizsgálatok. (Magyar with Engl. summ.: The beech forests of the Vértes-Mountain: II. An investigation of interspecific correlations.) *Bot. Közl.* 58: 47—52.

WHITTAKER, R. H., – 1952 – A study of summer foliage insect communities in the Great Smoky Mountains. *Ecol. Monogr.* 22: 1—44.

WHITTAKER, R. H., – 1967 – Gradient analysis of vegetation. *Biol. Rev.* 49: 207—264.

WILLIAMS, W. T., M. B. DALE, & P. MACNAUGHTON-SMITH, – 1964 – An objective method of weighting in similarity analysis. *Nature, Lond.* 201: 426.

YARRANTON, G. A., – 1967a – Principal components analysis of data from saxicolous bryophyte vegetation at Steps Bridge, Devon. I. A quantitative assessment of variation in the vegetation. *Can. J. Bot.* 45: 93—115.

YARRANTON, G. A., – 1967b – Principal components analysis of data from saxicolous bryophyte vegetation at Steps Bridge, Devon. II. An experiment with heterogeneity. *Can. J. Bot.* 45: 229—247.

YARRANTON, G. A., – 1967c – Principal components analysis of data from saxicolous bryophyte vegetation at Steps Bridge, Devon. III. Correlation of variation in vegetation with environmental variables. *Can. J. Bot.* 45: 249—258.

10 EVALUATION OF ORDINATION TECHNIQUES

Robert H. Whittaker and Hugh G. Gauch, Jr.

Contents

10 EVALUATION OF ORDINATION TECHNIQUES

10.1 Introduction

Historically, ordination techniques have had a number of sources (WHITTAKER 1967), from early work in direct gradient analysis (articles 2 and 4) and the use of similarity measurements in various schools (articles 5 and 6) to the development of Wisconsin polar ordination and its modifications (article 7) and the introduction of multivariate techniques from other fields (articles 8 and 9) to the testing of these techniques and development of new techniques appropriate to ecological data and research purposes (articles 10 and 11). The following article contains a brief summary of this history (11.2). The combinations of different ordination approaches, algorithms, similarity coefficients, and criteria for axis determination provide an uncountable number of possible techniques. Summarization of this range of techniques and evaluation of their usefulness for data from natural communities are needed. We consider here how ordination techniques are affected by the peculiarities of community data, how they can be evaluated, and what recommendations on their use can be offered.

10.2 Kinds of Ordination Techniques

Table I gives a classification. Ordinations can be classified by a number of criteria: (i) whether direct or indirect, (ii) the model or principle by which samples or species are arranged, (iii) the particular mathematical procedures, or algorithm, of the ordination, (iv) whether the technique ordinates samples, or species, or (as in dual techniques) both at the same time, (v) the kinds of data transformation and similarity measurements used (especially for groups 3-6 in Table I), (vi) the number of ordination dimensions or axes, (vii) whether these are required to be orthogonal or permitted to be oblique, (viii) whether the technique deals (as conventional indirect techniques do) with species composition of samples, or instead applies multivariate techniques to environmental variables or growth forms or other structural characteristics of communities, and (ix) the degree to which the technique is informal (by which we mean that it can be carried out, if necessary, by an investigator with paper and pencil and calculator) or formal (mathematically advanced and dependent on more complex calculations for which a computer is desirable or essential).

The first four of these criteria are used in Table I. The techniques are almost as difficult to classify as communities, and the reader may note omissions and overlaps. Ordination techniques are related not in a hierarchy but a multidimensional plexus, and Table I is an effort to express only some part of the relationships on a few major axes.

Table I. A Classification of Ordination Techniques

DIRECT	INDIRECT
1. Environmental measures or indices	
a. Direct environmental sample ordination	Compositional gradient analysis Primary matrix ordering Parametric mapping
b. Direct environmental species ordination	Gaussian ordination (dual)
2. Weighted averages	
a. Sample ordination	Reciprocal averaging (dual)
b. Species ordination	
3. Matrix and plexus arrangements	
a. Secondary sample matrix and plexus	Secondary reciprocal averaging (sample or species)*
b. Secondary species matrix and plexus	Multidimensional scaling
4. Polar ordinations	
a. Sample ordinations	
Two endpoints chosen (standard polar and importance projection)	Wisconsin technique (most dissimilar endpoint samples)
Similarity projection (samples or species)	
Single endpoint samples	Simple ordination Position vectors ordination
Multiple reference samples	
b. Species ordinations	Wisconsin inverse ordination
5. Correlative techniques	
a. Sample composition with environmental factors or scalars	Factor analysis (dual)* Canonical correlation analysis*
b. Species importance with environmental factors or scalars	
6. *Eigenvector techniques	Principal components analysis Principal coordinates analysis Correspondence analysis (= reciprocal averaging)

The first division in the table is that between direct and indirect ordination (chapter 1): direct ordinations that seek arrangement of samples (or species) on expressions of environmental gradients accepted as given, vs. indirect ordinations that seek to extract axes representing major directions of environmental and community variation from calculations on sample data. There is no discontinuity between these; mixed and intermediate approaches are often appropriate (11.5.1). It is also possible to distinguish environmental ordinations (employing measurements of environmental factors, or environmental indices such as elevation) and vegetational ordinations (employing calculations applied to sample composition) (Austin 1968). This division does not simply parallel that between direct and indirect approaches, for certain techniques (weighted averages and polar ordination) are vegetational but direct. These may be thought an intermediate group (Westman 1975). The most fundamental distinction in direction of research, however, would seem to be that between using a known or assumed environmental relationship for ordination, and seeking to derive axes of environmental relationships from the ordination.

The first two techniques in the table, ordination by an environmental index or by a weighted average, are part of direct gradient analysis as this developed in American ecology especially (articles 2 and 3). Environmental indices include measurements (such as elevation, soil pH, and insolation exposure) that are not assumed to be the critical factors for distribution of plant populations, but that express relative positions along complex-gradients involving numerous environmental variables (article 2.7.2). Samples may be ordinated by such measurements of their sites or biotopes; species, in contrast, are ordinated by their distributions along gradients. Species may, in particular, be arranged in sequence by positions of their population centers or modes along a gradient. Sets of species centered in different parts of a given gradient are ecological groups; and these may in turn be the basis of sample ordination by the second technique, that of weighted averages (article 2.2.3.1). Weighted averages are normally used to ordinate samples (Table I, 2a) by weights applied to species, but it is also possible to use weights for sample positions along a gradient to ordinate species (2b); weighted averages are in the latter case computed for each species with the weights applied to its importance values in different samples.

The right side of Table I gives indirect techniques that are roughly equivalent to some of these (although in many cases there is no real correspondence). Compositional gradient analysis or the technique of leading dominants (chapter 7.2, CURTIS & MCINTOSH 1951, BROWN & CURTIS 1952) is an informal technique that arranges samples in such a way as to produce a smooth rise and fall of species importance values. From this a major axis of community variation in response to environment may be derived (if the sample data are not complicated by additional axes). The ordering of primary matrices in the school of Braun-Blanquet (article 20) may have the same effect of revealing a major direction of community differences. Gaussian ordination (GAUCH et al. 1974, IHM & GROENEWOUD 1975, article 11) is a formal, indirect equivalent of compositional gradient analysis that proceeds by optimized sample arrangement and fitting of Gaussian curves to species data; like compositional gradient analysis, Gaussian ordination is largely limited to the recognition of a single major axis. Parametric mapping or continuity analysis (SHEPARD & CARROLL 1966, KRUSKAL & CARROLL 1969, NOY-MEIR 1974a, 1974b, article 11) is a formal technique that arranges samples in such a way as to maintain maximum continuity of species scores in one or more directions.

The techniques grouped under 3 and 4 (and some of those in 5 and 6) employ pairwise comparisons of samples by their composition (or of species by their distributions through samples) for ordination; the ordinations then depend on similarity measures as reviewed in article 5. Matrix and plexus techniques (article 6) are informal techniques of ordinating by rearranging samples (or species) in the matrices of similarity values, or arranging samples (or species) into graphical representations (plexuses) of their relative similarities. The arrangement need not assume a system of Cartesian coordinates, although the arrangement should express directions of relationship of samples and environment that might be treated as axes. In some cases matrix and plexus techniques can be employed for a simple indirect ordination (article 6). Multidimensional scaling (KRUSKAL 1964a, b, FASHAM 1977, article 11) is a mathematically advanced, indirect technique that produces an optimal arrangement of samples by their similarity measurements in a space with a specified number of axes. Index iteration or secondary reciprocal averaging (GOFF & COTTAM 1976, GAUCH et al. 1977, article 11) is an indirect technique that applies the algorithm of reciprocal averaging (HILL 1973, 1974) to extract a major axis of sample variation (and, potentially, additional axes); it is a formal analogue of secondary matrix arrangements as described in article 6. Polar ordination derives from BRAY & CURTIS (1957), but many variants have been proposed (articles 7, 11). These techniques have in common the arrangement of samples by relative similarity to one or more reference or endpoint samples; the choices of these reference samples define the axes of the ordination. The original technique of Bray and Curtis used a prior double standardization of the primary data matrix (7.3.1), a particular similarity index (article 4, equation 50), and choice of endpoint samples of maximum dissimilarity to each other as means of indirect ordination. Many applications prefer direct choice of endpoints to represent known or suspected environmental relations and employ other data treatments (10.4.3). Simple ordination (Orlóci 1966, 10.5.2) is a variant that uses Euclidean distance measures and defines the first axis by two endpoint samples of maximum distance; subsequent axes extend perpendicularly from the first axis (or preceding axes) to single reference samples. Position vectors ordination selects axes through actual samples, those samples the position vectors of which lie closest to principal components axes (Orlóci 1966, article 9.7.4.2). Similarity projection (Gauch & Whittaker 1972b) is a direct ordination that can use any sample similarity measurement to project sample positions onto axes defined by endpoint samples. Importance projection is a direct technique applied to the primary matrix of importance values in samples rather than to a secondary matrix (Gauch & Whittaker 1972b). It is identical in function to polar ordination using a Q-matrix of Euclidean distance values and the same endpoint samples. Direct ordinations can be based on a single reference sample (as in the study of retrogression, article 3), or may employ multiple reference samples (Ramensky 1930, article 4). Other variants of polar ordination have been proposed by Loucks (1962), Knight (1965) and Knight & Loucks (1969), Monk (1965), Orlóci (1966, 1974a), Swan & Dix (1966), Swan et al. (1969), Maarel (1969); see also article 11. Any of the polar techniques can, in principle, be applied to ordinate species by relative similarity of distributions and comparison with endpoint species.

The polar or comparative techniques are limited by their use of only a few samples to define axes. In principle techniques using the full range of sample information to determine axes would be preferable. Factor analysis (Dagnelie 1960, 1962, Greig-Smith 1964, Whittaker 1967, article 8) uses an R-matrix of SC or SA values and derives extracted factors as principal directions of species correlation. These extracted factors become axes (that need not be orthogonal) for ordination both of species in a loading space and of samples in a factor-value space. Factor analysis can be applied also to an R-matrix of correlations of environmental measures for samples, or to a mixed R-matrix of correlations of species distri-

nation. Reciprocal averaging = correspondence analysis is then an apparently optimal eigenvector technique for indirect ordination, difficult to place in Table I because it is a dual technique (producing simultaneous sample and species ordinations) and is related to principal components analysis, matrix arrangement, and weighted averages (Hill 1973, Gauch et al. 1977).

10.3 **Bases of Evaluation**

The central problem of ecological ordination is how to produce arrangements of samples and species that are realistic and informative – arrangements that effectively express, and consequently may permit recognition and interpretation of, major relationships of species populations and communities to environmental gradients. The behaviors of ordination techniques are profoundly affected by the peculiarities of ecological data. Mathematical techniques from outside the field, most of which embody linear assumptions, are not necessarily appropriate to ordination. Only by systematic testing with ecological data of known properties is choice among the range of techniques in Table I possible. The tests should be realistic in several ways that relate to ecological uses of ordination: (i) they should deal with sample sets embodying the kinds of curvilinearities characteristic of community data; (ii) they should take into account the wide range of sample set properties affecting ordinations (10.4.2); (iii) they should apply ordinations as full techniques from original data to final results, for an optimization of one criterion for one step may be detrimental to over-all performance; and (iv) they should permit evaluation in terms relevant to field ecological research.

As bases of testing ordinations we discuss first the theory of indirect ordination, second the criteria of evaluation, and third the simulated data that make controlled test and evaluation possible.

10.3.1 Ecological Space and Theory of Ordination

Understanding of ecological ordination starts with the properties of coenoclines as these have been shown by research in direct gradient analysis (Fig. 1 in article 2). The relationships of species populations to environmental gradients are complex, individualistic, predominantly continuous, nonlinear, and over sufficient ranges of gradients nonmonotonic. Direct gradient analysis is able to handle these peculiarities of ecological data because it uses a known environmental gradient or index as a standard for ordibutions and environmental measurements (DAGNELIE 1960). Inverse factor analysis employing a Q-matrix is possible (DAGNELIE 1960) but of more uncertain meaning. Canonical correlation analysis (AUSTIN 1968, PIELOU 1969, GAUCH & WENTWORTH 1976) seeks joint relationships in matrices of two types of variables such as environmental and community measurements; trial ecological applications seem not promising (AUSTIN 1968, GAUCH & WENTWORTH 1976). Factor analysis and canonical correlation analysis are, like many formal techniques including others of the eigenvector group, limited by the curvilinear, non-monotonic relationships of species scores to environmental gradients. These same curvilinear relationships have limited the use of conventional correlation techniques for direct ordination. Correlation of community measurements (weighted averages, life-form representation, productivity, etc.) with environmental measurements as independent variables, could be used to obtain regression equations predictive of community variables, and the equations could be used to ordinate samples. Because of the complex response surfaces formed by both species populations and community variables in relation to environmental variables (GITTINS 1968, AUSTIN 1972, article 2), such correlative techniques may have little usefulness for ordination.

A final group of techniques employ eigenvector calculations to derive axes onto which sample positions are projected. The most widely used technique, principal components analysis, projects samples from a multidimensional, Euclidean space onto axes that minimize variance of sample distances from a given axis, hence maximizing variance along the axis (ORLÓCI 1966, GITTINS 1969, article 9.7.3). Successive axes are chosen so that each maximizes along itself variance remaining from preceding axes, in a direction perpendicular to those axes. Principal components analysis thus can obtain ordinations of both samples and species (on differently scaled axes) from a primary matrix and either a Q or R secondary matrix using Euclidean distances (articles 9, 11). Principal coordinates analysis (GOWER 1966, 1969, PIELOU 1969, article 9.8.2) can produce comparable ordinations using non-Euclidean distance or similarity measurements. It is consequently not limited to use of covariance (as in principal components analysis) or correlation (as in factor analysis). Correspondence analysis (BENZÉCRI 1969, GUINOCHET 1973, HILL 1973, 1974, CHARDY et al. 1976) is closely related to principal components analysis in its calculations as an eigenvector technique, but is less affected by non-linearities (GAUCH et al. 1977, article 11). Correspondence analysis, as an eigenvector calculation, is identical in function to reciprocal averaging (HILL 1973, 1974) as a formal, indirect analog of weighted average ordi-

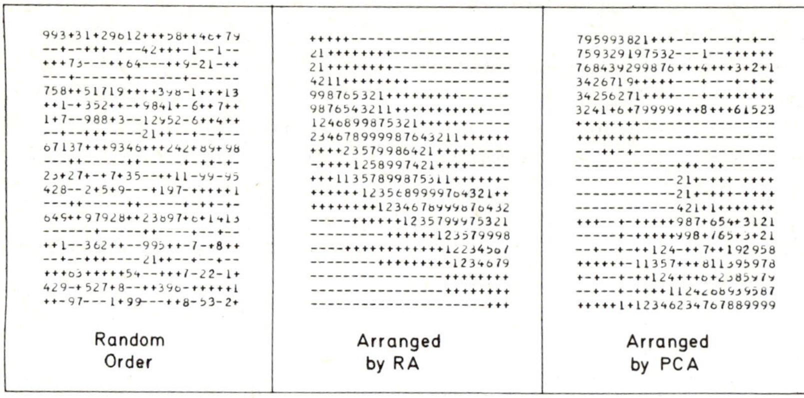

Fig. 1. A primary data matrix arranged at random (left) and according to species and sample ordination values from centered, non-standardized principal components analysis (right), and reciprocal averaging (middle). Note that RA arranges the matrix to concentrate higher values along the matrix diagonal. The data matrix is a simulated coenocline with Gaussian species distributions and a beta diversity of 5 half changes having 20 species and 24 sites. Absence is indicated by a dash (—) and scores are rounded into ten equal classes marked '+' and '1' to '9' (Gauch et al. 1977). Columns represent samples, and rows species.

nation and produces an arrangement in which the behavior of species populations can be observed. The challenge of indirect ordination is the extraction of a comparable arrangement from data embodying the complex nonlinearities of species distributions. Fig. 1 illustrates a successful indirect ordination of a coenocline by reciprocal averaging, and an unsuccessful ordination by principal components analysis.

Although a coenocline summarizes essential relationships affecting ordination in one dimension, actual problems are usually multidimensional. An individual ordination often must, given data embodying several directions of nonlinear variation in communities, seek to derive the major directions of that variation in response to environment. An ecological ordination deals with samples representing points in a multidimensional ecological space (Whittaker 1967, Beals 1973, Austin 1976a) and seeks to arrange the samples in a way that is realistic in relation to that space.

The properties of the ecological space are crucial to the problems of ordination. Consider first, as an abstraction from a landscape, a habitat-factor hyperspace of which the many environmental (and successional) variables affecting species populations are axes. This space, given the multiplicity of and differing partial correlations among the factors, is intractable as a basis of ordination;

concern is instead with a next-order abstraction, a habitat hyperspace the axes of which are complex-gradients (2.2.2), directions of concomitant change in a number of environmental variables. A space defined by a limited number of complex-gradients – such as elevation, topographic moisture, soil fertility, successional time – is the space in which a multidimensional ordination seeks to locate samples and species.

The habitat hyperspace is occupied by species populations. Along a given gradient species have bell-shaped, apparently Gaussian distributions; in relation to two gradients the populations form bell-shaped response surfaces. In a multidimensional space the populations have scattered atmospheric distributions, each with a center of maximum density away from which in all directions population density decreases in Gaussian functions (2.5.2). The habitat hyperspace is filled by variously dispersed and continuously overlapping populations that form a complex compositional space, the multidimensional equivalent of the coenocline. The compositional space is assumed to represent community response (by way of successional processes and the distributions and interactions of species populations) to the habitat space. Habitat and compositional space are coupled and together form the ecological space. In principle it is possible, given a successful ordination the axes of which have been interpreted, to predict (a) from composition of a sample that can be located in the ordination, the probable characteristics of its site, and (b) from position of a site on the ordination axes (as this might be independently determined from environmental measurements) the probable composition of a sample from that site (Austin 1976a).

Indirect ordinations seek to infer, from samples representing the compositional space, relative positions of these samples in the habitat space. In indirect ordination there need be no one-to-one correspondence between the number, orientation, and scaling of the axes and conventionally accepted complex-gradients (Greig-Smith 1971a). The axes of an indirect ordination are likely to be collectively rotated, or individually oblique and variously rescaled relative to conventional complex-gradients or the environmental measurements that are used as indices for these. Indirect ordination seeks, not a conventional reproduction of environmental factors, but an arrangement of samples that represents their positions relative to one another and major directions of environmental variation. Different scales and axes may be sufficiently correct much as maps on different scales with different projections and different orientations in relation to true north may all adequately represent relationships of geographic places to one another. Given a successful indirect ordination the ecological significance of the axes may be determined. Some of the structure of the ecological space, including the relationships of species populations and community characteristics to axes and the habitat space, can then be recognized. An ordination that most effectively represents ecological space will at the same time most effectively characterize species habitats in the sense of Whittaker et al. (1973), for it will show Gaussian habitat response surfaces of maximum concentration on ordination axes that express habitat factors (Hill 1976).

If it is successfully to infer structure of the ecological space, the ordination technique must be in some way adapted to deal with the nonlinear relationships in the compositional space. The ordination should not seek to summarize in its axes the full complexity of species distributions and sample composition. That complexity may be conceived in forms different from the compositional space – as a species-dimensional space (in which the species become axes in relation to which the samples are located as points) or sample-dimensional space (in which the samples become axes in relation to which the species are located as points). The species-dimensional and sample-dimensional spaces (Goodall 1954, Gittins 1969, article 9, Orlóci 1975, Austin 1976a) are mathematically attractive concepts: they bear a dual or converse relationship to one another, distances in them are Euclidean, and certain techniques, notably principal components analysis, provide optimal summarization of variation in them (article 9). Accurate representation of this Euclidean space is not, however, itself a purpose of ecological ordination (Beals 1973). Ordinations may be interpreted also in terms of 'delta' vegetational spaces (Beals 1973) in which samples (or species) are dispersed at distances defined by similarity measures; for a given sample set there are as many possible delta spaces as similarity measures. Correct representation of a delta space in ordination is desired only if that space is a useful approximation of the ecological space. Ecological, delta, and species- and sample-dimensional spaces are transformations of one another, but it is the ecological space on which the theory of indirect ordination is based.

An ordination technique that is compelled by its assumptions to treat as linear the nonlinear relations of species populations is thereby compelled to locate samples from a coenocline in a multidimensional meander in the ordination hyperspace. The ordination is then distorted; ecological relationships are obscured and the axes may be ecologically meaningless. Such is the failing of linear techniques, given ordination tasks to which they are not suited. Successful ordination must instead seek to derive coenoclines as optimal paths across the ecological space (Dale 1975) in despite of the curvilinear relationships of species populations along those paths. Successful ordination is possible by one of four approaches (see also AUSTIN 1976a, article 11.3.5): (i) Environmental complex-gradients are accepted as given, and samples (or species) are ordinated by an index (which may be an environmental measure or a weighted average) expressing relative positions along the gradient – direct ordination (article 2). (ii) An ordination technique embodying linear assumptions is applied to samples with a relatively narrow range of compositional variation, within which species distributions are predominantly monotonic (article 9.7.3.9., ORLÓCI 1975). Within such limits linear techniques (such as principal components analysis and factor analysis) may give satisfactory results (AUSTIN 1968, JEGLUM et al. 1971, article 8). (iii) The ordination technique is accommodated to nonlinearity of species distributions and either tolerates, or takes advantage of as a basis of ordination, the nonmonotonic rises and falls of species populations along environmental gradients. (iv) The ordination is accommodated to nonlinearity in similarity measures (and zero values in the secondary matrix) and employs these measures in such a way as to minimize distortion of sample positions (article 11.4.5.3). Among the formal techniques Gaussian ordination and parametric mapping represent approaches of type (iii), and multidimensional scaling and index iteration represent type (iv); these nonlinear techniques will be considered further in article 11. Our concern in this article is with evaluation of some widely used techniques – polar ordination, principal components analysis, and reciprocal averaging. We shall try to determine how these techniques are affected by properties of sample sets, the degrees to which they are successful in treatment of nonlinear data, and the circumstances in which they may be profitable tools of research.

10.3.2 CRITERIA OF EVALUATION

Among the desired properties of ecological ordination procedures are:

(i) Freedom from distortion – relative success of the ordination in representing the relations of samples and species in the ecological space, vs. representation that is distorted.

(ii) Range – the relative extent of community or environmental variation that the ordination technique can treat without undue distortion.

(iii) Lucidity – clarity to the investigator and others of the basis of the calculations and their relation to characteristics of the data, and hence of both the computational and ecological meaning of the ordination results.

(iv) Objectivity – degree to which results are free from influence by the choices of the investigator in the application of the technique.
(v) Efficiency – effectiveness in use of information, vs. either sacrifice of information that might help to clarify relationships or over-elaborate computation generating spurious information.
(vi) Heuristic function – success in revealing otherwise unrecognized relations of species and communities to environment, relations on which ecological interpretation and possible further research may be based.
(vii) Computational expense and dependence on access to computers.

The most critical of these for our tests is distortion. By distortion we mean misplacement of samples or species in an ordination, so that the relation of these to environmental gradients is obscured rather than clarified. Distortions of an ordinated coenocline may involve a number of effects (cf. Orlóci 1974b, 1975): (1) Distances along the axis can be transformed, with relative compression and expansion of different parts of the axis without changes in sample sequence. (2) Reversals of true sample sequence can occur with (a) errors that misplace individual samples, or (b) systematic misplacement of groups of samples. (3) Samples representing a linear sequence may be curved away from a first ordination axis, in a space defined by other axes, so that true relationships of these samples to the axes are obscured. (4) From such curvature apparent ordination axes that are spurious, themselves products of the distortion, can result. (5) Strong curvature can imply involution: turning inward of the ends of a coenocline so that its extreme samples are projected onto a given ordination axis at positions within those of less extreme samples. The same kinds of distortions apply to coenoplanes and multidimensional patterns but are less simply characterized for these. Interrelationships among the types of distortion will be noted. When a coenocline is strongly curved away from an axis in an arch effect (3), false higher axes (4) result, and the samples are likely to be spaced more closely toward the extremes than the middle of the first axis, hence a distance transformation of type (1) also occurs. When curvature is extreme with type (5) distortion systematic reversals of sample sequences, type (2b) result. The non-systematic reversals of type (2a) can result from sample error or noise in the data. The overriding problem of distortion, however, is curvature: the fact that indirect ordination techniques to varying degrees give linear community relationships curvilinear expression in ordination space, thereby altering or obscuring the relationships that ordination seeks to represent.

A number of approaches to measuring relative success of ordination, vs. degree of distortion or sample displacement, are possible. Simple tests of direct ordinations, based primarily on relative closeness of observed species distributions in transects to the ideal bell-shaped curve, have been described (article 2.2.4, WHITTAKER 1967). Some possible approaches to detecting or measuring distortions in indirect ordination are:

(i) Observation of the arrangement of samples in an ordination space as compared with known relationships of these samples along a simulated or actual coenocline (or in a simulated or actual community pattern). Plotting of simulated coenoclines and coenoplanes in the first two or three dimensions of ordination space can effectively illustrate the kind and degree of distortion (NOY-MEIR & AUSTIN 1970, AUSTIN & NOY-MEIR 1972, GAUCH & WHITTAKER 1972b, KESSELL & WHITTAKER 1976, GAUCH et al. 1977). When environmental relationships of real community samples are well known, plotting their positions on the first axes of an indirect ordination can show the distortion of the latter (JEGLUM et al. 1971, WESTMAN 1975).

(ii) Correlation of positions of samples along an ordination axis (or axes), with known positions in a coenocline (or pattern). For a coenocline PEARSON's coefficient of correlation, computed for sample position on the simulated coenocline axis versus sample location when projected onto the first axis of the ordination can provide an index of the extent to which the original spacing has been reproduced (without assuming normal distribution of the variates) (GAUCH & WHITTAKER 1972b, WHITTAKER & GAUCH 1973).

(iii) Number of reversals of true sample sequence as a measure of effects of sample error (and involution if any) on different ordinations of a coenocline (GAUCH et al. 1977). Rank order correlation of the true and ordinated sequence can be used in the same way (FASHAM 1977).

(iv) Mean sample displacement on a given axis, measured as the percent of the axis by which the positions of ordinated samples depart from their true positions in the coenocline ordinated (KESSELL & WHITTAKER 1976). This may be the most direct and effective measure. Like the preceding measures and (v), it is best adapted to the distortion of an ordinated coenocline but corresponding measures for coenoplanes or community patterns should be feasible.

(v) Average distance of samples from the ordination axis, divided by the length of the ordination axis (GAUCH & WHITTAKER 1972b). Since the coenocline has a single axis, a zero value for this distance is expected; values above zero indicate that sample positions have been displaced into nonexistent higher axes, representing spurious information in techniques other than polar ordination.

(vi) Correlation of distances between samples as ordinated, with known sample separation. This test may be ineffective because it fails to detect curvilinear distortion. It is not valid as used by AUSTIN & ORLÓCI (1966) to test polar ordination against principal components, accepting Euclidean distance as true sample separation. Eulcidean distance is a premise of principal components but not of ordination in general and it does not bear a linear relation to environmental separation (article 3).

(vii) Per cent of variance accounted for by a given number of axes is a mathematically attractive test of ordination efficacy (ORLÓCI 1966, BANNISTER 1968). This test also is prejudiced by its assumption that variance, which the axes of principal components analysis are defined to maximize, is a fair test applied to other ordinations. There is no demonstrated relationship between per cent extraction and ecological interpretability (AUSTIN & NOY-MEIR 1972), and no basis for assuming that an axis of maximum variance is of maximum ecological significance.

10.3.3 SIMULATED DATA

11.3.3.1 *Coenoclines*

Of the criteria of evaluation, it is freedom from distortion that can be most effectively tested and measured and that may in turn affect other criteria. Distortion is most easily tested by submitting synthetic or simulated coenoclines to ordination. When samples of known relations to one another along a linear coenocline axis are ordinated, distortion appears as altered spacing of the samples and curvature of the coenocline that displaces samples into dimensions or axes not present in the simulated data. A simple, simulated coenocline of bell-shaped curves, evenly spaced along the axis, was first used for testing ordination by SWAN (1970) and NOY-MEIR & AUSTIN (1970). Results from direct gradient analysis were studied by GAUCH & WHITTAKER (1972a) to develop a more realistic coenocline. The field data studied suggested these assumptions:

(i) The modes, or peak importance values, of minor species are distributed at random along the coenocline axis. The modes of major species may, however, be distributed more evenly because of selection toward the reduction of competition by the centering of their populations in different parts of a gradient (WHITTAKER 1967, article 2).

(ii) The heights of these modes, for the species of a coenocline, form either lograndom or lognormal distributions. When maximum importance values of tree species in forest coenoclines were grouped into octaves or doubling-units, the numbers of species in octaves were approximately even, implying lograndom distribution. (In this case a limit is imposed on the upper end of this distribution because maximum importance values for the dominants of different parts of the coenocline cannot exceed a certain range of values, and a lower limit is imposed by sample adequacy for rare species). When plant species not subject to an importance-value ceiling (forest undergrowth species, and desert species) were treated the same way, their importance values formed lognormal distributions. Lograndom and lognormal distributions were used as alternative assumptions for the coenocline program; lograndom distribution was used for the tests by GAUCH & WHITTAKER (1972b, WHITTAKER & GAUCH 1973).

(iii) Relative widths, or dispersions, of species populations in coenoclines were observed. Using as units the steps or intervals of composite transects, species whose distributions reached neither end of of a transect were arranged into frequency distributions by the numbers of steps through which their distributions extended. The results suggest a normal distribution for these population amplitudes in relation to a complex-gradient. The standard deviation for species amplitudes was around 0.3 times its average value (i.e. these amplitudes have a coefficient of variation of about 0.3). It is assumed that this coefficient of variation, computed for total numbers of transect steps, should apply also to amplitudes expressed as standard deviations.

(iv) Correlations between modal positions, peak importance values, and amplitudes for species populations in coenoclines were considered. No correlation between peak importance value and amplitude was found, despite the suggestion of McNAUGHTON & WOLF (1970) that a weak correlation might exist. When species diversity changes markedly along a complex-gradient, correlations of peak importance value and amplitude with modal position might be expected, but no such correlations were found in the forest and desert coenoclines tested. Noncorrelation of the three parameters of species populations was assumed in the program.

(v) Coenoclines must differ from one another, however, in richness in species of particular communities or samples (alpha diversity), in degree of compositional change or species turnover along the coenocline (beta diversity), and in total richness in species (gamma diversity). Field samples used for ordination must differ further in the degrees of sample error or population irregularity affecting them. The simulation program [1] was designed to produce coenoclines in which alpha diversity, beta diversity, and sample error could be independently varied. Representative simulated coenoclines are illustrated in Fig. 2.

10.3.3.2 *Coenoplanes*

While coenoclines permit a wide range of tests with ordination techniques, means of testing these techniques with two-dimensional patterns are also needed. A simple, two-dimensional model with evenly spaced Gaussian surfaces was used for tests by AUSTIN & NOY-MEIR (1972). More complex two-dimensional community patterns or coenoplanes were developed by GAUCH & WHITTAKER (1976) as a logical extension of their coenocline. The coenoplanes involve these assumptions:

(i) Species scores are assumed from published data (WHITTAKER 1956) to be Gaussian surfaces that are variously elliptic in outline, depending on the scaling of the two axes. These surfaces may be defined by six parameters: maximum value (a), mode on the X-axis (μ_x), mode on the Y-axis (μ_y), standard deviation on the X-axis, (σ_x), standard deviation on the Y-axis (σ_y), and an angle (θ) to be described below. For the coenoplane position (x, y), the species score z becomes:

$$z = ae^{-\left[\left(\frac{\mu_x - x}{\sigma_x}\right)^2 + \left(\frac{\mu_y - y}{\sigma_y}\right)^2\right]/2}$$

This surface may be rotated by the angle θ, by using standard equations for rotation of axes. The modes of minor speces are scattered at random, but modes of major species may be dispersed in a relatively uniform pattern that implies reduced competition among them. Sums of the scores of all species can be set at a ceiling value to simulate the effect of competition in limiting species; setting the ceiling at 100 relativizes the samples.

(ii) As in the coenocline, heights of modes (a) may form lognormal or lograndom distributions, and species dispersions (σ_x, σ_y) are distributed normally with a coefficient of variation of about 0.3 times the average value (GAUCH & WHITTAKER 1972a).

(iii) The angles of orientation of the longer axes of ellipses (θ) might be variously specified as uniform, random, or normally distributed. Distributional figures available (WHITTAKER 1956,

[1]) The program, termed 'Coenocline Simulation' developed in IBM Fortran IV and cataloged as CEP-1, is further described by GAUCH & WHITTAKER (1972a) and will be made available at cost on request to the authors.

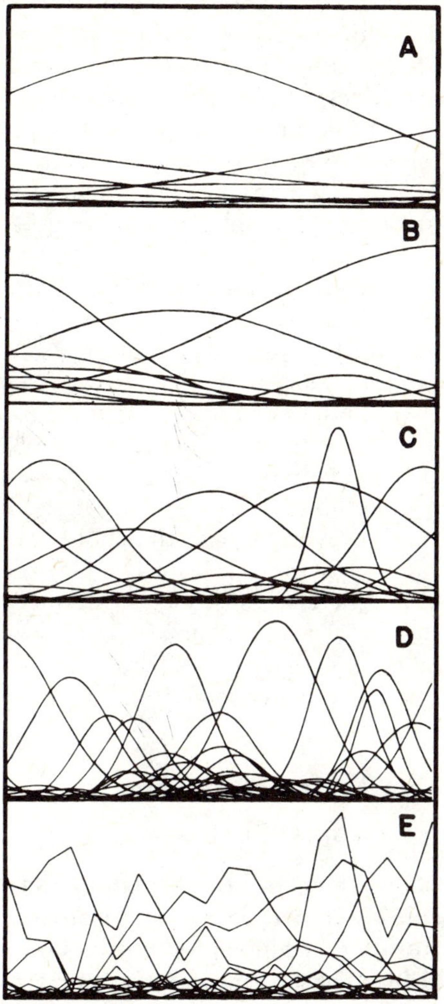

Fig. 2. Five simulated coenoclines of the type used for testing of ordination performance (GAUCH & WHITTAKER 1972a). Characteristics of the species distributions are designed to conform to those of natural communities along environmental gradients (cf. article 2, Fig. 1). Each curve represents a species population, with importance values on the vertical axis, and a gradient in relation to which the species is distributed on the horizontal axis. Alpha and beta species diversities double from A to B, B to C, and C to D. Coenocline C has a beta diversity of $z = 6.0$ (4.45 half-changes) and total species number of 52. E differs from C in the introduction of sampling error. The four diversity levels are intended to resemble conditions in actual forest coenoclines: A should resemble a taiga coenocline with only two dominant tree species, B and C warm-temperate coenoclines of lower and higher diversities, and D a subtropical forest coenocline.

WHITTAKER & NIERING 1965) show the long axes of species distributions correlated in direction (the effect may be a consequence of relative scaling of the elevation and topographic moisture gradients), but with some differences among species in orientation. A normal distribution of θ about a major direction may be a reasonable assumption, but for the test coenoplanes θ was set as zero. The two coenoplane axes were given arbitary ranges from 0 to 100, but were assigned species-turnover or beta-diversity lengths to give coenoplanes of different forms.
(iv) Correlations of the six parameters were again sought in field data, but none were found (apart from effects of axis scaling on θ), and none were assumed for the simulation.
(v) Coenoplanes, like coenoclines, will differ in alpha, beta, and gamma diversity (with beta diversity here defined in two directions). Simulation should provide for varying all three of these aspects of diversity, and also should make possible trends of changing alpha diversity across the coenoplane. Rare species may be excluded from simulated samples, however, by a threshold (e.g. 0.1% of the maximum value) at which the flanks of the Gaussian distributions are cut off. Sample error is a variable, as in the coenocline, and certain other properties of the sample sets can be varied; in particular, samples can be taken for a uniform, random, or variously clustered (or interrupted) arrangement in the coenoplane.

10.4 Ordination Tests and Results

10.4.1 REVIEW OF ORDINATION TESTS

Before discussing effects of sample set characteristics and data treatments on ordination, we summarize some of the studies of ordination using simulated coenoclines, coenoplanes, and field data. Abbreviations will be used for three techniques: polar ordination (PO), reciprocal averaging (RA), and principal components analysis (PCA).

10.4.1.1 *Coenoclines*

A simulated coenocline was first used by SWAN (1970) to observe the effect of bell-shaped species distributions and zero values on PO. NOY-MEIR & AUSTIN (1970) applied PCA to the same coenocline, which then appeared as a three-dimensional loop in the first three axes (Fig. 3). GAUCH & WHITTAKER (1972b) carried out a series of tests of ordination – PCA, similarity projection, impor-

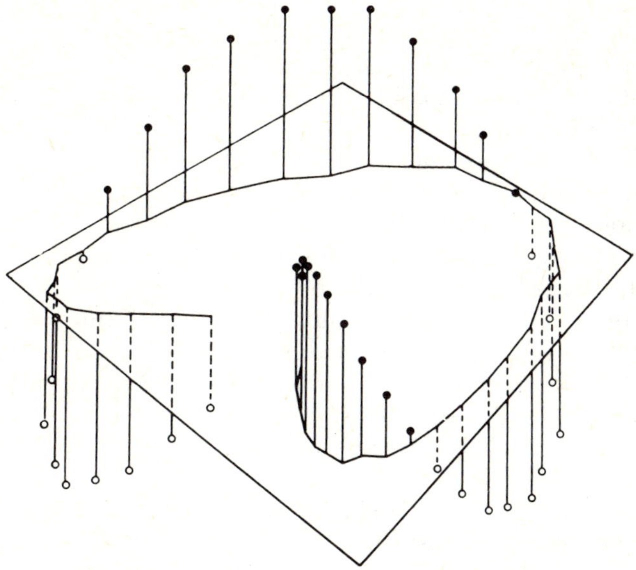

Fig. 3. Principal component ordination (species-centered, non-standardized) of SWAN's (1970) model V coenocline showing the first three axes. The original linear gradient has become a complex multidimensional curve in these three (and higher) axes. (NOY-MEIR & AUSTIN 1971).

tance projection, and PO with different similarity measures (CC, PS, and ED, see 5.5 and 10.4.3) with and without sample error. The over-all ranking of techniques suggested was (i) best, PO and similarity projection with CC or PS, (ii) the same techniques with ED, and importance projection, and (iii) worst, PCA (centered, nonstandardized). PCA was found subject to involution at about 2-3 half changes and the most vulnerable of the techniques to effects of increasing beta diversity and sample error. KESSELL & WHITTAKER (1976) compared PO, PCA, and discriminant function analysis for coenoclines of varying length. PO using PS or CC measures was found preferable to PO with ED, and PCA was found subject to distortion as beta diversity increased and vulnerable to effects of noise and sample clustering. Discriminant function analysis seemed generally unsuitable for ordination, but likely to be useful for sample classification. GAUCH et al. (1974, 1977) extended the tests with coenoclines by two other techniques – RA and Gaussian ordination. Both were found to give good ordinations of coenoclines, at least marginally superior to polar ordination, and to be effective up to high beta diversities, whereas effectiveness of PO breaks down at 5-6 HC. RA and PCA were shown to produce polynomial distortion curves of increasing order in higher axes (HILL 1973, GAUCH et al. 1977) and to be vulnerable to effects of outlier samples and partial sample set disjunctions (GAUCH et al. 1977). FASHAM (1977) obtained generally favorable results for coenocline ordination with multidimensional scaling (11.5.4). AUSTIN (1976b) has used the original coenocline of SWAN (1970), and coenoclines with species distributions departing from Gaussian curves toward skewed, shouldered, and weakly bimodal forms, to test Gaussian ordination, RA, multidimensional scaling, and parametric mapping. The first three of these were found effective to high beta diversities, given symmetrical Gaussian species distributions, but even these effective techniques were vulnerable to mislocation of sample positions when offered coenoclines with asymmetrical species distributions.

10.4.1.2 *Coenoplanes*

AUSTIN & NOY-MEIR (1972) subjected their coenoplane of evenly spaced, Gaussian species distributions to PCA with and without data standardization. Centered but nonstandardized PCA strongly involuted the corners of the coenoplane (Fig. 4); with standardization, and especially Wisconsin double standardization, the ordination was much less distorted and free from involution. In extensive tests by GAUCH et al. (1977) coenoplanes of different shapes and sizes with and without sample error, sample clusters, and outlier samples were submitted to PO, PCA, and RA. RA was shown much less subject to involution than PCA, but subject to curvatures into higher axes that could make interpretation difficult. AUSTIN (1976b) found RA and multidimensional scaling effective in ordinating the AUSTIN & NOY-MEIR (1972) coenoplane. GAUCH & WENTWORTH (1976) found canonical correlation analysis ineffective for relating simulated coenoplanes to simulated 'ecoplanes' or patterns of environmental gradients. FASHAM (1977) found multidimensional scaling always better than PCA and in most cases better than RA.

10.4.1.3 *Comparisons using Field Data*

Among many ordinations of field data, we mention only some that have sought to evaluate different techniques in relation to each other or in relation to recognized environmental gradients. Some tests of direct ordinations are discussed in article 2.2.4.

GROENEWOUD (1965), in an extensive study of Swiss and Canadian forest samples compared PCA of the covariance matrix,

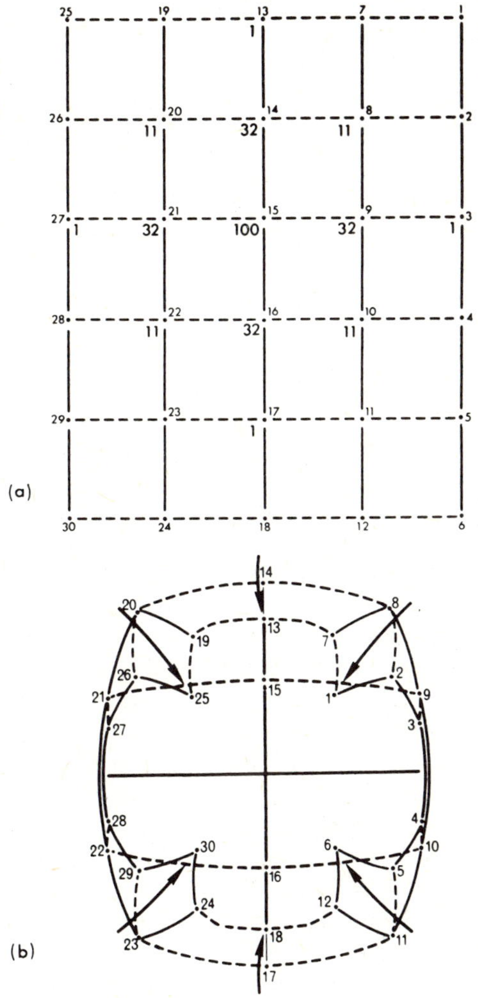

Fig. 4. Results of principal components ordination of a two-dimensional pattern of vegetation (Austin & Noy-Meir 1972). Above (a) is the simulated community pattern with a grid of samples as indicated by the small numbers. The larger numbers are importance values for one of the species, and these form a bell-shaped solid centered at sample 15. Below (b) is the result of principal components ordination of the samples, as projected onto a two-dimensional surface. Principal components ordination produces a complex curvilinear surface from the data, with the edges and corner samples turned within other samples.

PCA for transformed Mahalanobis D^2 values, and multidimensional scaling of D^2 values, with a cluster analysis and BRAUN-BLANQUET classification. Similar groupings of stands were produced by the different techniques, but the results gave no clear-cut evaluations of the ordinations. GROENEWOUD (1965) found that the first and third axes of the ordinations could be interpreted, but not the second axes, and he commented on the problems of nonlinearity, suggesting that the range of habitats sampled should be kept narrow to minimize the problems. AUSTIN & ORLÓCI (1966) and ANDERSON (1971) compared PO, PCA, and other treatments of field data with results they judged unfavorable to PO; AUSTIN (1968) showed useful results for PCA and unfavorable ones for canonical correlation analysis of chalk grasslands. JEGLUM et al. (1971) compared direct, environmental ordination and PCA for a wide range of communities from forest to marsh; the environmental gradients became loops in the ordination (Fig. 5). When the data were stratified into four sets with lower beta diversity, the PCA gave ordinations that were acceptable and seemed preferable to the environmental ordinations. WESTMAN (1971, 1975) applied several ordinations to a pattern of vegetation with a gradient of communities and soils from *Sequoia* forest to pygmy *Cupressus* forest. The sequence of ordinations from best to worst, as judged by representation of the gradient, appeared to be: similarity projection, PO with PS, importance projection, PCA, and factor analysis. Factor analysis gave consistently disappointing results while PCA (centered, nonstandardized) curled the major axis of the vegetation into a spiral (Fig. 6). CHARDY et al. (1976) compared several eigenvector ordinations of marine benthic samples and judged correspondence analysis (= RA) the best of these.

PEET (1975) and WENTWORTH (1976) have compared ordination techniques applied to sets of 40-70 vegetation samples representing elevation belts in mountains. RA was judged preferable to PO in PEET's application; RA seemed highly effective in extracting the first axis corresponding to the topographic moisture gradient, but it did not produce interpretable second axes for these data. WENTWORTH's (1976) study tested several direct ordinations (article 2.2.4) against PO, RA, PCA, and Gaussian ordination and (GAUCH & WENTWORTH 1976) canonical correlation analysis. RA gave consistently good results, somewhat better than PO and markedly better than direct ordination. PCA was least satisfactory; Gaussian ordination gave results as good as RA or marginally better. Performance of RA was improved by the removal of outliers and extreme samples from sets; RA did not produce useful second

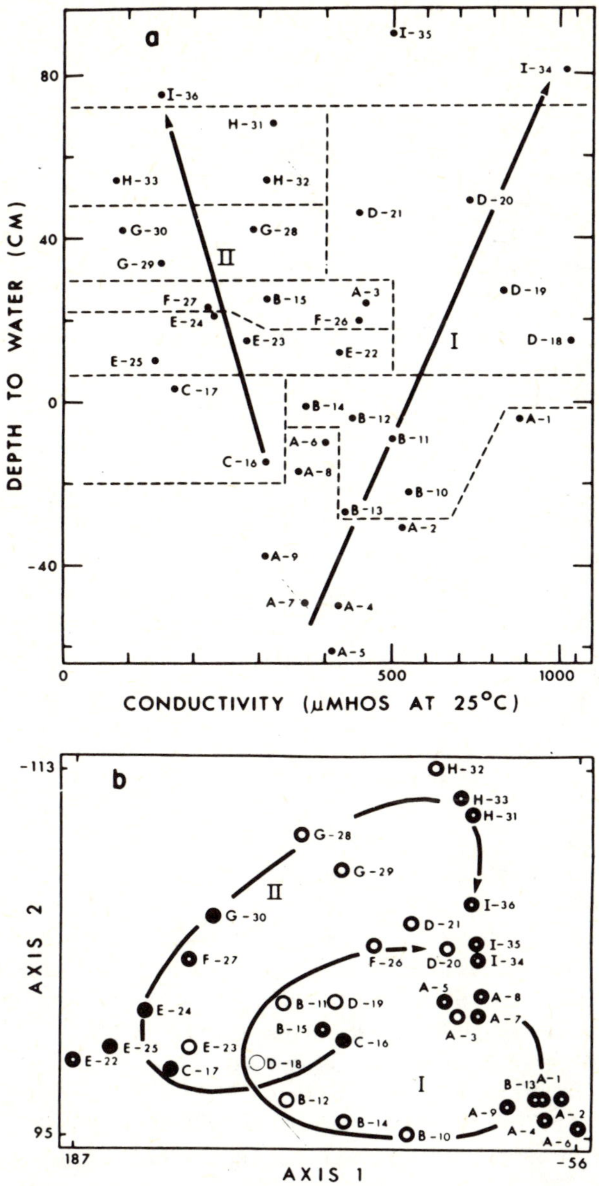

Fig. 5. Results of principal components analysis of field data representing a two-dimensional community pattern (JEGLUM et al. 1971, Fig. 1). Above is a direct, environmental ordination using gradients of water level and water conductivity. Two coenoclines or community-gradients are indicated by the arrows. Below, a principal components ordination projected onto the plane of the first two axes; the community-gradients have become overlapping, open loops.

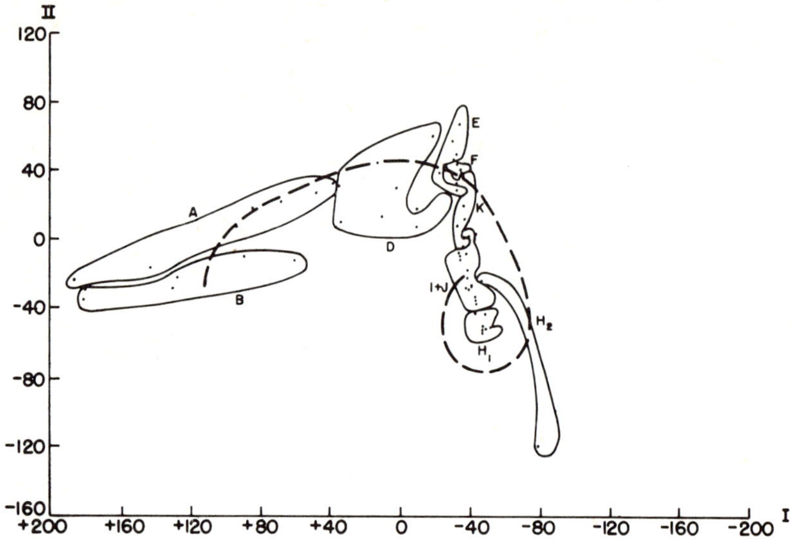

Fig. 6. Results of principal components ordination of field data representing an actual coenocline (WESTMAN 1971, 1975). The coenocline connects *Sequoia sempervirens* forests (A, B) through *Pinus muricata* heath (E, F) and other intermediate types, to pygmy forests of *Cupressus pygmaea* and *Pinus contorta* ssp. *bolanderi* (H2, I), on marine terraces of Mendocino County, California. The principal environmental factor is believed to be soil nutrient conditions, from the normal soils for the region under *Sequoia* forest to the intensely leached and sterile podzols under the pygmy forests. In principal components ordination the nutrient axis becomes a spiral as indicated by the broken line.

axes. Canonical correlation analysis was judged to have little value for ordination.

A further development is the comparison of ordinations and numerical or formal classifications of the same sample sets (GROENEWOUD 1965, GITTINS 1965, RAMSAY & DE LEEUW 1965, ELLENBERG 1967 and ROGERS 1970, YARRANTON 1967, ORLÓCI 1967, WEBB et al. 1967, GREIG-SMITH et al. 1967, CRAWFORD & WISHART 1967, 1968, KERSHAW 1968, FLENLEY 1969, TALLIS 1969, CRAWFORD et al. 1970, MOORE et al. 1970, NORRIS & BARKHAM 1970, EBELING et al. 1970, GRIGAL & GOLDSTEIN 1971 and GOLDSTEIN & GRIGAL 1972, ALLEN 1971, SINGH & WEST 1971, WARD 1971, AUSTIN et al. 1972, FREY & GROENEWOUD 1972, HINNERI 1972, JANSSEN 1972, RITTER 1972, HUGHES et al. 1972, WERGER 1973, WALI & KRAJINA 1973, GOLDSMITH 1973, HOWARD-WILLIAMS & WALKER 1974, HILL et al. 1975, BOUXIN 1975, MITCHELL & MOIR 1976, FONDA & BERNARDI 1976, PAKARINEN 1976, see also 11.4.4, 20.9 and 20.12). GREIG-

Smith et al. (1967) suggest that classification is likely to be more satisfactory on a higher, and ordination on a lower level of vegetational variation; Moore et al. (1970) judge the Braun-Blanquet classification (article 20) more efficient than numerical classification or ordination.

In principle the two approaches can serve not only to supplement but to evaluate one another: one may ask which classification is most consistent with ordinated relationships, and which ordination best expresses relationships of samples to community-types and environment. Work combining ordination and classification does not yet permit such conclusions but indicates that ordinations (and other continuous multivariate treatments) are useful adjuncts to classification and often means of relating the classified units to environment. Some of the multivariate techniques used (factor analysis, principal components analysis, canonical correlation analysis, discriminant function analysis) are subject to strong distortion when used to ordinate samples by species scores (see below). These techniques may, however, be aids to classification without being ecological ordinations in the sense of this article (see 10.6). A technique of classification of particular interest in this connection is the nodal component analysis of Noy-Meir (1973a, 1973b, Noy-Meir & Whittaker 1977, article 11) in which the axes of a principal components analysis, applied to noncentered data, define noda or community-types in a range of community variation accepted to be continuous (see chapter 11).

10.4.2 Sample Set Characteristics

These results do not indicate a clear preference for any one technique. Various characteristics of sample sets affect performance of different techniques; these characteristics include: sample set size and total species number, alpha and beta diversity, trends in these and other community properties, sample error or noise, number and relative extent of directions of sample variation, sample distribution through the range of variation including sample clusters and gaps, and presence of outlier samples, atypical species distributions, and disjunction or relative discontinuity in the sample set. These define a kind of hyperspace of sample-set characteristics of which Gauch et al. (1977) sought to explore some major implications for ordination. Some results from this and other studies will be summarized.

10.4.2.1 *Alpha Diversity and Sample Error*

Effects of alpha diversity and sample error or noise were investigated by GAUCH & WHITTAKER (1972b). Increase in alpha diversity of coenoclines, in the range from 20 to 50 species, improved ordination efficacy only slightly. A corollary is that, for vegetation fairly rich in species, reduction in the number of species used for ordination only slightly reduces ordination efficacy (AUSTIN & GREIG-SMITH 1968, GREIG-SMITH 1971b, AUSTIN et al. 1972, ORLÓCI & MUKKATTU 1973).

The general effect of sample error or noise is to scatter points from their true (noise-free) positions along a coenocline. This scatter, or reduction of ordination acuity, can be measured by correlation of true with ordinated sample positions (GAUCH & WHITTAKER 1972b) or, more simply, by means of per cent displacement of samples of a given coenocline along a first ordination axis (KESSELL & WHITTAKER 1976). The degree of sample error may be expressed as replicate similarity or 'internal association', the per cent similarity of replicate samples at a given position along a coenocline as these per cents decrease from 100 (noise-free data) to low (90), moderate (70-80), and high (50-60) noise levels. GAUCH & WHITTAKER (1972b), using noise levels from 100-55 % similarity, found a sequence of increasing vulnerability to deterioration, from PO through similarity projection and importance projection to PCA. This sequence is somewhat affected by the similarity measurements used; ordinations using Euclidean distance deteriorate more rapidly with increasing noise than those using percentage similarity or coefficient of community for distance measures. Marked vulnerability of PCA to increasing distortion with increasing sample error was shown also by KESSELL & WHITTAKER (1976). GAUCH et al. (1977) found observed scatter from noise was greatest for PCA and least for RA; it was only slightly greater for PO than for RA. Effects of noise are influenced by other characteristics of sample sets and by data treatments. For all the ordinations studied by GAUCH et al. (1977) scatter was strongly influenced by beta-diversity, with scatter worst at low beta. For PCA with beta diversity above 3 half changes, sequence reversals due to involution are admixed with those due to scatter, further limiting interpretation of axes. Toward high beta diversities RA has increasing advantage over PCA (and PO) in its resistance to effects of noise. Beta diversity influences the effect of noise because, as beta diversity becomes smaller, a greater part of the structure of the data is in the noise rather than in trends of sample composition. At very low beta and high noise, the axes of an indirect ordination will be directions of sample error.

Effects of noise on ordination of coenoplanes have not been investigated; they should not differ in principle from those observed with coenoclines. Relations to data treatment and similarity measures will be discussed below.

10.4.2.2 *Beta Diversity and Dimensionality*

The principal effect of beta diversity has been indicated; increasing distortion of ordinations (especially those with linear assumptions) with increasing beta diversity. Beta diversity, the degree of compositional change along a coenocline or compositional difference among the samples of a set, is conveniently measured in half-changes (HC, article 3.4). Despite the influence of noise and other variables on effects of increasing beta diversity, statements of the relative vulnerability of different techniques are possible. PCA, nonstandardized, is most vulnerable of the major techniques, with distortions of type (1)occurring below 2-3 half changes (Fig. 7) and involution and distortions of type (5) above that range. In the field ordinations of Wentworth (1976) with vegetation transects of low beta diversities (2-3 HC), PCA did not produce recognizable involution but gave markedly less effective ordinations than RA and PO. Standardization, and especially double standardization, can extend the effectiveness of PCA to higher beta diversities (Fig. 7, Austin & Noy-Meir 1972, article 11). RA can give effective ordination up to about 10 HC, a beta diversity much exceeding that of

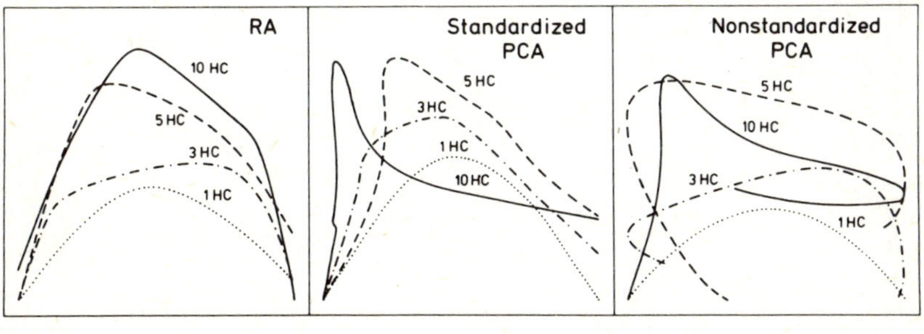

Fig. 7. Distortions of coenoclines (one-dimensional vegetation gradients) into arches by reciprocal averaging (RA), centered non-standardized principal components analysis (PCA), and centered species-standardized PCA, with sample ordinations for four levels of beta diversity (1, 3, 5, and 10 half-changes). In all cases the expected or true result is a straight line. Distortion increases as beta diversity increases, but PCA always has a greater arch distortion than RA. Note also that RA is distorted into a simple arch, whereas at beta diversities of 3 half-changes and more nonstandardized PCA involutes the gradient ends, resulting in sequence reversals (Gauch et al. 1977).

most field coenoclines (GAUCH et al. 1977). Among other techniques Gaussian ordination and multidimensional scaling can handle coenoclines of high beta diversity (GAUCH et al. 1977, FASHAM 1976) but may be dependent on adequate sample representation of the coenocline and a reasonable first guess ordination for success with such data.

PO gives generally good ordinations up to about 5 HC, it can give fair ordinations with correct sample sequences but with distorted sample spacing to somewhat higher diversities. In proper use (with beta diversity of 5 HC or less, percentage similarity or coefficient of community, and with endpoint samples representing the extremes of the coenocline) PO does not involute. It can, however, involute when Euclidean distance is used (Fig. 9), and it will involute outlier samples, including those beyond an endpoint sample of a coenocline of high beta diversity. The arch formed by samples in PO (Fig. 7) implies some displacement of sample positions toward the endpoints even at relatively low beta diversity (KESSELL & WHITTAKER 1976); such distortion (type 1) is greater in PO than in RA, but less than that in PCA (even when the latter is applied at low beta diversity and without involution).

PO and all eigenvector techniques are subject to an arch effect: curvature of ordinated coenocline positions away from the first axis. In PO the distance (*e*) off the axis is not used for further ordination, and the curvature on one axis has no effect on subsequent axes. In PCA, RA, and other eigenvector techniques the curvature off a given axis becomes part of the determination of ordinated positions of samples on the second and higher axes, with potentially vexing consequences for ordination. If the curvature off the first axis on the second forms an arch of quadratic form, that on the third axis is of sigmoid or cubic form, that on the fourth axis a fourth power function, and so on (HILL 1973, GAUCH et al. 1977). Thus the sample displacements take the forms of polynomial curves of increasing order but decreasing amplitudes from the second axis on (Fig. 8). Such effects are implicit, but sometimes unrecognizable, in eigenvector ordinations of field data. Sample error scatters the points off the curves, tending to obscure their forms. What the field investigator observes as ordinated positions on second or higher axes may then be complex combinations of distortion curvature and sample error with real relationships to environmental variables, and effects of species and sample amplitudes (GAUCH et al. 1977). The amplitude effects involve the fact that species of wide ranges in a sample set (and samples of relatively equitable species composition) are ordinated proximally (within an arch or other curve), whereas

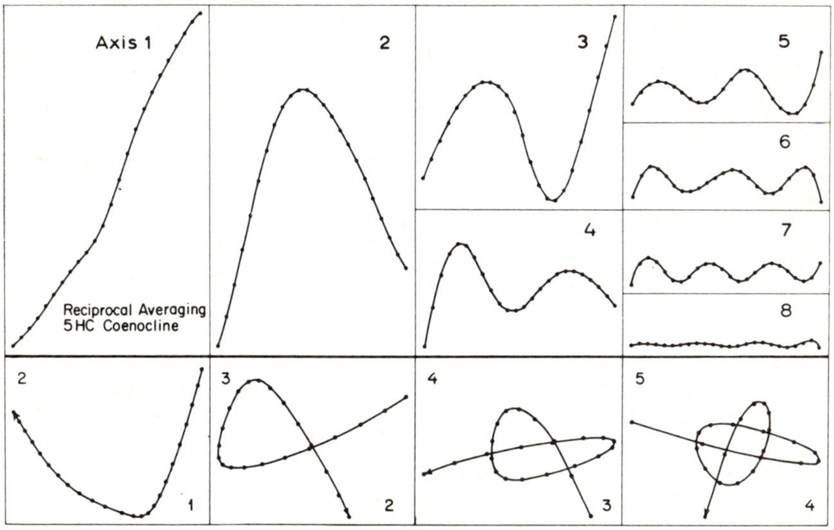

Fig. 8. Reciprocal averaging ordination of a 5 half-change coenocline. Above, results to 8 dimensions. Ordinates show ordination values, and abscissas show the correct or expected values for the original coenocline axis. Bottom, results for successive pairs of axes (1 and 2, 2 and 3, 3 and 4, and 4 and 5). Abscissas for the last three pairs are enlarged to the same length as axis 1 of the first pair, and ordinates are then scaled by the ratios of the square roots of the axes' eigenvalues.

species of narrow ranges (and samples with strong dominance) are ordinated distally to the curve. This mixture of effects implies that the second and higher axes of eigenvector techniques can be quite uninterpretable, even without involution. RA is subject to these effects, but the curvature into higher axes is less intense with RA than with PCA.

This distortion necessarily affects the way ordinations handle coenoplanes and higher-dimensional data. GAUCH et al. (1977) found that square coenoplanes (1.5 × 1.5 to 5 × 5 HC) were effectively ordinated by PO and RA; the outlines of the coenoplanes became mildly curved at higher HC values (Fig. 9). Nonstandardized PCA involuted the corners of the coenoplane at about 3 × 3 HC and higher, but doubly standardized PCA gave reasonable ordination to 5 × 5 HC (cf. AUSTIN & NOY-MEIR 1972). When no involution occurs, however, the coenoplane is arched into the third

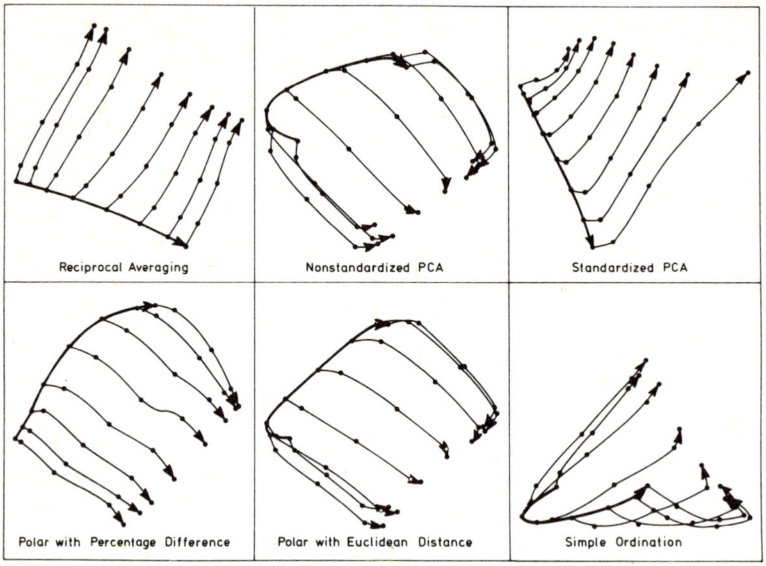

Fig. 9. Ordinations of a 4.5 × 4.5 half-change coenoplane by six techniques. The sampling pattern and the expected result is a square grid of points, eight rows in one direction and five in the other. Arrows are drawn to help show the configurations.

(and higher) axes; a spurious third axis is thus produced that with field data might be difficult to recognize as spurious.

With rectangular coenoplanes (e.g. 1.5 × 4.5 HC) an eigenvector ordination sometimes will and sometimes will not give interpretable results for the second ordination axis (Fig. 10). If the second axis of the original coenoplane is relatively narrow, the second axis of the ordination may be a quadratic distortion function of the first axis, whereas the second axis of the coenoplane appears in the third ordination axis (while the fourth and higher ordination axes represent distortion curvatures on the lower axes). Thus the second axis of a rectangular coenoplane may appear in the second, or third or higher, ordination axis depending on the relation of the variance on the second coenoplane axis to the variances of the distortion curves on the ordination axes. Such effects do not preclude successful multidimensional eigenvector ordinations in favorable cases, but they recommend caution in the interpretation of eigenvector axes. Not only are the effects of distortion compounded with those of sample error and sample equitability, but it may be very difficult with field data to recognize the relative variances, or components of beta diversity, represented by different directions of true sample variation in response to environmental gradients.

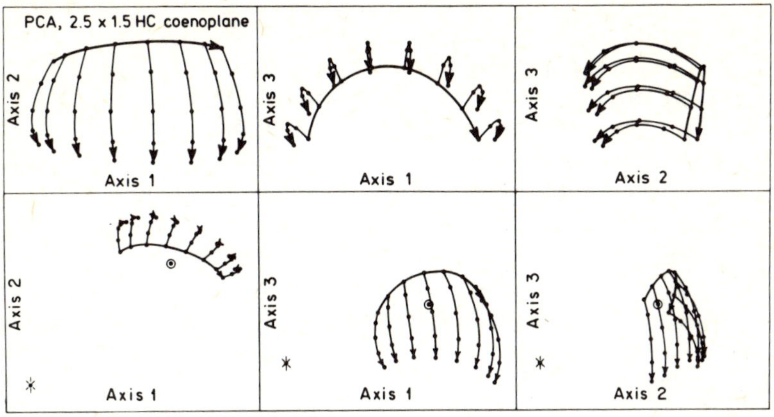

Fig. 10. Above, ordination of a simulated community pattern or coenoplane by centered, nonstandardized principal components analysis (PCA). The coenoplane had 40 samples in an 8x5 grid with a beta diversity of 2.5×1.5 half-changes. Arrows are drawn to help show the configurations. A perfect ordination would give a rectangular grid with sides 2.5 units on axis 1 to 1.5 units on axis 2 and no sample displacements from the origin on axis 3. Below, PCA ordination of the same coenoplane, but with two samples added: the circled point a type 'a' outlier, and the starred point a type 'b' outlier.

10.4.2.3 *Clusters and Outliers*

In many cases communities are sampled as they are encountered, or sample choice is determined by availability of undisturbed stands. Samples may then be irregularly distributed in the range of community variation, and the sample set may include clusters of similar samples, deviant or outlier samples, and gaps in sample representation. Because of the response of PCA to variance, clusters attract the axes of PCA (Kessell & Whittaker 1976). When the cluster is not at the extreme of the coenocline (or coenoplane), a spurious axis may result; and the distortion of the ordination (as it appears projected on the first two axes) may be much increased. Clusters of samples were found by Gauch et al. (1977) to have little effect on RA; in this technique clusters apparently do not attract the axes away from the extremes of a coenocline toward a group of samples within the range of variation sampled. Because only the endpoint samples determine positions in PO, sample clusters have no effect on ordinations with that technique.

Sample sets also may include deviant samples or outliers – samples with characteristics well outside the range of variation represented by other samples of the set. Gauch et al. (1977) simulated outliers in two ways (a) random selection, for each species, of one of its scores from among the samples already present, and (b) random selection of one species to be strongly dominant, about half the others to be of random but modest importance values, and half zeroes. The first was intended to simulate moderately deviant samples from atypical habitats, and the second strongly deviant samples from disturbed sites. These types of outliers have mathematical, as well as ecological implications. Type 'a' outliers locate themselves near or within the arch formed by samples as projected onto the first two axes (and the manifold formed by samples in a multidimensional, Euclidean space); whereas type 'b' outliers are located outside the arch and manifold. These simulated outliers were added to coenoclines without noise, a coenocline with 25 % noise, coenoplanes, and field data.

For polar ordination, outliers not chosen as endpoints automatically ordinate at or near the midpoint of the ordination axes. All outliers will ordinate at this position even if they are also very dissimilar to one another. As endpoint samples, outliers are particularly poor choices because of their low relative similarities to all other samples. (Unfortunately, the original technique of Bray & Curtis (1957) and the simple ordination of Orlóci (1966) choose outliers in sample sets including these.)

RA was robust against effects of type 'a' outliers – they tended to ordinate around the center of the ordination field, but caused negligible displacement of the other samples. Type 'b' outliers, in contrast, ordinate around the periphery of the ordination field and strongly affect positions of the other samples (Fig. 10). One such outlier from a coenocline ordinates at one end of the first axis and compresses the other samples in a tight cluster at the other end; the coenocline then emerges in the second axes. With 3 or 5 type 'b' outliers in a coenocline, the outliers are well scattered and other samples clustered. For a 2.5 × 1.5 HC coenoplane, one type 'b' outlier causes the first axis to have the outlier at one end and the remaining samples in a cluster at the other end; the 2.5 HC coenoplane axis then emerges in the second RA axis, and the 1.5 coenoplane axis emerges in the third RA axis. Three or five such outliers severely compress the original coenoplane samples, although their arrangement is still discernible. Outliers in field data thus have somewhat unpredictable but potentially severe effects on RA.

PCA ordinates type 'a' outliers centrally to the other samples. One such outlier has relatively little effect, but a few cause some displacement of sample positions in both coenoclines and coenoplanes. Effects of type 'b' outliers are more drastic; each outlier determines the direction of an axis, causing strong compression of the other samples into a small region of the ordination field, as well as distortion within this region. PCA is thus even more vulnerable to outliers than RA.

Outliers are detrimental to ordination and consequently should be identified and removed from sample sets. Identification may be on the basis of low mean similarity of certain samples with all other samples of a set. Identification may also be possible by an ordination (with, say, RA) that reveals the deviance of one or more samples from the set and implies the need for a second ordination with the outlier(s) removed.

10.4.2.4 *Disjunction*

When groups of samples have no species in common, the resulting species-by-samples data matrix is fully disjunct. Partial disjunction, in which groups of samples sharing many of their species are connected by few species to other groups, is more common in field data and is also more problematic for ordination. Even complete disjunction may not be obvious in an original data matrix, but becomes apparent after arrangement of rows and columns by the first-axis RA scores. Since unrelated subsets are involved, the simplest course is to separate these to avoid effects of the high intrinsic dimensionality of the full set that will complicate ordination results. PO endpoints will be nondiscriminating for samples not belonging to their subsets. Eigenvector ordinations, given a data set with N disjunct subsets, will first use $N—1$ axes to separate subsets. The following eigenvectors will differentiate within only one subset each, and in order of decreasing variance accounted for will switch from subset to subset as in a story with pages interleaved from several sources (see also 11.4.4 and 11.6).

Partial disjunction was tested by Gauch et al. (1977) using a 10 HC coenocline of evenly spaced Gaussian curves and samples. Two sections of 1, 2, or 3 HC were removed, thereby leaving three subsets related at three different levels of partial disjunction. PO ordinations were unaltered except for deletion of the missing samples. With different degrees of partial disjunction eigenvector ordinations (RA, nonstandardized PCA, and standardized PCA) produced various types of axes; (i) differentiating between two or more subsets, (ii) differentiating within one subset (with or without outlier effects of certain samples in other subsets), (iii) differentiating within two or more subsets, (iv) polynomial distortions of one or more lower axes, and (v) mixtures of these types. The polynomial nature of higher axes (Fig. 8) permits differentiation between more than one subset, possibility (iii), but also implies that polarity of differentiation may be reversed between these subsets. RA seemed to give a somewhat higher proportion of interpretable axes than standardized or nonstandardized PCA. Noy-Meir (1973a, b),

Noy-Meir & Whittaker (1977), and article 11 discuss disjunct sample sets and the application of noncentered PCA and classification to them.

10.4.3 Similarity Values and Transformations

Three major measures of sample similarity are used in ordination – coefficient of community (CC), percentage similarity (PS), and Euclidean distance (ED) (article 5, formulas 40, 50 or 51, and 62). For ordination CC and PS are normally subtracted from 1.0 to give distance values, CD and PD. It can be argued that the subtraction should be not from 1.0 but from a lower value that is obtained when replicate samples from a given stand are compared with one another. This replicate similarity or internal association (article 3) is often between 0.7 and 0.9, less commonly lower or higher. In ordination of field data a replicate similarity value is often not available, and the highest similarity in the matrix may differ from replicate similarity; many studies consequently subtract CC and PS from 1.0. Whether that value or a replicate similarity is used appears to make little difference in performance of polar ordination (Bannister 1968, Gauch 1973b).

The three measures represent a progression of increasing emphasis on quantitative species scores and on dominant species, from CC (which calculates for species presence and absence only) through PS (which calculates from absolute differences in species scores) to ED (calculated from squared differences in species scores). Tests of polar ordination with coenoclines indicate that in comparison with PS, ED increases distortion and vulnerability to sample error in ordination (Fig. 9, see also Gauch & Whittaker 1972b, Kessell & Whittaker 1976 and Gauch et al. 1977). The disadvantage of ED results from the squaring of species scores, which tends to exaggerate the importance of major species and the curvilinear relations and sample errors affecting their scores, beyond the comparable effects on PS. Contrasts in sample size also can have stronger distorting effects in ordination with ED than with PS, and can result in the anomaly that two samples sharing all or many of their species may have a larger ED than two samples with no species in common (Orlóci 1967, Bannister 1968, article 9). Orlóci (1967, article 9.7.1.2) suggests a standardized Euclidean distance as a solution. The present authors have used also a relative Euclidean distance scaled between 0 and 1.0: $ED_j' = \sqrt{\Sigma(p_{ij}-p_{ik})^2/2}$, in which p_{ij} and p_{ik} are relative importance values (totaling 1.0 for all species in a given sample), for species i in samples j and k

(Whittaker & Gauch 1973). Implications of the choice between ED and PS for eigenvector ordinations (e.g. secondary reciprocal averaging and principal coordinates analysis) have apparently not been tested. In general, however, PS seems the measure of preference for ordinations using species scores for distance measurement (see also article 9.7.1.2).

Tests have given conflicting results on PS and CC, some suggesting preference for the former (Kessell & Whittaker 1976) and some for the latter (Gauch & Whittaker 1972b); see also Austin & Greig-Smith (1968), Austin & Noy-Meir (1972). CC is less subject to the effects of complex and ditonic curves of species distribution; PS bears a more strongly curvilinear relation to separation along environmental gradients than CC (Gauch 1973a). Thus in some ordination circumstances more may be gained by the use of the less strongly curvilinear measure, CC, than is lost by sacrifice of information. For a sample set representing a limited range of communities, in which most species have monotonic distributions along a given axis, the advantage is probably reversed. Also, in sample sets with contrasting alpha diversities of samples, use of CC tends to spread apart species-rich samples and to press together species-poor samples in ordination space (Orlóci 1966). When samples are of low alpha diversity, PS may be preferred, for CC fails to use much of the information available in the species scores; when species scores are strongly affected by disturbance (whereas floristic composition is less affected) CC may be preferable because less vulnerable to distortion by sample error. The advantage is probably with PS (with or without species standardization) for the majority of ordinations, with CC having the advantage in some cases of high alpha and beta diversity and sample error affecting dominant species.

Sample standardization removes effects of different sample sizes on distance calculations; relativization (such that each sample and column of the primary matrix totals 1.0) before ordination is probably generally desirable. Calculations of similarity values are thereby simplified: $PS = \Sigma \min (p_{ij}, p_{ik})$. Species standardizations (division of a species' scores by the total, maximum, mean, or standard deviation of the scores for that species in that row of the primary matrix) tend to equalize the effects of species on the similarity values and the ordination (article 11). Standardization by the standard deviation is often employed for eigenvector ordinations, and this standardization reduces distortion in PCA (Noy-Meir et al. 1975, Gauch et al. 1977). Standardization by row (species) totals can be employed for polar species ordination and permits ready calculation of percentage similarity of species distribution (article 5, formula 23). For PO of samples, species standardization reduces the effect of major species while retaining the effects of differences in scores for a given species; its effect may be a compromise between that of PS without species standardization and CC.

The Wisconsin school used a double standardization (of rows first and then of columns, article 7.3.1); AUSTIN & GREIG-SMITH (1968) and AUSTIN & NOY-MEIR (1972) found that some ordinations are improved by such treatment of the data. This treatment and the simultaneous double standardization of reciprocal averaging have compromise characteristics that are optimal in some respects for ordination (article 11). Species standardization may not be desired, however, it if is assumed that major species are more effectively measured and their data consequently more significant for ordination than rare species.

Various transformations of scores before ordination have been tried (article 11, NOY-MEIR et al. 1975, CHARDY et al. 1976, SMITH 1976). Square root and log transformations of scores, by damping the effects of major species on ED, can produce some improvement in ordinations employing this measure (GAUCH et al. 1977). Log transformation is often thought appropriate for the treatment of animal population data (CHARDY et al. 1976). WENTWORTH'S (1976) tests suggested that log transformation of coverages before PS calculation gave a modest improvement in ordination of field vegetation data. Use of the phytosociological coverage scales – the 5-point scale of BRAUN-BLANQUET or the 10-point scale of DOMIN (CAIN & OLIVIERA 1959, BANNISTER 1966, SHIMWELL 1971: 111, MUELLER-DOMBOIS & ELLENBERG 1974: 62, article 20.4.5) – should have a similar effect in damping without removing contrasts between species in abundance. Efforts have also been made to improve ordinations by the use of transforms of percentage similarity – the arcsine (LOUCKS 1962), reciprocal (GIMMINGHAM et al. 1966), and logarithm (FRYDMAN & WHITTAKER 1968). The arcsine transformation seems to produce little or no improvement in ordination (ORLÓCI 1966). None of these transformations fits the relationship of percentage similarity to environmental separation, which is a nonstandardized complement of an error function (GAUCH 1973a, article 3). Use of the inverse of the error function as a transformation gave modest improvement of ordinations with noiseless, simulated data (GAUCH 1973b), but apparently not with field data. SWAN (1970) has commented on the distorting effect of zero importance values and possible corrections.

No transformations are known that fully solve the problems of curvilinear relations and zero values. Because of the bell-shaped form of species distributions, effective measurement of similarities of species distribution seems more problematic than measurement of sample similarity (see also article 5). A number of measurements are possible if beta diversities are not too high and are discussed in article 5 (see also articles 7 and 8). Reciprocal averaging seems especially effective for its indirect species ordinations (Fig. 11, HILL 1973, GAUCH et al. 1977). Article 5 gives a broader survey of similarity measurements, while articles 9.8.1 and 11.3 consider further the major measures used in ordination and effects of transformations and standardizations.

10.4.4 TESTS OF VARIANT TECHNIQUES

10.4.4.1 *Principal Components Analysis Variants*

Several PCA variants were compared by GAUCH et al. (1977) with coenoclines of lengths from 1 to 10 HC. Among these variants, noncentering appeared disadvantageous for ordination (although it may be advantageous for classification) (NOY-MEIR 1973 a, b, NOY-MEIR & WHITTAKER 1977, article 11). Sample standardization gave little or no improvement in these tests (but may do so if samples are of different sizes). Wisconsin double standardization improved ordinations substantially in some cases (cf. AUSTIN & NOY-MEIR 1972) with results intermediate to RA and nonstandardized PCA. Logarithm, square root, and cube root transformations gave some improvement, but there was no clear basis of preference among them. Presence/absence transformation gave various but usually poorer results. The effects of these variations on PCA are dependent on sample set properties. PCA ordinations may be improved, however, by data treatments that reduce the effects of dominant species on the ordination, whether by species standardization or by logarithmic or other transformations. These treatments extend the range of beta diversities that can be ordinated without severe distortion. Reasons for these effects of standardization are discussed in article 11.4.3 (see also CASSIE 1969, NOY-MEIR et al. 1975, and CHARDY et al. 1976).

10.4.4.2 *Polar Ordination Variants*

The essential approach of polar or comparative ordination – arrangement of samples (or species) on axes by relative similarity to reference samples (or species) – admits many variations from the original technique of BRAY & CURTIS (7.3.4, 10.2, 11.4.1).

Results of some of the tests with variations were summarized by GAUCH et al. (1977). Comparisons of PO and PCA (GAUCH & WHITTAKER 1972b, KESSELL & WHITTAKER 1976) have applied these to the same primary matrix without Wisconsin double standardization. Although double standardization can much improve the performance of PCA, it gave modest or no improvement in PO in the tests by GAUCH et al. (1977). (Whether double standardization is advantageous depends on sample set characteristics and the weighting that should be given to species – see 11.4.2.) Corrections in PO for curvilinearity of sample similarity values (GAUCH 1973b) and nonorthogonality of axes (ORLÓCI 1966, 1974a, BANNISTER 1968) gave little or no improvement in ordination of simulated data with sample error. Use of Euclidean distance rather than percentage similarity for PO (ORLÓCI 1966, 1974a, BANNISTER 1968) usually increases sample displacement along the axis (GAUCH & WHITTAKER 1972b, KESSELL & WHITTAKER 1976) and can produce involution of the extremes where percentage difference and the complement of community coefficient do not so involute (GAUCH et al. 1977, Fig. 9). The simple ordination of ORLÓCI (1966) uses Euclidean distance, and the first axis goes through the least similar pair of samples. For the second (and higher) axes single endpoint samples define axes orthogonal to the first axis (or earlier axes). Applied to a coenocline simple ordination creates a spurious second axis of relative distance from one sample that is highest on the arch off the first axis. Applied to coenoplanes exceeding a few HC simple ordination produces severely distorted configurations (e.g. Fig. 9).

We have recently experimented further with algorithms and data treatments. Alternative algorithms include (i) the Pythagorean projection customary in polar ordination (percentage distance in the equation given in 7.3.1), (ii) one-ended proportionality (distance of sample from one endpoint divided by base-line distance), (iii) two-ended proportionality (distance of sample to one endpoint, divided by sum of distances to that and the other endpoint, SCRUGGS 1976), and (iv) two-ended squared proportionality (same as preceding, but all distances squared). These were tested, using both percentage difference and Euclidean distance, as applied to data sets that were (i) untransformed, (ii) relativized by sample totals, (iii) Wisconsin double standardized, and (iv) arcsine of square roots of similarity values (LOUCKS 1962), using coenoclines of 1.5, 3, 5 and 7 halfchanges. Results were assessed by average displacement of samples along the ordination axis (KESSELL & WHITTAKER 1976). The algorithms performed best to worst in the order: original Pythagorean equation of BRAY & CURTIS (1957) and BEALS (1960), two-ended squared proportionality, two-ended proportionality, and (much the worst) one-ended proportionality. (Complications involving different advantage at different beta diversities affect some of these statements; the two-ended proportionalities have some advantage over the Pythagorean projection at low beta diversity – SCRUGGS 1976.) Transformations made less difference than change of algorithm; but relativization produced modest improvement over untransformed data, and double standardization produced somewhat greater improvement in these experiments. Percentage difference was preferable to Euclidean distance, as in other comparisons of these. The arcsine transformation gave modest improvement (comparable to that of the inverse error function transformation of GAUCH 1973a) at intermediate beta diversities (3 HC) with untransformed data. It appeared to give no significant improvement with doubly standardized data (or, presumably, field data with sample errors). We found no combination clearly preferable to that used by BRAY & CURTIS (1957). Their technique embodies some effective (if approximate) compensations for problems of nonlinearity – a fact often not recognized. In particular, efforts to modify the technique on the basis of Euclidean assumptions reduce its effectiveness for ecological ordination.

10.4.4.3 *Secondary Reciprocal Averaging*

GOFF & COTTAM (1967) applied an RA algorithm to a secondary matrix of sample similarities, instead of the usual application to a primary matrix of species scores. This technique of index iteration (GOFF & COTTAM 1967) or secondary reciprocal averaging (SRA) may have advantage over PO for objective recognition of a first axis of vegetational variation because it employs the full set of similarity values rather than comparisons with chosen endpoints. SRA was applied by GAUCH et al. (1977) to matrices of percentage similarity values, by samples and by species, for coenoclines of 1 to 10 HC and several coenoplanes. Results of SRA of samples were much like those of RA for the 1 to 5 HC coenoclines, but the 10 HC coenocline was distorted less (about as much as a 5 HC coenocline with regular RA). Results of SRA of species (by percentage similarities of species distributions) were, in contrast, less satisfactory than RA species ordination because of problems caused by rare species. SRA applied to coenoplanes showed greater curvilinear distortion than RA. SRA may be useful for sample ordinations with high beta diversity and may also be useful for analyzing data in papers where only the secondary matrix was published, for SRA can be applied to any matrix of similarity or distance values. Its performance in that use has not been compared with principal coordinates analysis (GOWER 1966). For most uses SRA should have no advantage over

RA with its effective, simultaneous ordinations of samples and species.

10.5 Summary Comments on Major Techniques

10.5.1 Direct Gradient Analysis

Direct ordination with samples arranged into transects, and patterns or mosiacs, can give particularly clear results in favorable circumstances. Direct ordination can also handle longer axes of community variation (by either environmental indices or weighted averages) without distortion than many other techniques. It is thus of wide applicability, for sample sets ranging from small to very large. For problems of small scale involving a single environmental complex-gradient direct ordination is a simple and lucid technique. For such larger projects as transect analysis of the vegetation of mountain ranges (Whittaker 1956, 1960, Whittaker & Niering 1965, 1968, Peet 1975) it is demanding of a rather large effort in field sampling and data treatment. Given such effort large sample sets of wide community variation – that would be difficult to handle with any other ordination technique – can be treated to give results including detailed knowledge of distributions of species and community-types scarcely obtainable by other techniques. Effectiveness of direct gradient analysis may be increased in some studies by scalar approaches to environmental factors (Loucks 1962, Wentworth 1976, article 2), and combinations of direct and indirect approaches. Among the latter possibilities are: (i) stratification of a sample set on a direct first axis, followed by indirect ordination within the resulting subsets (e.g. stratification of mountain vegetation into elevation belts, followed by RA or Gaussian ordinations of these) (Peet 1975, Wentworth 1976), (ii) use of an indirect technique such as RA to derive a first axis, and of direct techniques for additional axes not clarified by the indirect technique, (iii) joint use of direct and indirect ordinations of a sample set with the indirect technique(s) indicating major directions of variation and the direct techniques aiding in the interpretation of indirect axes, or providing the basis of rotation of these for more effective interpretation.

10.5.2 Polar Ordination

Polar ordination is a simple, adaptable, relatively effective technique. Its simplicity makes it easy to use as a trial ordination, or as a comparison with and check on more formal techniques. PO will in some cases give results as good as those of any other technique. Polar ordinations also have advantage over more formal techniques for some purposes. The calculations are not only simple, they are lucid: The investigator can more easily see the computational reasons samples are arranged as they are than in PCA and other formal techniques. PO is less subject to distortion and the production of ecologically meaningless axes than PCA; and PO is capable of dealing with moderately high beta diversity (about 5 HC) on the first axis with little distortion. PO also may permit recognition of a second axis, narrower than the first, that PCA and other eigenvector techniques obscure with distortion curvature. PO is relatively robust as an ordination technique for several reasons: (i) Because the ends of the axes are anchored by reference samples, the axes will not involute with proper use of PO as they may in PCA. (Proper use implies beta diversity not over 5 HC, and endpoint samples representing the extremes of the gradient sampled. Only in some cases will PO project samples more extreme than an endpoint sample onto the axis beyond that endpoint sample.) (ii) Because the curvature off a given axis between extreme samples (the arch effect, sample displacement off the axis measured by *e*) does not affect other axes, PO is free of the problems of curvilinear distortion on lower axes affecting sample positions on higher axes that afflict the eigenvector techniques. (iii) The Pythagorean projection between two endpoints reduces, though it does not remove, effects of the nonlinear relation of sample similarities to the coenocline distance in locating samples on the axis (Gauch 1973b). (iv) The definition of axes by endpoints only, though it sacrifices information, also protects the ordination against effects of sample clusters and outliers not used as endpoints. PO consequently has some characteristics of nonlinear ordination (Gauch et al. 1977). The technique's merits should not be underestimated because of its simplicity (Beals 1973).

We regard PO as a useful technique despite its limitations: the necessity of choosing endpoint samples, the heavy emphasis of a relatively few endpoints in determining the ordination, and the vulnerability of the ordination to effects of different endpoint choices. Among PO variants: (i) In most cases we employ PO for direct ordination, choosing the endpoints for known or suspected major directions of environmental relationships in the sample set. (ii) We use PS for either relativized samples or doubly standardized data as a first similarity calculation, and CC as a possible further calculation for comparison with PS. For some data log transformation (and relativization) of species scores before calculation of PS is probably desirable. (iii) We do not use error function, arcsine, and other transformations of similarity values. (iv) Endpoints for higher axes are chosen that are close together on lower axes, and which consequently define axes that are nearly orthogonal. True positions of these endpoints (of, say, the second axis on the first axis) are retained in the plot as in Fig. 4 in article 7, but relative positions of endpoints and other samples on the second axis are determined by projection onto a perpendicular second axis. The use of single endpoints to define second and higher axes in the simple ordination of ORLÓCI (1966) increases distortion (GAUCH et al. 1977), and the perpendicular technique of ORLÓCI (1974a) is unnecessarily complex and does not improve ecological ordinations; we do not recommend these modifications.

10.5.3 PRINCIPAL COMPONENTS ANALYSIS

PCA is appropriate for a multivariate normal sample distribution, which implies a hyperellipsoidal cloud of points in a multidimensional space as illustrated in some of the papers that discuss the rationale of PCA (GROENEWOUD 1965, GITTINS 1969, PIELOU 1969). The problem is that in a Euclidean, species-dimensional space, sample positions form curvilinear clouds or manifolds, not hyperellipsoidal clouds. For sample sets not of very low beta diversity sample positions in the manifold are curved by the overlapping, nonmonotonic distributions of species scores, around an interior that is hollow because some species combinations are infrequent or absent. Projections of such manifolds onto PCA axes are necessarily curvilinear: in ecological terms they are distorted. In its most familiar form (species-centered, nonstandardized) PCA is highly vulnerable to involution and other distortions, and it is vulnerable to effects of sample error or noise, outlier samples, and sample clusters (GAUCH et al. 1977). Performance in ordination, and the beta-diversity of sample sets that can be ordinated without severe distortion, can be markedly improved by species standardization and particularly double standardization (NOY-MEIR et al. 1975, GAUCH et al. 1977, article 11). (Species standardization shrinks the manifold's extremes and reduces its curvature in species-dimensional space.)

Given this improvement and the easy availability of the programs, PCA will continue to be used for ordination. We recommend caution in its use, care regarding beta diversity and standardization, and testing of the results and their interpretation with other ordinations. We are not encouraging one possible approach to ecological ordination: application of PCA (or other formal technique) to a data set the beta diversity and other characteristic of which are unknown to extract axes subject to distortion effects that cannot be tested, which axes are labeled as ecological factors and the results presented as mathematical ecology. Practices for which we imply no criticism are: (i) application of PCA and other linear techniques to monotonic variables other than species scores, such as environmental factors, structural characteristics of communities (KNIGHT 1965, KNIGHT & LOUCKS 1969, JAMES 1971), and species groups or higher taxa (MAAREL 1969, WHITTAKER 1972), (ii) noncentered application of PCA as an approach to classification (nodal component analysis, NOY-MEIR 1973a, b, NOY-MEIR & WHITTAKER 1977, chapter 11.5), and (iii) other uses of multivariate techniques as aids to classification (10.4.1.3, 10.6, 11.6.3).

10.5.4 RECIPROCAL AVERAGING

RA is closely related as an eigenvector technique to PCA (11.4) but has distinctive properties (HILL 1973, 1974, CHARDY et al. 1976, GAUCH et al. 1977). RA appears an optimal eigenvector technique for ordination, less vulnerable to distortion than singly standarized and much less vulnerable than unstandardized PCA. RA is highly resistant to involution up to high beta diversities (about 10 HC). It is a reliable technique for extracting the first, major axis of community and environmental variation from a sample set; it can extract the first axis over a larger range of beta diversity and with less sample displacement along the axis than PO. In favorable cases RA will give efficient ordination in second or higher axes, but RA is (like PCA) subject to the problems of spurious higher axes produced by curvilinear distortion. Such distortion is not a serious problem in ordination of sets that embody a single major direction of variation (or can, as in the separate ordination of elevation belts in mountains, be stratified so that they do). RA is resistant to the effects of clusters and other peculiarities of sample sets except outliers, which should be removed from an ordinated set. A point of special attraction for RA is its production of simultaneous species and sample ordinations on axes that can be treated as on the same scales (Fig. 11, GAUCH et al. 1977).

10.6 Conclusion

Other techniques remain to be evaluated; a few of these (multidimensional scaling, parametric mapping, Gaussian ordination) are discussed along with RA in the following article. The

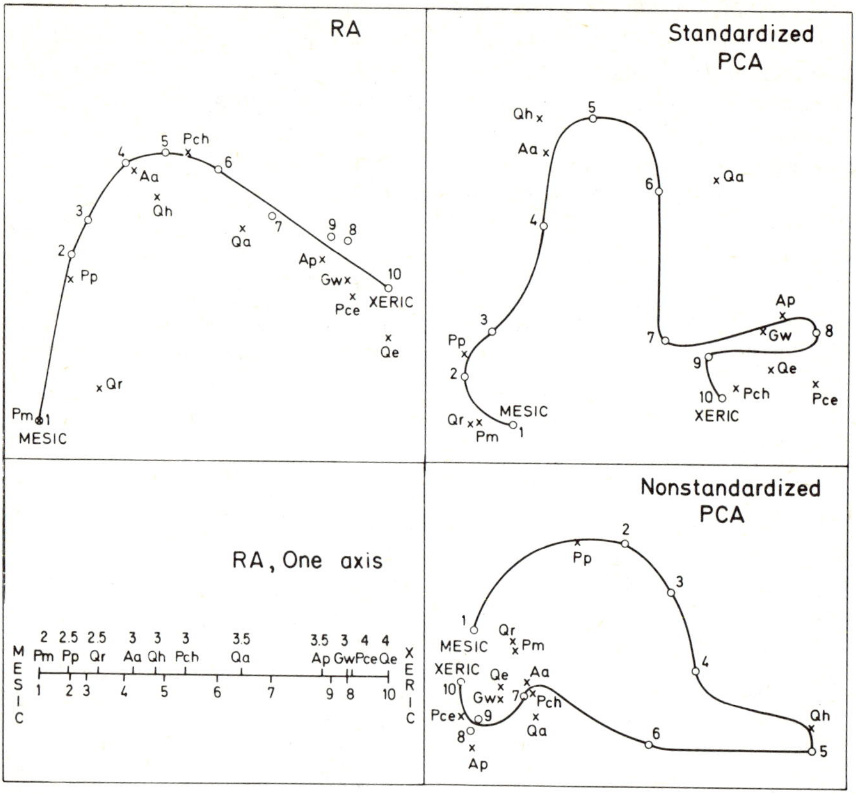

Fig. 11. Ten composite samples numbered in sequence from a topographic moisture gradient at 1,830-2,140 meters in the Santa Catalina Mountains, Arizona (Whittaker & Niering 1965: data tabulated and graphed in Gauch et al. 1974), ordinated in two dimensions by reciprocal averaging (RA), centered, species-standardized principal components analysis (PCA), and centered, nonstandardized PCA. The samples are indicated by circles numbered from 1 (most mesic) to 10 (most xeric); ordinated positions of the species are indicated by X's and initials. RA ordinates the samples into an arch (left) in the first two axes; PCA ordinates the samples into much more complex curves. In the lower left panel both samples and species positions are indicated along the first axis of RA; the arrangement is considered realistic from field experience. (Both techniques reverse samples 8 and 9 from their sequence, presumably because of sample errors.) The tree and arborescent shrub species ordinated are: *Pseudotsuga menziesii*, *Pinus ponderosa*, *Quercus rugosa*, *Arbutus arizonica*, *Q. hypoleucoides*, *Pinus chihuahuana*, *Q. arizonica*, *Arctostaphylos pringlei*, *Garrya wrightii*, *Pinus cembroides*, and *Q. emoryi*. Numbers above the species initials in the one-dimensional RA graph are weights used for direct weighted-averages ordination by Whittaker & Niering (1964, 1965).

history of gradient analysis has seen an apparently exponential proliferation of techniques available, and a progression of interest in four stages from (i) direct ordination, through (ii) indirect but informal techniques, and (iii) formal techniques little suited to ecological data, to (iv) formal techniques adapted to the nonlinearity of ecological data. Meanwhile, the implications of different sample set properties have become apparent; some of these properties – notably beta diversity, dimensionality, clusters, outliers, and disjunctions – can strongly influence ordinations (GAUCH et al. 1977). We think that the diversity of sample set characteristics, together with the problems of nonlinearity, imply that no fully satisfactory, all-purpose ordination technique will be found. Ordination may become an area of expertise, employing a number of techniques of different advantages that are widely used by investigators, in a context including also specialty techniques and understanding of the qualifications and complications affecting different techniques. We offer here an interim report that we hope clarifies some of the major techniques; the following article offers comments on nonlinearity and some of the stage (iv) techniques appropriate to it.

One point that emerges from the study is the implication of interrelations among sample set characteristics for ordination problems (GAUCH et al. 1977). Data set characteristics influence ordination but may be difficult to recognize before ordination. Effects of different combinations of these on eigenvector ordinations are relatively unpredictable. Sample error in particular tends to obscure distortion consequent on other sample set characteristics. Given moderate scatter in sample positions because of sample error, the curvilinear distortion in the second axes (Figs. 8-11) and higher axes may be difficult to recognize. When effects of distortion and error are combined with possible effects of clusters, deviants, partial disjunction, and sample equitability, interpretations of axes may be tenuous. In general, consideration of sample set properties and careful use of other information (knowledge of environmental relationships, which may be represented in direct or polar ordinations) are needed for realistic use of eigenvector techniques as tools of research in gradient analysis.

This need for care in interpretation bears on some of the desired properties of ordinations (10.3.2). The different techniques offer different balances of these. PO represents a relatively favorable combination of freedom from distortion, range of applicability, lucidity, and efficiency, but it is short on objectivity. PCA is more objective and more efficient in use of data but weak in other respects. RA combines the objectivity and efficiency of PCA with greater range and freedom from distortion for the first axis, but is subject to difficulties in higher axes. Gaussian ordination (11.5.2) may maximize efficiency in data used and freedom from distortion for a single axis but is limited to a single axis. None of these techniques escapes the need for careful use of supplementary information for the interpretation of axes. None can meet in a simple and automatic way the desired heuristic properties of ordinations – capacity to reveal otherwise unrecognized relations of species and communities to environment, on which further research may be based. The heuristic function is not served by the uncritical production and labeling of PCA axes. The heuristic function may be served instead by thoughtful application of field knowledge to the results of an ordination, or of different ordinations of the same sample set. Particular, unsuspected relationships of interest are most likely to be recognized in a context of other, better known relationships. Rewards of ecological research are likely to be directly proportional to the investigator's investment of time, effort, interest, and understanding, and often to willingness to explore alternative techniques that may supplement and inform one another. Scientific value of ordination research can be inversely related to the degree of uncritical reliance on a given mathematical technique.

Related to this are two concepts that have influenced discussions of ordination: objectivity and mathematical appropriateness (GAUCH et al. 1977). Objectivity refers to the production by a technique of information that is to the least degree influenced by the investigator's choices in applying that technique. PO is subjective in its requirement for endpoint choices, and the eigenvector techniques are attractive because axes are objectively produced by calculations employing the full range of sample data. The contrast is not, however, one-sided. For PCA a range of variants employing different combinations of transformation, standardization, centering, and rotation give different ordination results among which an investigator can choose. In the chosen ordination the significant axes are recognized and interpreted by subjective means, mainly observed correlations and intuitive perceptions of relationships. Supplementary information is needed, and the ideal is not a simple objectivity, but an effective interaction of information and investigator. The process is one of progressive approximation (POORE 1956). Ecological ordination itself may be thought a kind of reciprocal averaging of field information against ecologists' understanding toward the optimal correspondence of these.

Mathematical appropriateness is necessarily determined in part in relation to the curvilinearities of ecological data. For ecological ordination principal components analysis is simplistic. Search for techniques and problems comparable to indirect ordination in other fields leads not to PCA but to multidimensional scaling and parametric mapping. PCA should be used with due sense of both its serviceability and its relative crudity for ecological ordination. RA and PO are in different ways better adapted to the nonlinearity of ecological data, but both also are subject to limitations. Other techniques that may be appropriate to indirect ordination are under test: Gaussian ordination (GAUCH & CHASE 1974, GAUCH et al. 1974, IHM & GROENEWOUD 1975, AUSTIN 1976b), parametric mapping or continuity analysis (SHEPARD & CARROLL 1966, KRUSKAL & CARROLL 1966, NOY-MEIR 1974a, b), and nonmetric multidimensional scaling (KRUSKAL 1964a, b, SHEPARD 1974, FASHAM 1977). Results so far (AUSTIN 1976b, NOY-MEIR & WHITTAKER 1977, article 11) suggest that these will share places in the field with PO, PCA, and RA but will not solve all ordination problems and will not escape the need for careful and informed use of any technique for the ordination of community data.

The question of mathematical appropriateness relates also to the concept of ordination. The term 'ordination,' by its Latin base and derivation from RAMENSKY (1930) and GOODALL (1954a, b), denotes an ordering or arrangement of entities. We have assumed throughout this article that in ecology the arrangement sought is in relation to environmental gradients, or abstract axes that should represent directions of community change in response to environmental gradients (WHITTAKER 1967, 1973, AUSTIN 1976a). (The environmental gradients may include successional or disturbance gradients.) This restricted definition is the tradition among ecologists and the purpose of most ecological investigators for whom ordination is a primary research method. An alternative, broader definition is also current, however, using ordination as a collective term for techniques in which ecological (ORLÓCI 1975, DALE 1975) or taxonomic (SNEATH & SOKAL 1973) entities are arranged. The relation to environmental gradients is not necessarily the goal of ordination in this sense. In ecology the broader definition offers both evident convenience, and problems. The restricted definition of ordination is important for understanding the major role of this approach in field research. It also defines the basis for techniques and criteria by which the efficacy of different ordination techniques can be evaluated and the success of ordinations can be judged. The broad definition has tended to confuse different functions of multivariate arrangements – inquiry into environmental relationships of communities, and expression of similarity between samples as an adjunct to classification (10.4.1.3). Both purposes are justified, but it is sometimes unclear that an ordination by the broad definition serves either purpose well. Without a well-defined purpose the broad definition can, at worst, lead to triviality in ordination: production not of understanding but of publishable figures that serve as a multivariate dressing for ineffective research.

The dual definition has contributed to the controversy regarding PO and PCA. The limitations of linear techniques for ecological ordination were known (GOODALL 1954, BRAY & CURTIS 1957, GROENEWOUD 1965, WHITTAKER 1967). However, the broader definition was attractive as linear techniques for ordination were more extensively considered (ORLÓCI 1966, GITTINS 1969, article 9). Beyond examination of these techniques, which was appropriate, failure to recognize the different implications of the two definitions is suggested by advocacy of linear techniques and Euclidean criteria for ecological ordination in the restricted sense, and criticism of PO for non-Euclidean properties. Ordination in the restricted sense is one of the most important concepts of ecological method, and this concept deserves a term of clear meaning. Multivariate arrangements for other purposes are also justified, but should be so labeled that the restricted meaning of ordination is not assumed when the purpose and choice of technique do not justify that assumption. If 'ordination' can no longer be assumed to have its restricted sense in ecology, the basis and use of a given ordination should be clearly indicated. The two major uses appear to be:

(i) Ecological ordination that seeks to express or reveal relationships of species and communities to environment. 'Ecological' is here used in its Continental sense of relationship to environment, and it is understood that ecological ordination must deal with the curvilinear relations of species populations and community composition to environmental gradients. Direct ecological ordination has also been called environmental ordering and direct gradient analysis, and indirect ecological ordination may also be termed catenation (NOY-MEIR 1974b, 11.4.5.3). Ecological ordinations are dual-space ordinations that seek to relate to one another the variables in two kinds of spaces (habitat, and compositional) with complex, nonlinear relations to one another.

(ii) Taxonomic or taxometric ordination that seeks to express similarities or affinities of entities within a single-space ordination. Correspondence of axes and environmental gradients is not assumed, curvilinearity of data structure is not necessarily a concern, and linear techniques such as PCA may provide an optimal summarization of the relationships (ORLÓCI 1975 and article 9). 'Taxonomic' or 'taxometric' is here used in a general sense (of

Greek *taxis*) relating to any arrangement, division, or classification. Taxometric ordinations are often used as aids to classification but are not necessarily so used, and taxometric ordinations can be used in synecology and other fields as well as for the systematics of organisms. We suggest, however, that in ecology there might be advantage in designating figures representing taxometric ordinations scatter diagrams (article 9) or 'taxograms' to distinguish these from figures representing ecological ordinations.

This article is a study of ecological ordination; our recommendations are not intended for taxometric ordination. This volume is primarily about ecological ordination, but much of the treatment in article 9 is appropriate for taxometric ordination.

In some cases (especially at low beta) a given ordination may serve both functions, but in many cases it will not. Both taxometric and ecological ordinations may be appropriate to a study of community classification; the former may aid in the classification itself, but the latter may provide additional information on environmental relations that is welcome. Ordination in support of classification is one of the links between major approaches in synecology – that through study of gradient relationships, for which ordination is the means, and that through characterization of classes or types, involving classification in the broad sense. It is difficult to frame a general statement on the relationship of these two approaches (see also article 11). Classification is (in some forms at least) less demanding, and it may be the natural basis for preliminary work and for mapping and other practical applications. Ordinations are in many cases more demanding of research effort, but for this effort they may tell much about relationships of communities and species to one another and environment. Penetrating study of some of these relations through classification is possible, but gradient relationships and continuities are often lost sight of when vegetation is divided into classes. Choice may have little to do with the character of the vegetation: Classification may seem the more natural approach to vegetation made mosaic-like by human disturbance, yet ordination is by no means inappropriate to such vegetation; ordination may seem most appropriate for extensively intergrading, undisturbed communities, yet classification of such vegetation is also feasible and usually also desired. For either choice of primary method, the other method may be a valuable complement. Most gradient analyses use classification to summarize some relationships; and many studies based on classification gain by using ordination to interrelate, as parts of a landscape pattern, the community-types of one of the approaches to classification described in the second volume of this book.

10.7 **Summary**

A wide variety of ordination techniques are now available, and evaluation of their relative usefulness is needed. Ecological ordination is the arrangement of community samples (and species) in relation to environmental gradients, or abstract axes that are related to and interpretable in terms of environmental gradients. The community response to environmental gradients is a population continuum or coenocline in one dimension, a population pattern or compositional space in several dimensions. The compositional space of dispersed, overlapping species populations corresponds to a habitat space, the axes of which are complex-gradients. The challenge of indirect ordination is the derivation of arrangements that are realistic and interpretable in relation to habitat space, from measurements affected by the curvilinear and nonmonotonic relationships of species populations in the compositional space.

Ordination techniques are evaluated by testing their relative success in representing known relationships. Effective tests are based on (i) simulated coenoclines or community gradients of Gaussian species populations, (ii) simulated coenoplanes or two-dimensional community patterns of Gaussian species response surfaces, and (iii) field data for which direct gradient analysis has shown species and community relationships to environmental gradients. Ordinations are much affected by characteristics of sample sets – especially beta diversity, sample error, dimensionality, and presence of sample clusters, outliers, and disjunctions – and simulated data sets are varied in these characteristics to determine their effects on different ordination techniques. Relative success of ordination may be determined by plotting sample positions on ordination axes to observe distortion, and by measurements such as mean displacement of samples along the first ordination axis from true positions in the coenocline ordinated.

Three major ordination techniques – polar ordination (PO), reciprocal averaging (RA), and principal components analysis (PCA) – have been subjected to systematic test. Alpha diversity has limited effect and sample error significant effect on these; PCA is most and RA least subject to displacements by sample error. Nonstandardized PCA involutes with severe distortion at beta diversities above 2-3 half-changes, whereas PO and RA are generally free from involution and can successfully ordinate coenoclines of 5 half-changes or more. In PCA and RA a linear coenocline becomes an arch in relation to the second axis and a polynomial curve of increasing order in higher axes, with consequent distortions of sample positions in higher axes. PO is not subject to this distortion. Square coenoplanes up to 5 × 5 half-changes can be ordinated by RA, PO, and standardized PCA; but in RA and PCA rectangular coenoplanes (e.g. 4.5 × 1.5 HC) may be distorted in the second axis while the second (1.5 HC) coenoplane axis appears in the third ordination axis. Sample clusters attract axes of PCA and can increase distortion on a given axis; clusters have less effect on RA and none on PO. Deviant or outlier samples can increase distortion in PCA and RA but have no effect on PO (unless chosen as endpoints). Sample sets with disjunctions or relative discontinuities should generally be divided into subsets for ordination; RA will reveal disjunctions.

PO is a simple and useful technique, relatively well adapted to the nonlinearity of ecological data, but requires subjective choice of endpoints. Percentage similarity seems the best similarity measurement in general for PO; coefficient of community is appropriate for some sample sets; Euclidean distance increases distortion and is less desirable for ordination than percentage similarity. RA has advantage over PO for objective recognition of a first axis of community variation, but is subject to difficulties in higher axes. Nonstandardized PCA is highly vulnerable to distortion; but PCA gives ordinations that are better (but not so good as RA) with species-standardized or doubly standardized data. Because of the general problem of curvilinearity, and the special problems of different sample set characteristics, there is probably no fully satisfactory ordination technique. Careful use of these and other techniques with sample set characteristics in view, thoughtful checking of the axes of indirect ordinations with environmental data and distortion problems in mind, and comparison of different techniques including direct ordination of the same sample set, are suggested.

ACKNOWLEDGEMENT

Contribution from a research project supported by the National Science Foundation. We thank M. P. AUSTIN, M. O. HILL, I. NOY-MEIR, L. ORLÓCI, R. K. PEET, and T. R. WENTWORTH for discussions relating to this work.

REFERENCES

ALLEN, T. F. H., – 1971 – Multivariate approaches to the ecology of algae on terrestrial rock surfaces in North Wales. *J. Ecol.* 59: 803-826.

ANDERSON, A. J. B., – 1971 – Ordination methods in ecology. *J. Ecol.* 59: 713-726.

AUSTIN, M. P., – 1968 – An ordination study of a chalk grassland community. *J. Ecol.* 56: 739–757.

AUSTIN, M. P., – 1972 – Models and analysis of descriptive vegetation data. In 'Mathematical Models in Ecology,' ed. J. N. R. JEFFERS. *Symp. Brit. ecol. Soc.* 12: 61–86.

AUSTIN, M. P., – 1976a – On non-linear species response models in ordination. *Vegetatio* 33: 33–41.

AUSTIN, M. P., – 1976b – Performance of four ordination techniques assuming three different non-linear species response models. *Vegetatio* 33: 43–49.

AUSTIN, M. P., P. S. ASHTON & P. GREIG-SMITH, – 1972 – The application of quantitative methods to vegetation survey. III. A re-examination of rain forest data from Brunei. *J. Ecol.* 60: 305-324.

AUSTIN, M. P. & P. GREIG-SMITH, – 1968 – The application of quantitative methods to vegetation survey. II. Some methodological problems of data from rain forest. *J. Ecol.* 56: 827-844.

AUSTIN, M. P. & I. NOY-MEIR, – 1972 – The problem of non-linearity in ordination: experiments with two-gradient models. *J. Ecol.* 59: 763-773.

AUSTIN, M. P. & O. ORLÓCI, – 1966 – Geometric models in ecology. II. An evaluation of some ordination techniques. *J. Ecol.* 54: 217-227.

BANNISTER, P., – 1966 – The use of subjective estimates of cover-abundance as the basis for ordination. *J. Ecol.* 54: 665-674.

BANNISTER, P., – 1968 – An evaluation of some procedures used in simple ordinations. *J. Ecol.* 56: 27-34.

BEALS, E., – 1960 – Forest bird communities in the Apostle Islands of Wisconsin. *Wilson Bull.* 72: 156-181.

BEALS, E. W., – 1973 – Ordination: mathematical elegance and ecological naïveté. *J. Ecol.* 61: 23–35.

BENZÉCRI, J. P., – 1969 – Statistical analysis as a tool to make patterns emerge from data. In 'Methodologies of Pattern Recognition,' ed. S. WATONABE, pp. 35-74. Academic Press, London.

BOUXIN, G., – 1975 – Ordination and classification in the savanna vegetation of the Akagera Park (Rwanda, Central Africa). (French summ.) *Vegetatio* 29: 155-167.

BRAY, J. R. & J. T. CURTIS, – 1957 – An ordination of the upland forest communities of southern Wisconsin. *Ecol. Monogr.* 27: 325-349.

BROWN, R. T. & J. T. CURTIS, – 1952 – The upland conifer-hardwood forests of northern Wisconsin. *Ecol. Monogr.* 22: 217-234.

CAIN, S. A. & G. M. DE OLIVEIRA CASTRO, – 1959 – Manual of Vegetation Analysis. Harper, New York. 325 pp.

CASSIE, R. M., – 1969 – Multivariate analysis in ecology. *Proc. N.Z. ecol. Soc.* 16: 53-57.

CHARDY, P., M. GLEMAREC & A. LAUREC, – 1976 – Application of inertia methods to benthic marine ecology: practical implications of the basic options. *Estuar. coast. mar. Sci.* 4: 179-205.

CRAWFORD, R. M. M. & D. WISHART, – 1967 – A rapid multivariate method for the detection and classification of groups of ecologically related species. *J. Ecol.* 55: 505-524.

CRAWFORD, R. M. M. & D. WISHART, – 1968 – A rapid classification and ordination method and its application to vegetation mapping. *J. Ecol.* 56: 385-404.

Crawford, R. M. M., D. Wishart & R. M. Campbell, – 1970 – A numerical analysis of high altitude scrub vegetation in relation to soil erosion in the Eastern Cordillera of Peru. *J. Ecol.* 58: 173-191.

Curtis, J. T. & R. P. McIntosh, – 1951 – An upland forest continuum in the prairie-forest border region of Wisconsin. *Ecology* 32: 476-496.

Dagnelie, P., – 1960 – Contribution à l'étude des communautés végétales par l'analyse factorielle. (Engl. summ.) *Bull. Serv. Carte phytogéogr.*, Paris, Sér. B, 5: 7-71, 93-195.

Dagnelie, P., – 1962 – L'étude des communautés végétales par l'analyse des liaisons entre les espèces et les variables écologiques. Inst. Agron. de l'Etat, Gembloux. 135 pp.

Dale, M. B., – 1975 – On objectives of methods of ordination. *Vegetatio* 30: 15-32.

Ebeling, A. W., R. M. Ibara, R. J. Lavenberg & F. J. Rohlf, – 1970 – Ecological groups of deep-sea animals off southern California. *Bull. Los Angeles County Mus. Nat. Hist.* 6: 1-43.

Ellenberg, H., ed., – 1967 – Vegetations- und bodenkundliche Methoden der forstlichen Standortskartierung: Ergebnisse eines internationalen Methodenvergleichs im Schweitzer Mittelland. (French & Engl. Summs.) *Veröff. geobot. Inst. ETH, Stiftg, Rübel*, Zürich 39: 1-296.

Fasham, M. J. R., – 1977 – A comparison of nonmetric multidimensional scaling, principal components and reciprocal averaging for the ordination of simulated coenoclines, and coenoplanes. *Ecology* 58: 551-561.

Flenley, J. R., – 1969 – The vegetation of the Wabag region, New Guinea Highlands: a numerical study. *J. Ecol.* 57: 465-490.

Fonda, R. W. & J. A. Bernardi, – 1976 – Vegetation of Sucia Island in Puget Sound, Washington. *Bull. Torrey bot. Club* 103: 99-109.

Frey, T. & H. van Groenewoud, – 1972 – A cluster analysis of the D^2 matrix of white spruce stands in Saskatchewan based on the maximum-minimum principle. *J. Ecol.* 60: 873-886.

Frydman, I. & R. H. Whittaker, – 1968 – Forest associations of southeast Lublin Province, Poland. (Germ. summ.). *Ecology* 49: 896-908.

Gauch, H. G. Jr., – 1973a – The relationship between sample similarity and ecological distance. *Ecology* 54: 618-622.

Gauch, H. G. Jr., – 1973b – A quantitative evaluation of the Bray-Curtis ordination. *Ecology* 54: 829-836.

Gauch, H. G. Jr. & G. B. Chase, – 1974 – Fitting the Gaussian curve to ecological data. *Ecology* 55: 1377-1381.

Gauch, H. G. Jr., G. B. Chase & R. H. Whittaker, – 1974 – Ordination of vegetation samples by Gaussian species distributions. *Ecology* 55: 1382-1390.

Gauch, H. G. Jr. & T. R. Wentworth, – 1976 – Canonical correlation analysis as an ordination technique. *Vegetatio* 33: 17-22.

Gauch, H. G. Jr. & R. H. Whittaker, – 1972a – Coenocline simulation. *Ecology* 53: 446-451.

Gauch, H. G. Jr. & R. H. Whittaker, – 1972b – Comparison of ordination techniques. *Ecology* 53: 868-875.

Gauch, H. G. Jr. & R. H. Whittaker, – 1976 – Simulation of community patterns. *Vegetatio* 33: 13-16.

Gauch, H. G., R. H. Whittaker & T. R. Wentworth, – 1977 – A comparative study of reciprocal averaging and other ordination techniques. *J. Ecol.* 65: 157-174.

Gimingham, C. H., N. M. Pritchard & R. M. Cormack, – 1966 – Interpretation of a vegetational mosaic on limestone in the island of Gotland. *J. Ecol.* 54: 481-502.

GITTINS, R., – 1965 – Multivariate approaches to a limestone grassland community. III. A comparative study of ordination and association-analysis. *J. Ecol.* 53: 411-425.
GITTINS, R., – 1968 – Trend-surface analysis of ecological data. *J. Ecol.* 56: 845-869.
GITTINS, R., – 1969 – The application of ordination techniques. In 'Ecological Aspects of the Mineral Nutrition of Plants,' ed. I. H. RORISON. *Symp. brit. Ecol. Soc.*, 1968, 9: 37-66.
GOFF, F. G. & G. COTTAM, – 1967 – Gradient analysis: the use of species and synthetic indices. *Ecology* 48: 793-806.
GOLDSMITH, F. B., – 1973 – The vegetation of exposed sea cliffs at South Stack, Anglesey. I. The multivariate approach. *J. Ecol.* 61: 787-829.
GOLDSTEIN, R. A. & D. F. GRIGAL, – 1972 – Definition of vegetation structure by canonical analysis. *J. Ecol.* 60: 277-284.
GOODALL, D. W., – 1954a – Objective methods for the classification of vegetation. III. An essay in the use of factor analysis. *Aust. J. Bot.* 2: 304-324.
GOODALL, D. W., – 1954b – Vegetational classification and vegetational continua. (Germ. summ.) *Angew. PflSoziol.*, Wien, Festschr. Aichinger 1: 168-182.
GOWER, J. C., – 1966 – Some distance properties of latent root and vector methods used in multivariate analysis. *Biometrika* 53: 325-338.
GOWER, J. C., – 1969 – Multivariate analysis and multidimensional geometry. *Statistician* 17: 13-28.
GREIG-SMITH, P., – 1964 – Quantitative Plant Ecology. 2nd ed. Butterworths, London. 256 pp.
GREIG-SMITH, P., – 1971a – Analysis of vegetation data: the user viewpoint. In 'Statistical Ecology,' ed. G. P. PATIL, E. C. PIELOU & W. E. WATERS, 3: 149-166. Pennsylvania State University, University Park, Pa.
GREIG-SMITH, P., – 1971b – Application of numerical methods to tropical forests. In 'Statistical Ecology,' ed. G. P. PATIL, E. C. PIELOU & W. E. WATERS, 3: 195-206. Pennsylvania State University, University Park, Pa.
GREIG-SMITH, P., M. P. AUSTIN & T. C. WHITMORE, – 1967 – The application of quantitative methods to vegetation survey. I. Association-analysis and principal component ordination of rain forest. *J. Ecol.* 55: 483-503.
GRIGAL, D. F. & R. A. GOLDSTEIN, – 1971 – An integrated ordination-classification analysis of an intensively sampled oak-hickory forest. *J. Ecol.* 59: 481-492.
GROENEWOUD, H. VAN, – 1965 – Ordination and classification of Swiss and Canadian coniferous forests by various biometric and other methods. (Germ. summ.) *Ber. geobot. Inst. ETH, Stiftg. Rübel,* Zürich, 1964, 36: 28-102.
GUINOCHET, M., – 1973 – Phytosociologie. Masson, Paris. 227 pp.
HILL, M. O., – 1973 – Reciprocal averaging: an eigenvector method of ordination. *J. Ecol.* 61: 237-249.
HILL, M. O., – 1974 – Correspondence analysis: a neglected multivariate method. *J. R. statist. Soc.*, Ser. C, 23: 340-354.
HILL, M. O., – 1976 – Correspondence analysis and habitat width. (manuscript).
HILL, M. O., R. G. H. BUNCE & M. W. SHAW, – 1975 – Indicator species analysis, a divisive polythetic method of classification, and its application to a survey of native pinewoods in Scotland. *J. Ecol.* 63: 597-613.
HINNERI, S., – 1972 – An ecological monograph on eutrophic deciduous woods in the SW archipelago of Finland. *Annls Univ. åbo (turku)*, (*Turun Yliop. Julk.*), Ser. A, II (Biol.-Geog.-Geol.) 50: 1-131.
HOWARD-WILLIAMS, C. & B. H. WALKER, – 1974 – The vegetation at a tropical African lake: classification and ordination of the vegetation of Lake Chilwa (Malawi), *J. Ecol.* 62: 831-854.

JGHES, R. N., D. L. PEER & K. H. MANN, – 1972 – Use of multivariate analysis to identify functional components of the benthos in St. Margaret's Bay, Nova Scotia. *Limnol. Oceanogr.* 17: 111-121.

M, P. & H. VAN GROENEWOUD, – 1975 – A multivariate ordering of vegetation data based on Gaussian type gradient response curves. *J. Ecol.* 63: 767-777.

MES, F. C., – 1971 – Ordinations of habitat relationships among breeding birds. *Wilson Bull.* 83: 215-236.

NSSEN, J. G. M., – 1972 – Detection of some micropatterns of winter annuals in pioneer communities of dry sandy soils. *Acta bot. neerl.* 21: 609-616.

LUM, J. K., C. F. WEHRHAHN & J. M. A. SWAN, – 1971 – Comparisons of environmental ordinations with principal component vegetational ordinations for sets of data having different degrees of complexity. (French summ.) *Can. J. For. Res.* 1: 99-112.

RSHAW, K. A., – 1968 – Classification and ordination of Nigerian savanna vegetation. *J. Ecol.* 56: 467-482.

SSELL, S. R. & R. H. WHITTAKER, – 1976 – Comparisons of three ordination techniques. *Vegetatio* 32: 21-29.

IIGHT, D. H., – 1965 – A gradient analysis of Wisconsin prairie vegetation on the basis of plant structure and function. *Ecology* 46: 744-747.

IIGHT, D. H. & O. L. LOUCKS, – 1969 – A quantitative analysis of Wisconsin forest vegetation on the basis of plant function and gross morphology. *Ecology* 50: 219-234.

RUSKAL, J. B., – 1964a – Multidimensional scaling by optimizing goodness of fit to a nonmetric hypothesis. *Psychometrika* 29: 1-27.

RUSKAL, J. B., – 1964b – Nonmetric multidimensional scaling: a numerical method. *Psychometrika* 29: 115-129.

RUSKAL, J. B. & J. D. CARROLL, – 1969 – Geometric models and badness-of-fit functions. In 'Multivariate Analysis,' ed. P. R. KRISHNAIAH 2: 639-671. Academic Press, New York.

UCKS, O. L., – 1962 – Ordinating forest communities by means of environmental scalars and phytosociological indices. *Ecol. Monogr.* 32: 137-166.

AAREL, E. VAN DER, 1969 – On the use of ordination models in phytosociology. (Germ. summ.) *Vegetatio* 32: 21-29.

CNAUGHTON, S. J. & L. L. WOLF, – 1970 – Dominance and the niche in ecological systems. *Science, N.Y.* 167: 131-139.

ITCHELL, R. & W. MOIR, – 1976 – Vegetation of the Abbott Creek Research Natural Area, Oregon. *NW Sci.* 50: 42-58.

ONK, C. D., – 1965 – Southern mixed hardwood forest of northcentral Florida. *Ecol. Monogr.* 35: 335-354.

OORE, J. J., P. FITZSIMONS, E. LAMBE & J. WHITE, – 1970 – A comparison and evaluation of some phytosociological techniques. (Germ. summ.) *Vegetatio* 20: 1-20.

UELLER-DOMBOIS, D. & H. ELLENBERG, – 1974 – Aims and Methods of Vegetation Ecology. Wiley, New York. 547 pp.

ORRIS, J. M. & J. P. BARKHAM, – 1970 – A comparison of some Cotswold beachwoods using multiple-discriminant analysis. *J. Ecol.* 58: 603-619.

OY-MEIR, I., – 1973a – Data transformations in ecological ordination. I. Some advantages of non-centering. *J. Ecol.* 61: 329-341.

OY-MEIR, I., – 1973b – Divisive polythetic classification of vegetation by optimized division on ordination components. *J. Ecol.* 61: 753-760.

OY-MEIR, I., – 1974a – Multivariate analysis of the semiarid vegetation in south-eastern Australia. II. Vegetation catenae and environmental gradients. *Aust. J. Bot.* 22: 115-140.

Noy-Meir, I., – 1974b – Catenation: quantitative methods for the definition of coenoclines. *Vegetatio* 29: 89-99.

Noy-Meir, I. & M. P. Austin, – 1970 – Principal component ordination and simulated vegetational data. *Ecology* 51: 551-552.

Noy-Meir, I., D. Walker & W. T. Williams, – 1975 – Data transformation in ecological ordination. II. On the meaning of data standardization. *J. Ecol.* 63: 779-800.

Noy-Meir, I. & R. H. Whittaker, – 1977 – Continuous multivariate methods in community analysis: some problems and developments. *Vegetatio* 33: 79-98.

Orloci, L., – 1966 – Geometric models in ecology. I. The theory and application of some ordination methods. *J. Ecol.* 54: 193-215.

Orloci, L., – 1967 – An agglomerative method for classification of plant communities. *J. Ecol.* 55: 193-206.

Orlóci, L., – 1974a – Revisions for the Bray and Curtis ordination. (French summ.) *Can. J. Bot.* 52: 1773-1776.

Orlóci, L., – 1974b – On information flow in ordination. *Vegetatio* 29: 11-16.

Orlóci, L., – 1975 – Multivariate Analysis in Vegetation Research. Junk, The Hague. 276 pp.

Orlóci L. & M. M. Mukkattu, – 1973 – The effect of species number and type of data on the resemblance structure of a phytosociological collection. *J. Ecol.* 61: 37-46.

Pakarinen, P., – 1976 – Agglomerative clustering and factor analysis of south Finnish mire types. *Ann. bot. fenn.* 13: 35-41.

Peet, R. K., – 1975 – Forest vegetation of the east slope of the Colorado Front Range. Ph.D. Thesis, Cornell University.

Pielou, E. C., – 1969 – An Introduction to Mathematical Ecology. Wiley-Interscience, New York. 286 pp.

Poore, M. E. D., – 1956 – The use of phytosociological methods in ecological investigations. IV. General discussion of phytosociological problems. *J. Ecol.* 44: 28-50.

Ramensky, L. G., – 1930 – Zur Methodik der vergleichenden Bearbeitung und Ordnung von Pflanzenlisten und anderen Objekten, die durch mehrere, verschiedenartig wirkende Faktoren bestimmt werden. *Beitr. Biol. Pfl.* 18: 269-304.

Ramsay, D. McC. & P. N. de Leeuw, – 1965 – An analysis of Nigerian savanna. IV. Ordination of vegetation developed on different parent materials. *J. Ecol.* 53: 661-677.

Ritter, J., – 1972 – Les groupements végétaux des étages subalpin et alpin du Vercors méridional. Essai d'interprétation statistique. *Vegetatio* 24: 313-403.

Rogers, D. J., – 1970 – A preliminary ordination study of forest vegetation in the Kirchleerau area of the Swiss Midland. (Germ. summ.) *Ber. geobot. Inst. ETH, Stiftg Rübel*, Zürich, 1969, 40: 28-78.

Scruggs, W. M., – 1976 – Revised coordinate determinations for the Bray and Curtis ordination. *Ecology* (in press).

Shepard, R. N., – 1974 – Representation of structure in similarity data: problems and prospects. *Psychometrika* 39: 373-421.

Shepard, R. N. & J. D. Carroll, – 1966 – Parametric representation of nonlinear data structures. In 'Multivariate Analysis,' ed. P. R. Krishnaiah 1: 561-592. Academic Press, New York.

Shimwell, D. W., – 1971 – The Description and Classification of Vegetation. Sidgwick & Jackson, London. 322 pp.

Singh, T. & N. E. West, – 1971 – Comparison of some multivariate analyses of perennial Atriplex vegetation in southeastern Utah. (Germ. summ.) *Vegetatio* 23: 289-313.

SMITH, R. W., – 1976 – Numerical analysis of ecological survey data. Ph.D. Thesis, Univ. Southern California.
SNEATH, P. H. A. & R. R. SOKAL, – 1973 – Numerical Taxonomy: The Principles and Practice of Numerical Classification. Freeman, San Francisco. 573 pp.
SWAN, J. M. A., – 1970 – An examination of some ordination problems by use of simulated vegetational data. *Ecology* 51: 89-102.
SWAN, J. M. A. & R. L. DIX, – 1966 – The phytosociological structure of upland forest at Candle Lake, Saskatchewan. *J. Ecol.* 54: 13-40.
SWAN, J. M. A., R. L. DIX & C. F. WEHRHAHN, – 1969 – An ordination technique based on the best possible stand-defined axes and its application to vegetational analysis. *Ecology* 50: 206-212.
TALLIS, J. H., – 1969 – The blanket bog vegetation of the Berwyn Mountains, North Wales. *J. Ecol.* 57: 765-787.
WALI, M. K. & V. J. KRAJINA, – 1973 – Vegetation-environment relationships of some sub-boreal spruce zone ecosystems in British Columbia. *Vegetatio* 26: 237-381.
WARD, S. D., – 1971 – The phytosociology of *Calluna-Arctostaphylos* heaths in Scotland and Scandinavia. III. A critical examination of the Arctostaphyleto-Callunetum. *J. Ecol.* 59: 697-712.
WEBB, L. J., J. G. TRACEY, W. T. WILLIAMS & G. N. LANCE, – 1967 – Studies in the numerical analysis of complex rain-forest communities. I. A comparison of methods applicable to site/species data. *J. Ecol.* 55: 171-191.
WENTWORTH, T. R., – 1976 – The vegetation of limestone and granite soils in the mountains of southeastern Arizona. Ph.D. Thesis, Cornell University.
WERGER, M. J. A., – 1973 – On the use of association-analysis and principal component analysis in interpreting a Braun-Blanquet phytosociological table of a Dutch grassland. (Germ. summ.) *Vegetatio* 28: 129-144.
WESTMAN, W. E., – 1971 – Production, nutrient circulation, and vegetation-soil relations of the pygmy forest region of Northern California. Ph.D. Thesis, Cornell Univ., Ithaca, N.Y.
WESTMAN, W. E., – 1975 – Edaphic climax pattern of the pygmy forest region of California. *Ecol. Monogr.* 45: 109-135.
WHITTAKER, R. H., – 1956 – Vegetation of the Great Smoky Mountains. *Ecol. Monogr.* 26: 1-80.
WHITTAKER, R. H., – 1960 – Vegetation of the Siskiyou Mountains, Oregon and California. *Ecol. Monogr.* 30: 279-338.
WHITTAKER, R. H., – 1967 – Gradient analysis of vegetation. *Biol. Rev.* 42: 207-264.
WHITTAKER, R. H., – 1972 – Convergences of ordination and classification. (Germ. summ.) In 'Grundfragen und Methoden in der Pflanzensoziologie,' ed. E. VAN DER MAAREL & R. TÜXEN. *Ber. Symp. int. Ver. Vegetationskunde*, Rinteln 1970: 39-57.
WHITTAKER, R. H. & H. G. GAUCH, Jr., – 1973 – Evaluation of ordination techniques. (Germ. summ.) In 'Ordination and Classification of Communities,' ed. R. H. WHITTAKER. *Handb. Veget. Sci.* 5: 287-321.
WHITTAKER, R. H., S. A. LEVIN & R. B. ROOT, – 1973 – Niche, habitat, and ecotope. *Am. Nat.* 107: 321-338.
WHITTAKER, R. H. & W. A. NIERING, – 1964 – Vegetation of the Santa Catalina Mountains, Arizona. I. Ecological classification and distribution of species. *J. Ariz. Acad. Sci.* 3: 9-34.
WHITTAKER, R. H. & W. A. NIERING, – 1965 – Vegetation of the Santa Catalina Mountains, Arizona (II). A gradient analysis of the south slope. *Ecology* 46: 429-452.

Whittaker, R. H. & W. A. Niering, – 1968 – Vegetation of the Santa Catalina Mountains, Arizona. IV. Limestone and acid soils. *J. Ecol.* 56: 523-544.
Yarranton, G. A., – 1967 – A quantitative study of the bryophyte and macrolichen vegetation of the Dartmoor granite. *Lichenologist* 3: 392-408.

11 RECENT DEVELOPMENTS IN CONTINUOUS MULTIVARIATE TECHNIQUES

IMANUEL NOY-MEIR and ROBERT H. WHITTAKER

Contents *Page*

11 RECENT DEVELOPMENTS IN CONTINUOUS MULTIVARIATE TECHNIQUES

11.1 Introduction

Our purpose is to consider the present state of multivariate techniques for description and interpretation of communities in terms of continuous variables. Mainly, we attempt to define some major problems of concept and method that came to light in the period from 1965 to 1970, and to examine the contribution of recent developments towards solution of these problems.

Multivariate techniques, both discrete and continuous, are applied by ecologists to a community composition matrix with scores for species in samples or sites, in the hope that these techniques will facilitate analysis, description, and interpretation of a structure that is implicit in the data but not easily recognized in the matrix. Most classificatory techniques assign samples and species to units that are assumed to be discrete or discontinuous with one another. By contrast, all the methods discussed here arrange samples and species in relation to one or several continuous variables (axes, components, dimensions, gradients). By this process, the compositional similarities of samples and species are represented in a simplified and condensed form. This representation facilitates the identification and interpretation of the main trends or patterns of variation, which is the purpose of multivariate analysis in the most general terms (Orlóci 1975).

Two major types of patterns are of interest, and accordingly alternative major objectives of the analysis can be distinguished (cf. Dale 1975):

(i) Identification and definition of sequences or directions of community compositional variation (coenoclines, Whittaker 1960, 1967), that can be directly related to gradients of environmental factors (gradient analysis);

(ii) Identification and typification of groups of related samples and species (without necessarily drawing discrete boundaries between them). Some users of continuous multivariate techniques have explicitly emphasized one or the other objective. Others have done so implicitly, or have not clearly distinguished between the two objectives. Some of the controversy about the relative merits of different techniques may reflect implicit differences in objectives. Confusion has resulted also from different definitions of 'ordination' (Dale 1975, article 10.6): (i) a restricted sense, for the arrangement of ecological samples (or species) in relation to one or more environmental gradients, or in relation to abstract axes that should represent major directions of community change in response to environment (WHITTAKER 1967, 1973, AUSTIN 1976a, b), or (ii) a broader sense, as a collective term for multivariate techniques by which samples or other entities are arranged and metric information extracted (ORLOCI 1973, NOY-MEIR 1973a, b, DALE 1975). In this review we use 'ordination' in its restricted sense in the context of gradient analysis.

11.2 A Brief History

Methods of classification have for a long time served as convenient and conceptually simple tools for the description of community variation. But any ecologist who tries to classify vegetation in the field can hardly fail to notice the limitations and difficulties that are inherent in the discrete structure (of boxlike, discontinuous sets) assumed in most hierarchical classifications. The need for adequate representation of continuous aspects of community variation began to be felt in the 1920's (e.g. GLEASON 1926, RAMENSKY 1926; reviews by WHITTAKER 1962, 1967), and first attempts were made to develop techniques for this purpose (RAMENSKY 1930; see article 6). Early studies of continuous change in plant communities along environmental gradients were carried out by MØLHOLM HANSEN (HANSEN 1930, 1932). The development and acceptance of these approaches was slow at first, perhaps because both the concepts and the techniques required by continuous models are somewhat more sophisticated than those of discrete models. Another reason may have been the dominant position of the great classificatory schools in the phytosociological establishment.

A new era of multivariate treatment was pioneered by TUOMIKOSKI (1942), who grouped species, and then ordered samples, on the basis of between-species correlations. Techniques based on continuous models gained momentum thereafter, when several quantitative methods were developed independently by plant ecologists: use of similarity measures for community classification (SØRENSEN 1948), matrix and plexus arrangement (MOTYKA 1947, MATUSZKIEWICZ 1948, VRIES 1953, FALINKSI 1960, article 6), direct gradient analysis (WHITTAKER 1951, 1952, 1956; also ELLENBERG 1950, 1952, 1956, article 2), compositional gradient analysis (CURTIS & MCINTOSH 1951, BROWN & CURTIS 1952, article 7), and polar ordination (BRAY & CURTIS 1957, LOUCKS 1962, articles 7, 10). The use of quantitative measures and objective procedures in these techniques was a major innovation in community ecology.

But the objective procedures usually did not cover all stages of the process; they were supplemented by informal and more or less intuitive procedures. The initial success of the new techniques in clarifying community variation and its relation to environment led to rapidly expanding application, as well as to further work on their improvement; in particular, several modifications of polar ordination (PO) were suggested in efforts to make it more objective.

At about the same time ecologists realized that some techniques of multivariate analysis that had been developed in other sciences were relevant to community description (GOODALL 1954). Particularly promising was the family of eigenvector methods, primarily principal component analysis (PCA) and factor analysis [later also canonical correlation analysis (AUSTIN 1968, 1972) and multiple discriminant (canonical variate) analysis (NORRIS & BARKHAM 1970)]. These methods were characterized by a higher level of formality and mathematical sophistication than the ordination techniques developed by ecologists; and their use was facilitated by the development of computers. The first applications to ecological data gave promising results (GOODALL 1954, DAGNELIE 1960, CASSIE 1963, GROENEWOUD 1965). The relation of these techniques to the concepts of Euclidean spaces (with samples as points in a space of species as dimensions, or vice versa) and to geometrical operations in these spaces was clarified (GOODALL 1963, ORLÓCI 1966, 1975, GITTINS 1969). An extension to accommodate non-Euclidean measures was also suggested (principal coordinate analysis, GOWER 1966). Since 1966, principal component analysis was increasingly applied to vegetation data. At some stage it may have seemed to be the definitive solution to the problem of community analysis.

However, between 1965 and 1970, as continuous multivariate methods (and particularly PCA and PO) gained ever-wider application in ecology, some methodological problems caused increasing concern and controversy among their users. In the following section we attempt to define these problems as they appeared then.

11.3 Some Emerging Problems

11.3.1 Axis Definition

For some time, a kind of contest between polar ordination (PO) and principal component analysis (PCA) appeared in the ecological literature. A major difference between them is that in the former the direction of the ordination axes is defined by pairs of most different endpoint samples, while in the latter it is defined to account maximally for differences between all samples. Proponents of PCA claimed that the reliance on extreme stands in PO tended to lead to loss of information and to distortion (Orlóci 1966, Austin & Orlóci 1966, Anderson 1971) and felt that PO was at best an approximation, now superseded by PCA. The defendants of PO disputed this, stressing that PO was a simpler and clearer technique for community data and that it gave less distorted, and more realistically interpretable, results than PCA (Gauch & Whittaker 1972b, Beals 1973).

11.3.2 The R/Q Problem

Another division among ordinations was between *R*-methods, which derive axes from a matrix of similarities between species (e.g. factor analysis as normally used), and *Q*-methods, which derive them from a matrix of similarities between samples (PO, Orlóci's 1966 technique, and many applications of PCA). In *R*-methods the direct result is an ordination of species, from which an ordination of samples can be computed secondarily; in *Q*-methods the latter is obtained directly. The relative merits of these two apparent alternatives were a subject of discussion (e.g. Dagnelie 1960, Greig-Smith 1964). The issue was confused because it was not always clear whether *R*/*Q* referred to the matrix used (between species/between sites), to the standardization applied (by species/by site), or to the ordination obtained (of species/of sites) (see Ivimey-Cook et al. 1969).

11.3.3 Similarity Measures and Data Transformations

These can be considered together, for any transformation of the original data is equivalent to a redefinition of the similarity or distance measure. The two most common types of transformation are centering (Dagnelie 1960) and standardization (Austin & Greig-Smith 1968). Centering is the subtraction of the mean for the species or the sample. Standardizations are divisions by totals, norms ($\sqrt{\sum x^2}$), maxima, standard deviations, or ranges for a given species in all samples, or for all species in a given sample, or (double standardization) for both. Relativization is standardization by sample totals so that the scores in a sample total 1 or 100 % (Gauch & Whittaker 1972b); normalization is standardization by norms so that the sum of the squares of the scores for a sample (or a species) equals 1 (Orlóci 1967a).

The conventional *R*-measures were between-species covariance (Euclidean, centered by species mean, not standardized) and correlation (Euclidean, centered by species and standardized by species standard deviation). DAGNELIE (1960, article 8) tried several non-centered and differently standardized measures, but finally preferred the correlation coefficient. In *Q*-analyses a much wider range of between-site measures was used: Euclidean (ORLÓCI 1966) and non-Euclidean (BRAY & CURTIS 1957), without standardization, or standardized by sample, by species, or both. Systematic comparisons showed that quite different configurations (and hence interpretations) could result from analysis of the same data with different data transformations and similarity measures (DAGNELIE 1960, AUSTIN & ORLÓCI 1966, GOFF & COTTAM 1967, WEBB et al. 1967, AUSTIN & GREIG-SMITH 1968). This raised the question of which was the best measure to use for ecological data. In particular, this became another point of argument between PO-followers and PCA-followers (ORLÓCI 1966, 1974a, BANNISTER 1968, GAUCH & WHITTAKER 1972b, article 9), as the traditional BRAY & CURTIS package for PO used a particular non-Euclidean measure and double standardization (first by species, then by sample), while conventional PCA used a Euclidean measure, centered by species and unstandardized (or standardized by species) (articles 7.3.1 and 9.7.1.2). Alternatively, the question was put in a more general form: how to choose a similarity measure and transformation in relation to the nature of the data and the purpose of the analysis (AUSTIN & GREIG-SMITH 1968). The multiplicity of possibilities, among which the user had to choose subjectively, caused some ecologists to wonder how objective the objective methods really were.

11.3.4 CONTINUITY AND DISCONTINUITY

The use of continuous multivariate methods was often associated with a belief in the essentially continuous nature of community variation. However, it could hardly be denied that marked clustering and discontinuity in composition did sometimes occur in data sets from real vegetation. Were continuous methods applicable to the analysis and representation of compositional variation in such situations, or were they doomed to failure? Several ecologists (WILLIAMS & LAMBERT 1961, GREIG-SMITH 1964, 1971) reasoned that if the sample consisted (for instance) of two discontinuous clusters, the first axis would separate them, but on subsequent axes the main directions of variation within each cluster (presumably different) would be confounded with each other. DALE (1964) studied in detail the problem of dealing with discontinuity or 'between axis heterogeneity' and in general found conventional multivariate methods (in particular factor analysis) to be quite ineffective. It thus seemed that ordination could be usefully applied only after an initial classification had removed major discontinuities in the data (GREIG-SMITH 1964, LAMBERT & DALE 1964). This limitation might set ordination at a disadvantage compared with classification as a generally useful method of data analysis.

11.3.5 NONLINEARITY AND AXIS INTERPRETATION

In general, users of ordination expected to interpret each of the resulting axes directly as a compositional gradient (coenocline) corresponding to an environmental gradient. This interpretation involves some assumptions that are problematic with real vegetation data.

A first problem arose because the axes resulting from indirect ordinations are usually orthogonal by definition, whereas the effects of environmental factors on populations are often not independent, and the factors themselves may be correlated in the sample set. In principle, these problems may be solved by rotating the axes, if necessary to oblique (non-orthogonal) positions (DAGNELIE 1960, AART 1973) and by interpreting axes also in terms of factor-complexes rather than single factors.

A second problem was the fact that most techniques for indirect ordination assumed a linear relation between the axes and the original species variables. By contrast, the abundances of species are known to be related to environmental gradients by functions which are nonlinear, often also nonmonotonic and with zero values over part of the range (WHITTAKER 1967, articles 2, 10). Thus a compositional gradient (coenocline) which is closely related to a single environmental gradient (over a range that is not very narrow) will be mapped by the linear transformations of ordination into a nonlinear, multidimensional configuration in ordination space. This was appreciated by some of the first users of indirect ordination in ecology (GOODALL 1954, GROENEWOUD 1965; see also WHITTAKER 1967), but was often forgotten in the following rush of applications. Published ordinations of real vegetation data included many curvilinear configurations of points and some probably spurious interpretations of axes. The significance of these results was not generally understood, however, until it was shown that similar configurations resulted from PO and PCA ordinations of perfect artificial gradient data (SWAN 1970, NOY-MEIR & AUSTIN 1970, AUSTIN & NOY-MEIR 1971, AUSTIN 1972, GAUCH & WHITTAKER 1972a, LAFRANCE 1972, GROENEWOUD 1973). Problems of distortion and ineffectiveness for ordination affect not only PCA but other multivariate techniques: factor analysis (WESTMAN 1975), canonical correlation analysis (AUSTIN 1968, GAUCH & WENTWORTH

1976, Wentworth 1976), and discriminant function analysis (Kessell & Whittaker 1976), the last of which is not clearly appropriate for ordination but gives distance measures that can be used for ordination (James 1971, Norris & Barkham 1970, Green 1974).

Once the problem had been clearly defined, several remedies seemed possible (Austin & Noy-Meir 1971, article 10.3.1):
(a) To limit the use of linear indirect ordination techniques to sample sets in which the range of environmental and compositional variation was narrow enough for species/factor responses to be nearly linear, and to use classification or direct ordination for wider ranges. The problem then remains how to define and recognize the sufficiently narrow range (Groenewoud 1973).
(b) To find data transformations which would correct for linear relations and minimize the curvilinear distortion of coenoclines in ordination space (Swan 1970).
(c) To develop methods of ordination which would not be restricted to linear transformations and which could detect and unfold also nonlinear trends in compositional data. Compositional gradient analysis (Curtis & McIntosh 1951) was recognized to be a simple nonlinear ordination (Austin & Noy-Meir 1971), but more objective methods seemed desirable.
(d) To give up applying linear ordination techniques to species composition of samples for the detection of coenoclines and community patterns (but possibly to continue to use them for other purposes). Linear techniques might still be appropriate for variables that are monotonic over broader ranges than species scores, variables such as environmental factors, community structure (James 1971), and composition in terms of higher-level taxa (Maarel 1972, article 20.9.2).

11.4 **Recent Developments Towards Solutions**

11.4.1 Axis Definition

The original polar ordination method of Bray & Curtis (1957) was quite different from PCA in the criteria defining the axes. Several techniques presented either as improvements of PO or as simplifications of PCA (Table I) have sought to bridge this gap.

Beals (1960) replaced the graphical technique of projecting points onto axes in PO by an equivalent mathematical formula (article 7.3.1, see also 'importance projection,' Gauch & Whittaker 1972b). Goff & Cottam (1967), in a study on polar ordination of species, modified the endpoint criterion. They also

Table I. Comparison between axis selection techniques in Bray and Curtis polar ordination, principal component analysis, and some intermediate methods.

Method	Reference	Number of Stands Through Which Axis has to Pass	Criterion to be Maximized	Axis Orthogonality Required
polar ordination importance projection	Bray & Curtis 1957 Gauch & Whittaker 1972b	two	pair dissimilarity (or known environmental contrast)	no
polar ordination (modified)	Swan & Dix 1966	two	pair dissimilarity (+similarity with other stands)	no
simple ordination	Orlóci 1966	first axis two, further axes one	pair dissimilarity and sample deviance (distance from centroid)	yes
principal axes ordination	Maarel 1969	two	negative correlation tendency	no
optimized polar ordination	Swan et al. 1969	first axis two, further axes one	sum of squared projections	yes
position vector ordination	Orlóci 1966	one	sum of squared projections	yes
principal component analysis	Groenewoud 1965	none	sum of squared projections	yes

suggested a new iterative method (similar to reciprocal averaging, HILL 1973, see 10.4.4.3) for fixing the position of points along the axis, which could correct the initial choice of endpoints. ORLÓCI's (1966, see also 1974b) 'simple ordination' method made the second (and subsequent) axis orthogonal to the first, and defined it by the single sample farthest from the first axis, rather than by a pair of distant samples. SWAN & DIX (1966) modified the maximum pair distance criterion for selecting the reference stands for the first axes, by a constraint intended to avoid outlier stands (see 10.4.2.3), which had little in common with the bulk of the sample. MAAREL (1969) replaced the simple pair distance by a new criterion to be maximized: the 'negative correlation tendency' between two stands, which took into account their similarities with all stands in the sample. Finally, the 'best possible stand-defined ordination' of SWAN et al. (1969) maximized the sum of squared projections on the first axes defined by any two stands, not necessarily dissimilar ones. The procedure for selecting axes after the first seems to be identical to that in position vectors ordination (PVO), which was proposed by ORLÓCI (1966) as a simple and close approximation of PCA. The difference between optimized PO and PVO is that in PVO the first axis passes through one stand and the centroid, rather than through two stands. All these techniques attempted to make axis selection in PO dependent on more information about the assemblage of stands, instead of relying completely on a few extreme stands (article 7).

There seems to be little doubt that such improvement was needed, and that the dependence on pairs of most dissimilar stands was the major weakness of BRAY & CURTIS' original version of PO, which made it unduly sensitive to sampling accidents and to 'outlier' stands. This conclusion is not contradicted by the results of tests on simulated coenoclines, in which PO produced less distortion than PCA even when the same (Euclidean) distance measure was used (GAUCH & WHITTAKER 1972b, WHITTAKER & GAUCH 1973, GAUCH et al. 1977, KESSELL & WHITTAKER 1976). In these tests the first axis had been defined by the true endpoints of the coenocline, i.e. by the pair of ecologically most distant stands rather than the pair of compositionally most dissimilar stands. Above a certain range, compositional dissimilarity is practically unrelated to coenocline distance (GAUCH 1973a). In coenocline data with noise and outliers, the two criteria may select quite different pairs of stands. In fact, WHITTAKER & GAUCH (1973, article 10) and GAUCH et al. (1977) replace BRAY & CURTIS' orginal 'maximum dissimilarity pair' criterion by sample pairs that represent the extremes of recognized environmental gradients (while excluding from such choice samples that are outliers for the sample set). This criterion depends on the investigator's subjective interpretation of environmental causes of compositional variation, but may be successful with sample sets in which one or more major ecological gradients can be clearly recognized.

11.4.2 The R/Q Problem: Duality

The theoretical basis for resolving the R/Q problem was already present in the papers by Orlóci (1966, 1967b) and Gower (1966) (see also Pielou 1969, Noy-Meir 1973a). However, it seemed to have taken some time for the full implications to sink in.

In so far as cross-product (generalized Euclidean) similarity measures of the same data matrix $\mathbf{Y}$ (with the same transformation, if any) are used, the $m \times m$ between-species ($\mathbf{R}$) similarity matrix and the $n \times n$ between-site ($\mathbf{Q}$) similarity matrix are dual matrices that contain essentially the same information. They have the same p eigenroots ($\mathbf{\Lambda}$); and m-eigenvector (species values) and n-eigenvector (sample values) associated with each eigenroot (each PCA axis) are related. The species ordination ($m \times p$ eigenvectors $\mathbf{A}$) resulting from PCA of the $\mathbf{R}$-matrix can be used to compute indirectly a sample ordination ($n \times p$ eigenvectors $\mathbf{Z}$),

$$\mathbf{Z} = \mathbf{YA} \,.\, \mathbf{\Lambda}^{-\frac{1}{2}} \qquad (1)$$

identical with that obtained directly from PCA of the $\mathbf{Q}$-matrix (Orlóci 1973, article 9). Symmetrically, the latter can be used to compute an indirect ordination of species A, identical with the direct one from the R-analysis. Since either R- or Q-analysis can produce the same ordinations of both sites and species, the decision between them becomes a merely technical one. If the number of species m is smaller than the number of samples n, it is computationally more expedient to proceed through the $\mathbf{R}$-matrix; and conversely if n is smaller than m.

It must be noted that dual similarity measures are not always also analogous. Orlóci (1966) showed that the Q-dual of between-species covariance was not between-sites covariance (centered by site) but a between-sites measure centered by species. Gower (1966) showed also that the Q-dual of between-species correlation had to be standardized by species standard deviation; therefore it is quite different from between-sites correlation (see also Noy-Meir 1973a, Table 2). Only when the data transformation is symmetrical with regard to sites and species (no transformation, Crawford & Wishart 1967; or simultaneous double transformation, Benzécri 1969) have the dual R- and Q-similarities also analogous meanings.

Table II. Similarity (or distance) measures which are (A) special cases of a generalized Euclidean measure, with different data transformations, or (B) Manhattan-type measures (see also DAGNELIE 1960, article 5, NOY-MEIR et al. 1975).

In parentheses are initials of references (listed below) and numbers for equations in article 5.

* indicates that the measure was used only for presence data.

Centering	Standardization	R-similarity	Q-similarity
A. Euclidean measures			
none	none	co-occurrence* (D, CW)	Euclidean distance (Or, 62), presence form*, $\sqrt{b+c}$ (Or)
by species	none	covariance (D, Gr, Y)	weighted similarity (Or)
by species	by species	correlation (G, D, Y, 24) point correlation* (V, D, 15)	standard distance (GS, We*)
none	by species	congruence (D*)	
none	by sample		congruence, relative distance (Or), relativized Euclidean distance (WG)
by sample	by sample		correlation (We*)
none	by species and sample	X^2 distance, correspondence (B, H)	X^2 distance, correspondence (B, H)
B. Manhattan measures			
none	by sample		coeffient of community* (S, 40) percentage similarity–Czekanowski form (50) or relativized (Wh, 51)
none	by species and sample		Wisconsin similarity (BC, 50)
none	by species	percentage co-occurrence* (WF, 4) percentage similarity of distribution (WF, 23)	

Abbrevations for references: B - BENZÉCRI 1969, BC - BRAY & CURTIS 1957, CW - CRAWFORD & WISHART 1967, D - DAGNELIE 1960, G - GOODALL 1954, Gr - GROENEWOUD 1965, GS - GREIG-SMITH et al. 1967, H - HILL 1973, Or - ORLÓCI 1966, 1967a, 1973, S - SØRENSEN 1948, V - VRIES 1953, 1954, We - WEBB et al. 1967, Wh - WHITTAKER 1952, WF - WHITTAKER & FAIRBANKS 1958, WG - WHITTAKER & GAUCH 1973, Y - YARRANTON 1967.

In any case the choice of data transformation is a basic decision (see below) which should be completely independent of the technical decision between *R*- and *Q*-algorithms (or alternative algorithms proceeding directly from the primary data matrix; WOLD 1966, HILL 1973).

The simple *R/Q* duality relation is exactly valid only for PCA of generalized Euclidean measures. It is probably still nearly true for methods of axis selection which closely approximate PCA. To what extent duality of **R** and **Q** matrices can still be approximated, when both are of non-Euclidean measures, needs yet to be examined.

11.4.3 CHOICE OF SIMILARITY MEASURE: RATIONALIZATION

The number of similarity (and dissimilarity) measures used in ecology has continued to increase (see article 5 for a review), but some order seems to be emerging from this multiplicity. A systematic approach shows that many of the measures are related and can be classified by a few simple criteria; and rational, rather than arbitrary or trial-and-error, choices among them are possible (WILLIAMSON 1972, ORLÓCI 1972, 1975, article 9.7, NOY-MEIR et al. 1975, CHARDY et al. 1976).

Many measures of between-sample and between-species similarity and dissimilarity, which were used under many different names (Table II), belong to the family of Euclidean (cross-product) similarities, of the general form:

$$s_{ij} = \sum_{k=1}^{M} y_{ik} y_{jk} \text{ or } s_{ij} = \sum_{k=1}^{S} y_{ik} y_{jk} \tag{2}$$

and the corresponding Euclidean distances or dissimilarities:

$$d_{ij}^2 = \sum_{k=1}^{M} (y_{ik} - y_{jk})^2 \text{ or } d_{ij}^2 = \sum_{k=1}^{S} (y_{ik} - y_{jk})^2 \tag{3}$$

(i, j refer to two samples containing species k, or two species in sample k, in a sample set of S species and M samples.) The y-scores are either the original species-in-sample scores (x), or some specified transformation of these. The variants of these generalized Euclidean forms differ only in the data type (quantity or presence) and data transformation (usually centering and/or standardization). As mentioned above, the picture is further simplified by duality relations that exist between some of the *Q*-measures used and some *R*-measures. Moreover, the corresponding similarity and distance measures are simply related by the equation (GOWER 1966):

$$d_{ij}^2 = \sum_{k=1}^{S} y_{ik}^2 + \sum_{k=1}^{S} y_{jk}^2 - 2s_{ij} \qquad (4)$$

(or with summations over M samples, for species distances).

An analogous relation exists between non-Euclidean similarity measures of the general form:

$$s_{ij}' = \sum_{k=1}^{M} \min\,[y_{ik}, y_{jk}] \qquad (5)$$

(e.g. relativized percentage similarity, WHITTAKER 1952), and a 'city-block' or 'Manhattan' distance (LANCE & WILLIAMS 1967, BEALS 1973).

$$d_{ij}' = \sum_{k=1}^{M} |y_{ik} - y_{jk}| \qquad (6)$$

Both Euclidean and Manhattan distances are metrics of the Minkowski type, which differ only in the form in which the ones dimensional distances are pooled (as squared values or as absolute values). The percentage similarity measure used by BRAY & CURTIS (1957; and many other studies) is derived from a Manhattan measure; since the standardization is different for every pair of sites, it is a semimetric (ORLÓCI 1974a) and not a true metric (unless the original data are previously relativized, i.e. standardized by site total). Its presence-data analog is SØRENSEN's (1948) coefficient of community (article 9.7.1.2).

The conflict between Euclidean metrics (as a group) and Manhattan metrics and semimetrics in ordination has not yet been resolved. GOWER's (1966) extension of PCA to 'principal coordinate analysis' and a further extension by WILLIAMS et al. (1971) now allow one to apply eigenvector ordination to practically any non-Euclidean and even nonmetric measure (after suitable transformations) while minimizing the distortion of the original distances. Euclidean measures have some theoretical advantages (article 9.7, ORLÓCI 1974a, 1975) and make possible the R/Q duality in PCA. On the other hand, in several empirical tests, a Manhattan-derived semimetric (BRAY & CURTIS percentage similarity without standardization) was shown to be less sensitive than Euclidean distance measures to the distortion due to nonlinearity and zero values (GAUCH & WHITTAKER 1972b, BEALS 1973, KESSELL & WHITTAKER 1976, GAUCH et al. 1977; but in contrast ORLÓCI 1974a), which is now recognized to be a major problem in ordination. The fact that a Euclidean measure (that is not species-standardized) squares the differences in scores for dominant species implies heavy weighting of these species and may make the ordination more vulnerable to distortion and effects of sample error in measuring dominant species (KESSELL & WHITTAKER 1976; see also NOY-MEIR et al. 1975 and section 11.4.5.2).

The decision between centered and noncentered similarity measures in PCA has been reexamined theoretically and empirically (NOY-MEIR 1973a). It was confirmed that centering by species mean is effective in concentrating information on compositional variation on the first axis (DAGNELIE 1960, ORLÓCI 1966). But it was also shown that the information retained in the ordination by refraining from centering is not necessarily trivial and is relevant to the problem of discontinuity (see 11.4.4).

An attempt has been made (NOY-MEIR 1970, NOY-MEIR et al. 1975) to rationalize the decision about data standardization in PCA (none, by site, by species, or both). Theoretical and empirical results indicate that this is essentially a decision on the effective weight or importance that should be given to each sample and each species in the over-all multivariate analysis, in relation to sample equitability and species abundance (see also BURR 1968, WILLIAMS 1971, CHARDY et al. 1976). The effective weight of a sample or species in a multivariate analysis using Euclidean measures is the sum of the squares of scores (as transformed); this weight defines the (squared) distance of the sample or species from the origin, and the contribution to covariances and hence to the determination of axes. For data not centered by samples, the sums of squares of scores in individual samples, $\sum_{i=1}^{S} x_{ij}^2$, are affected by sample sizes (if there is any contrast in these) and, more significantly, by sample equitability vs. dominance. Equitability refers to relative closeness of species scores when arranged in sequence by those scores, or relatively gentle slope of the importance-value curve formed by those scores (LLOYD & GHELARDI 1964, WHITTAKER 1972b). An equitable sample, with low contrast among its species scores, will have a smaller sum of squares than a sample with one or more species strongly dominant. For sample-centered data the sum of squares, $\sum_{i=1}^{S} y_{ij}^2$, changes meaning and indicates the relative deviance, or distance of a sample from the centroid.

For data not centered and not standardized, the 'weight' of a species is the sum of squares of its scores in different samples, $\sum_{i=1}^{M} y_{ij}^2$; but the species' 'leverage' (effectiveness in determining sample positions in relation to one another through its effect on sample similarity measurements) is related to the range of its scores, and hence to the variance of those scores relative to one another,

$\sum_{i=1}^{M} (y_{ij}-\bar{y}_i)^2/M$. Sample centering removes the effects of contrasts between species in mean abundance, while retaining the effects of contrasts in ranges of scores. With species centering (which is more commonly used for PCA in ecology than sample centering) species weight and leverage are the same, and the effect of the species is expressed by its variance, and hence its range of scores. Standardization by species norms equalizes the weights of species, but not the effects of variances on leverages; the latter are equalized by standardization by species standard deviation. Standardization by species totals increases the weights of rare species with high relative variances, $\sum_{i=1}^{M} (y_{ij}-\bar{y}_i)^2/ \sum_{i=1}^{M} y_{ij}$. Standardizations by species maxima, standard deviations, or norms tend to more effectively equalize the effects of species on the ordination. Species with similar scores (as transformed) in all samples will have relatively little effect on similarity measures and will (like equitable samples) appear near the origin (DAGNELIE 1960, GAUCH et al. 1977). Species standardization increases the weight and leverage of samples with a concentration of generally rare species and (with centering) also of samples with high scores for species with high variances. This heavy weighting of samples that are extreme in this sense may reduce the involution of simulated coenoclines when treated by species-standardized PCA (GAUCH et al. 1977), but may also cause axes to be dominated by single unusual stands that are not necessarily at the extremes of the main over-all gradient (in real data; NOY-MEIR et al. 1975).

Standardization by sample (total, norm, etc.), in contrast to standardization by species, tends to equalize the weights and leverages of samples in the ordination, irrespective of species-richness or total biomass (cover, etc.). The differences between equitable samples and those with pronounced dominance will be removed only if standardization is by a stand-centered parameter. In general, standardization by sample will give prominence in the ordination to species that are relative dominants in many samples.

Double standardization intensifies the effects of both samples with concentrations of extreme values and species concentrated in a few (extreme) samples. The over-all weighting is then in favor of both rare species and species-poor (and inequitable) samples, and particularly coincidences of such species and samples (Noy-Meir et al. 1975). From this strong upweighting of the extreme corners of the primary matrix (as arranged by ordination) may result the improvement of PCA by Wisconsin double standardization (Austin & Noy-Meir 1971) and the strong resistance of RA to involution of a coenocline on its first axis (Gauch et al. 1977). On the other hand, doubly standardized ordinations are sensitive to outliers, i.e. samples or small groups of samples with unusual species combinations (Noy-Meir 1970, Hill et al. 1975, Noy-Meir et al. 1975, Gauch et al. 1977).

For a polar ordination of samples, samples are normally standardized (by prior relativization of samples, or by double standardization). Even without relativization or double standardization, the commonly used similarity measures PS and CC (article 5, formulas 50 and 39 or 40) are partially standardized by sample, although the standardization is applied at the level of each pair of samples compared, rather than to each sample relative to all samples. For polar ordination of samples, samples are 'weighted' only by their use as endpoints. Total scores for species have little meaning as weights for these species; species 'leverages' in determining sample positions relate primarily to the mean deviations of their scores, $\sum_{i=1}^{M} |y_{ij} - \bar{y}_i| / M$.

The prior allocation of importance by standardization (and centering) should be related to the interests of the investigator in a given study. Sample standardization (e.g. relativization) is appropriate to remove effects of different sample sizes, whatever the ordination technique, unless there is a special reason for weighting differences in sample totals in the ordination. Standardization by species tends to equalize the weights and leverages of species and implies that the ecologist is as much interested in rare as in dominant species. Nonstandardization by species, in contrast, implies heavy weightings on the scores for dominant species. Logarithmic transformation of species scores sometimes gives an advantageous, intermediate weighting – one that overemphasizes neither the dominants nor the rarities (Chardy et al. 1976, Gauch et al. 1977, Wentworth 1976). Certain standardizations can be associated with particular purposes of vegetation study and with one of the classical phytosociological schools (see also Dahl 1957, Noy-Meir et al. 1975, Noy-Meir & Whittaker 1976). The use of species standardization (or presence and absence) corresponds to the perspective of the school of BRAUN-BLANQUET (article 20), non-standardization by species to the emphasis of dominant species in British and American ecology (articles 13 and 14) and some Scandinavian phytosociology (article 18). Unsatisfactory results of a published ordination (or classification) can sometimes be explained by the use of a standardization inconsistent with the interests of the user.

11.4.4 CONTINUOUS TECHNIQUES AND DISCONTINUOUS DATA

The question 'ordination or classification?' has lost much of its sharpness in recent years. Most ecologists now accept the idea that both models can be usefully applied to the same data and may complement each other in various ways (WEBB 1954, WHITTAKER 1962, 1967, 1972a, WILLIAMS & LANCE 1968). Some possibilities are: parallel ordination and classification of sites, with group membership plotted in the ordination (DAGNELIE 1960, GROENEWOUD 1965, GREIG-SMITH et al. 1967, KERSHAW 1968, FLENLEY 1969); initial classification followed by ordination of samples within each group (GREIG-SMITH et al. 1967, JOHNSON 1973, CHARDY et al. 1976) or ordination of groups (ORLÓCI 1967a, CRAWFORD & WISHART 1968, FRYDMAN & WHITTAKER 1968, MAAREL 1969, WILLIAMS et al. 1971, WHITTAKER 1972a, article 20); initial classification revised by ordination (MORAL 1975); initial ordination from which a divisive polythetic classification is derived (LAMBERT 1971, LAMBERT et al. 1973, NOY-MEIR 1973b, HILL et al. 1975).

The problem of how continuous multivariate methods could deal with strong discontinuities has usually been avoided by trying to ordinate only sample sets that were reasonably continuous. However, it has been shown that, even in situations with complete discontinuity (disjunction), patterns of compositional variation can be usefully represented by noncentered PCA (NOY-MEIR 1973a). Moreover, such component analyses give an indication of the number and sharpness of discontinuities in the data set. If the sample consists of one continuous group with a strong common element, only the first principal component is unipolar (has only positive values for samples and species) and all subsequent components are bipolar (+ and — values). If the sample consists of two or more disjunct groups, which have no species and samples in common, there will be one unipolar component for each such group; each of the bipolar components will then represent variation within one group only, without interference with other groups.

A similar pattern emerges when a disjunct data matrix is analyzed by reciprocal averaging or correspondence analysis (BENZÉCRI 1969, HILL 1974), which may be regarded as a special case (doubly standardized) of noncentered eigenvector analysis. The additional advantage is that in this case all disjunct group components have equal eigenroots ($\lambda = 1$) and are extracted before any within-group components. Discontinuities that are not absolute can be detected by asymmetry of + and — values on components after the first. As groups become more connected, components become bipolar and interference increases; in these intermediate cases, rotation to simple structure helps to clarify the picture (see below, section 11.6.2). The PCA model, though essentially continuous, is thus capable of detecting and representing qualitative discontinuities in composition, as a limit case of continuous variation.

11.4.5 Nonlinearity and Axis Interpretation: Divergence

11.4.5.1 *Defining the Limits of Ordination*

The easiest solution to the nonlinearity problem would limit ordination to sample sets with a narrow range of environmental and compositional variation, i.e. with low beta-diversity (Whittaker 1956, article 3). But how is one to measure and define the range beyond which the level of 'type A distortion' (Orlóci 1974a) becomes unacceptable? Gauch & Whittaker (1972a, 1972b) applied PO and PCA ordinations to simulated coenocline data with varying beta-diversity and measured parameters expressing distortion in the results (10.3.2). Beta-diversity was measured in z units (length of gradient in units of average width of species response curves) or in units of 'half-change' (HC; gradient length over which between-sample similarity decreases by a factor of two; Whittaker 1956, 1960). The results (see also Gauch 1973b, Whittaker & Gauch 1973, Gauch et al. 1977. Kessell & Whittaker 1976) showed that PO (with percentage similarity) gave reasonable representations when beta-diversity was less than 4 or 5 HC (5-7 z); above this the distortion became severe. PCA ordinations often showed marked distortion and involution already at 2-3 HC (see also Jeglum et al. 1971, Westman 1975). Gauch (1973a, 1973b, article 3) also derived theoretical functions relating compositional similarity between two stands (s) to their 'ecological separation' along the gradient, Δz, assuming Gaussian-type species response curves to a single gradient. In the predicted functions, s approached zero as Δz exceeded 4 HC and was poorly sensitive to further increase in separation. Thus in any data set spanning 5 or more HC, a large proportion of between-stand similarities is expected to be near zero, irrespective of the 'true' ecological separation; any linear ordination based on these similarities will give a distorted picture of the coenocline.

Ordination techniques thus differ widely in their tolerance of beta-diversity from the low tolerance of nonstandardized PCA to the intermediate tolerance of PO and the high tolerance of techniques to be discussed below. In real vegetation samples beta-diversity is difficult to measure until the data have been successfully ordinated. However, a prior estimate could be obtained by considering the distribution of between-sample similarity values, and in particular the proportion of zero values.

10.4.5.2 *Reducing Distortion by Transformations*

Various transformations of the original data, or of the similarity values, reduce the distortion of coenoclines in ordination space. Swan (1970) recognized that the high proportion of zero values in data matrices with high beta-diversity was a major source of distortion. He suggested to transform each zero value to a value expressing the degree of absence of the species in the sample, calculated from the values of the 'association index' (over the whole sample) between the absent species and the species present in the site. In simulated coenoclines with high beta-diversity this transformation reduced distortion from a severe to a moderate level. In tests with real vegetation data (Walker 1974), it improved the results in some cases, but not in others. In his study of the similarity function Gauch (1973a, 1973b) used its inverse to transform between-sample similarity values into a measure linearly related to separation along the coenocline. This transformation reduces the distortion in ordinations of simulated data without sample error, but may give little or no improvement in ordination of simulated data with sample error or of field data. The form of the similarity function for the latter (article 3, Figs. 6 and 7) differs from the ideal error function, and no transformation is known that fits well the similarity function for real data.

Austin & Noy-Meir (1971) found that normalization (11.3.3) reduced the distortion of simulated two-dimensional coenoplanes in PCA ordination space in most cases (see also Gauch 1973b). They also found that stepwise double standardization (by species maximum, then by site total; Bray & Curtis 1957) and simultaneous double standardization (by sample and species total) were usually even more effective than simple standardization by sample, apparently because of the upweighting of samples (and/or species) at the extremes of the coenocline (Noy-Meir et al. 1975, 11.4.3). Gauch et al. (1977, article 10) found that Hill's (1973) reciprocal averaging was much less prone to distortion than normal PCA. Reciprocal averaging involves a simultaneous double standardization similar to that found successful by AUSTIN & NOY-MEIR (1971). In the same tests, polar ordination with a Manhattan distance (complement of percentage similarity) was consistently less distorted than the same technique with Euclidean distance (article 10).

To summarize, distortions in ordination from nonlinearity can be partially controlled by choices of data transformation and similarity measure. In general the worst types of distortion – notably involution of coenoclines and reversals of order – can be avoided either by these means or by limiting the range of sample variation (beta-diversity) in the set to be ordinated. The residual nonlinearity has a simple curvature (arch, dome, horseshoe, boomerang) in relation to the second axis. Thus, the between-sample distances may be somewhat compressed toward the extremes of the first axis, and the ordination may produce other, spurious axes separating the middle from the extremes. In dealing with field data, interpretation of second (and higher) axes may be obscured by unknown combinations of these curvature effects with real separation of sites along secondary ecological gradients and sample error (GAUCH et al. 1977, article 10).

11.4.5.3 *Nonlinear Techniques* (*Catenation*)

The basic problem is that linear ordinations (PCA, factor analysis) imply linear relationships between axes and species variables, whereas the axes that many ecologists hope to find in ordinations of real vegetation (environmental gradients, coenoclines) have strongly nonlinear relations with species variables. Transformations may alleviate the symptoms of this problem, but they do not solve it. The nature of the problem may be clarified by the concept of ecological space in which an indirect ordination seeks to use nonlinear data on species distributions (or similarity measurements) to recognize linear directions of environmental change to which the directions of community change are responses (article 10.3.1). For this purpose the ordination technique should not be constrained by linearity but should allow for, or take advantage of, the curvilinear relationships of the compositional continuum (NOY-MEIR 1974a, AUSTIN 1976a).

Several such techniques have recently begun to be applied to phytosociological data, either by adaptation from other sciences or by independent development. As a collective name for techniques of indirect ordination that are free of the linearity constraint the term 'catenation' has been suggested (NOY-MEIR 1971, 1974a, 1974b, cf. seriation in the sense of archeologists and DALE 1975). A survey of nonlinear techniques which have been used (or suggested) for ecological data is given in Table III. Two main approaches may be distinguished: (a) those which maximize fit to explicit nonlinear species/axis functions over the entire range (e.g. polynomial ordination, Gaussian ordination); (b) those which maximize some measure of local ordering which is consistent with similarity values (e.g. proximity analysis, multidimensional scaling, continuity analysis, curve-seeking).

Some early techniques (direct ordination by environmental measures, weighted averages, compositional gradient analysis) were in fact nonlinear ordinations in this sense. Their main deficiency was that they involved subjective decisions and therefore did not give unique solutions (except in very simple situations). However, when they were (for this reason) replaced by PO and PCA, the linear constraint slipped in unnoticed (AUSTIN & NOY-MEIR 1971). Also, the rearrangement of phytosociological tables and similarity matrices (article 6, DALE 1975) can be regarded as a kind of catenation when samples are thereby arranged along a primary axis of community and environmental variation. Recently several objective rearrangement procedures based on RA and other techniques have been suggested (e.g. ČEŠKA & ROEMER 1971, LIETH & MOORE 1971, MAAREL unpublished, and others).

Other methods are mathematically more complex. Each of the main aproaches has already been applied to ecological data, but experience with them is still limited. We shall, however, characterize and comment on some techniques of promise.

11.5 Recent Techniques of Indirect Ordination

Among ordination techniques more recently introduced and tested, four deserve special comment. We draw on articles by AUSTIN (1976b), NOY-MEIR & WHITTAKER (1977), GAUCH et al. (1977), and FASHAM (1977) for the following.

11.5.1 RECIPROCAL AVERAGING

Reciprocal averaging has had most testing in ecology among the four techniques and is discussed by HILL (1973, 1974), GAUCH et al. (1977, article 10), AUSTIN (1976b), and CHARDY et al. (1976). RA originated in work of HIRSCHFELD (1935) and FISHER (1940); it has since then been rediscovered by others (GUTTMAN 1959, HILL 1973, 1974) and applied by French phytosociologists as 'correspondence analysis' (analyse factorielle des correspondances)

Table III. Some methods of nonlinear ordination (catenation) that could be used, or have been used (*) in ecology.

Method	References	Principle	Problems or Limitations
compositional gradient analysis	*Curtis & McIntosh 1951, Brown & Curtis 1952	unimodal species/axis curves	one dimension; no unique solution
plexus diagrams	*Vries 1953, 1954, article 6	adjacency consistent with similarity	two dimensions; no well-defined procedure
direct gradient analysis (with weighted averages)	*Whittaker 1951, 1956, Curtis 1955, Whittaker & Niering 1965	species modes positioned on factor gradient; site positions derived as weighted averages	one dimension at a time; species positions and choice of factor not unique
table rearrangement	*Lieth & Moore 1971	minimum dissimilarities between adjacent points	presence data; no unique solution
curve-seeking	Sneath 1966	pull together and link adjacent points	very sensitive and problematic algorithm
polynomial ordination	McDonald 1962, 1966	fit polynomials to points in PCA-space	choice of polynomial
	Wood et al. 1964	fit polynomials to species/axes relations	rotation problems
proximity analysis	Shepard 1962		
multidimensional scaling	Kruskal 1964a, 1964b (*Anderson 1971)	between-point distance on axes	choice of 'distortion' measure?
smallest space analysis	Guttman 1968	monotonically related to similarity	algorithm for finding best solution?
continuity analysis (parametric mapping)	Shepard & Carroll 1966 (*Noy-Meir 1974a)	minimum 'discontinuity' (sum of ratios dissimilarity/distance, weighted by adjacency) in species/axes curves	choice of weighting parameters? algorithm problems
Gaussian ordination	*Gauch et al. 1974, Johnson 1973, Ihm & Groenewoud 1975	fit Gaussian curves to species/axes relations	good algorithm for one dimension only

(BENZÉCRI 1969, LACOSTE & ROUX 1971, ROMANE 1972, GUINOCHET 1973). Its general use is to reveal correspondences, for a number of observations, between two kinds of information (BENZÉCRI 1969) such as, in our application, species and samples. HILL (1973) describes RA as a weighted average ordination effected by successive approximations. In one of its algorithms (HILL 1973, appendix 2) species are weighted by positions along a rough initial gradient and the weights are used to calculate sample scores. These sample scores as weights are then used to calculate a new and improved calibration of the species. The new species weights provide further improvement in sample calibration and so on. Back and forth, iterative calculations lead to a stable, optimal solution that does not depend on the initial arrangement. RA is thus an indirect ordination related to techniques of direct ordination – weighted averages (GOFF & COTTAM 1967, WHITTAKER 1967, article 2) and elective means (RAMENSKY 1930, article 4) (see also article 10). RA is also an eigenvector technique related to principal components analysis but differing in (i) simultaneous double standardization of the primary data matrix, (ii) an additional division (as ξ, versus x, see HILL 1973, appendix 1), and (iii) employement of chi-square (rather than covariance or correlation) distances (CHARDY et al. 1976).

RA relates also to traditional phytosociological matrix arrangement, and it can be used to arrange phytosociological tables into community-types (ČEŠKA & ROEMER 1971, GUINOCHET 1973), but RA also produces an optimal sequence of both samples and species in relation to each other. The production of simultaneous ordinations of samples and species, approximately coordinate with each other (on scales of 0-100 on the first axis) is an especially attractive feature of RA. Finally, RA can be applied as the 'index iteration' of GOFF & COTTAM (1967) to secondary data matrices (of sample, or species similarity values). This, which might also be termed secondary reciprocal averaging (GAUCH et al. 1977, article 10.4.4.3) is then a mathematical optimization of another classic technique of phytosociology – maneuvering the high similarity values towards the secondary matrix diagonal (MOTYKA 1947, MATUSZKIEWICZ 1948, article 6).

Extensive tests of RA are reported by GAUCH et al. (1977) and article 10, from which we summarize: (i) RA is a technique of preference for derivation of a first, major axis of community variation, effective over a wide range of beta diversity with simulated and field data; (ii) for this use RA is generally less vulnerable to effects of sample error and clusters than principal components ordination, but RA ordinations may be strongly affected by outlier samples; (iii) RA is highly tolerant of nonmonotonic species distributions along a first axis of variation, but like other eigenvector techniques is subject to curvilinear distortions into second and higher axes. AUSTIN (1976b) found that RA was successful with simulated coenoclines and coenoplanes based on symmetrical Gaussian distributions, but that it was subject to involution when offered data with skewed and other distributions differing from the Gaussian.

11.5.2 GAUSSIAN ORDINATION

Gaussian ordinations are distinctive among those discussed, in that these techniques have been developed by ecologists themselves to fit the particular characteristics of ecological data. These ordinations seek to solve the problems of nonlinear ordination with algorithms based on Gaussian distributions of species populations. The attempt to base ordination on fitting species populations with polynomial functions was suggested by McDONALD (1966). An explicitly Gaussian ordination was developed by GAUCH et al. (1974, GAUCH & CHASE 1974). An initial estimate of sample positions is provided, and iterative calculations fit a Gaussian function to each of the species in turn; further calculations produce, by means of sample rearrangements, an optimal arrangement of samples to fit Gaussian functions for all species considered together. JOHNSON (1973) developed a maximum likelihood method, which takes account also of information from records of absence. IHM & GROENEWOUD (1975) have developed a third technique which is based on the Gaussian response function, with some additional assumptions. These assumptions allow species and sample positions to be estimated by a relatively simple eigenvector solution. First tests on artificial coenocline data gave fairly good results. At present the Gaussian ordination of GAUCH et al. (1974) is coded for a single ordination axis and is appropriate only for sample sets having a single major direction of community and environmental variation. For such use the technique is effective to very high beta diversities. JOHNSON (1973) and IHM & GROENEWOUD (1975) suggest extension to further axes, but make restrictive assumptions that may make such extension ineffective with field data.

The technique of GAUCH et al. (1974) is the most extensively tested, and apparently the only one yet effectively used with field data. The numerical routines seem robust and efficient; and the technique can recover the true coenocline from randomly ordered data, though its performance is improved by a reasonable first-guess ordination. Given a reasonable first guess, it has been quite successful with simulated coenoclines including sample error, and with field data with one predominant direction of variation (GAUCH et al. 1974, AUSTIN 1976b, WENTWORTH 1976). With field data it gives a somewhat larger percentage of variance accounted for than reciprocal averaging (WENTWORTH 1976). (Given the Gaussian assumptions of the technique, variance accounted for is an uncertain indication of ordination improvement over reciprocal averaging.) AUSTIN (1976) found Gaussian ordination unsuccessful with species distributions departing widely from the bell-shaped form; given simulated data with skew distributions, Gaussian ordination altered 'true' sample positions to make the distributions symmetrical and by this means accounted for a remarkable fraction of variance (99 %). Skew distributions are common in published transect data (AUSTIN 1976a). For such data, it is uncertain whether Gaussian ordination's symmetrical forms are a correction for sample error or an improvement on nature. Like reciprocal averaging, Gaussian ordination gives simultaneous ordinations of species and samples; unlike other techniques Gaussian ordination provides estimates of heights of modes and standard deviations (i.e. habitat amplitudes on the gradient) for species populations. The technique can thus be used to give a refinement of an initial single-axis ordination by another technique and to obtain in addition these species parameters. Reciprocal averaging has wider utility; but for data with high beta diversity, a well-defined major axis, and species distributions reasonably assumed to fit the bell-shaped form, Gaussian ordination may give a best and most informative ordination.

11.5.3 PARAMETRIC MAPPING

Parametric mapping or continuity analysis was developed in psychology by CARROLL (KRUSKAL & CARROLL 1969, SHEPARD & CARROLL 1966) and introduced into plant ecology by NOY-MEIR (1974a, 1974b); NOY-MEIR (1974b) describes the technique, which has been applied to both simulated and real vegetation data (NOY-MEIR 1970, 1974a, 1974b, and unpublished, SELIGMAN 1973, AUSTIN 1976b). In this technique, the species response curves are undefined, but the species data are fitted to underlying (environmental) axes in such a way that the species curves are as smoothly continuous as possible (i.e. no sharp inflections). An index of continuity is defined for this purpose.

In several tests with simulated data parametric mapping recovered the coenoclines (or the two dimensions of the coenoplane) accurately. In some real sample sets it detected coenoclines that were highly correlated with environmental variables, and in which species response curves had unimodal bell-shaped forms. However, in some other cases it was only partially successful (some distortion at the extremes, spurious dimensions with residual nonlinearity) or failed completely (no convergence to a true or plausible solution). It seems that the latter tends to occur in data sets with discontinuity and/or high random variation (not enough information about continuous links) and in data sets where no one or two dimensions of compositional variation are dominant (SELIGMAN 1973). AUSTIN (1976b) found continuity analysis able to deal with simulated Gaussian data; like reciprocal averaging and unlike Gaussian ordination it could recover a square coenoplane (with only slight distortion at the corners). Continuity analysis performed poorly, however, in tests with data departing from Gaussian forms. KRUSKAL & CARROLL (1969) report that unsatisfactory solutions can occur where the samples are randomly dispersed over the gradients; success may be dependent on having sample positions evenly distributed in the ecological space. Continuity analysis is limited in the size of sample sets to which it can be applied. There are other as yet unsolved problems, e.g. the choice of the parameters α and β in the model (NOY-MEIR 1974a) and computational difficulties with large matrices. Present indications are that the technique can be successful in favorable cases, but is more limited in its potential in ecology than the three other techniques discussed in this section.

11.5.4 MULTIDIMENSIONAL SCALING

Nonmetric multidimensional scaling (MDS) was developed by KRUSKAL (1964a, 1964b, KRUSKAL & CARMONE 1971). It was used for a biogeographic study by HOLLOWAY & JARDINE (1968) and for plant ecology by ANDERSON (1971); it has had more extensive use in palaeoecology, particularly for pollen diagram analysis (WHITTINGTON & HUGHES 1972, BIRKS 1974, GORDON & BIRKS 1974). Given a matrix of similarities, the technique seeks to arrange the objects in a space of a given number of dimensions so that monotonic relationships between similarity values and distances between objects are maintained in that space. In nonmetric MDS (in contrast to metric MDS, TORGERSON 1958) the form of the relationship of similarity and distance need not be specified. A measure of stress (KRUSKAL 1964b) is defined to measure departure from the monotonic relationship, and the technique seeks to minimize this stress measure (9.8.3).

GORDON & BIRKS (1974) and FASHAM (1977) comment that with single-gradient solutions, the technique may become easily trapped in local minima and not provide an optimal result. ANDERSON (1971) used two variants of MDS (KRUSKAL 1964a, 1964b, and a new method) on a vegetational sample set, and found that they compared favorably with PO and PCA. ANDERSON (1971) suggests that, with fewer than 30 samples, rank order preservation may mean considerable loss of information. More recent tests of MDS with simulated data indicate its ability to ordinate sample sets of high beta diversity with less distortion than other techniques (FASHAM 1977, AUSTIN 1976b). FASHAM (1977) notes that the stress measure may not be effective as an expression of ordination success, and he suggests that MDS can be used to improve on an ordination by another technique. MDS requires prior specification of the number of axes to be derived, and the solution for n axes is not the same as the projection of the solution for $n+m$ axes onto the first n axes of the $n+m$ space. MDS is a complex and demanding computation, but it seems promising as an effective nonlinear ordination. FASHAM (1977) found it in most cases better than RA and suggests use of the cos theta similarity coefficient.

11.6 Continuous Multivariate Classification

11.6.1 Between Ordination and Classification

Nonlinearity is a problem if linear multivariate methods are expected (as most users expect) to show coenoclines and reveal environmental gradients (objective i). However, this problem is by and large irrelevant if the purpose of the analysis is to give a simplified (low-dimensional) picture of phytosociological variation, and in particular to show and typify groups of related samples (or communities) and species (objective ii). The two objectives have not always been clearly distinguished, but some users have sought the second objective (e.g. DAGNELIE 1960, IVIMEY-COOK & PROCTOR 1967, MAAREL 1969, 1972). If the groups are derived from a continuum and are understood to intergrade or overlap, the result is not classification in its most familiar, hierarchical sense. It is, however, classification in the sense of typification (article 13, WHITTAKER 1975): recognition of types as foci within a range of variation, centers around which samples may be grouped into classes that represent parts of the range of variation but are not assumed to be discontinuous.

How effective are current linear ordination methods (PO, PCA) for the purpose of detecting phytosociological groupings? Usually such clusters or groupings (associations, vegetation-types, etc.) can be identified with areas in ordination space; but often they are not related in any simple way to the individual axis. This is not a major problem if one keeps to two (or three) dimensions, on which all the information can be displayed graphically; but it does make phytosociological interpretation of axes beyond the third cumbersome or impossible. In large samples of diverse vegetation there may be many axes beyond the third which contain information that is both significant and interesting (e.g. 12-20 axes in NOY-MEIR 1971).

11.6.2 NODAL COMPONENT ANALYSIS

Phytosociological interpretation of axes would be easiest, if each (unipolar) axis would pass from the origin through a single group of similar samples and/or species. When PCA (centered, nonstandardized) is used for ordination, the axes are drawn toward groups of samples in which particular species are strongly dominant (KESSELL & WHITTAKER 1976). When the dominant species are centered within the coenocline, the resulting axes are spurious and the effect is adverse for ordination. If recognition of groups of samples is the purpose, however, then this characteristic of PCA may be turned to advantage. Recent work shows that a structure of unipolar group axes can be obtained through PCA with two modifications; (a) noncentered data and similarities (i.e. leaving the origin at the absolute zero, rather than moving it to the centroid); (b) rotation of axes to simple structure. IVIMEY-COOK & PROCTOR (1967) applied varimax rotation to a four-dimensional centered PCA of species. [Varimax rotation (KAISER 1958) maximizes the variance of squared normalized projections on orthogonal axes, and in effect tends to orient axes towards clusters.] They found that the rotated solution was more readily interpretable than the original PCA: the axes were nearly unipolar into clusters and each could be associated with one vegetation-type. However, in other data sets, rotated axes derived from centered PCA are often bipolar, passing between two clusters at the negative and positive poles (WEBB et al. 1967, NOY-MEIR 1970); this again complicates the interpretation. In noncentered PCA without rotation, the first axis for each disjunct group is unipolar (NOY-MEIR 1973a; see above), but within-group axes are bipolar.

When simple structure (e.g. varimax) rotation is applied after PCA of a noncentered similarity matrix, all or most of the resulting axes are practically unipolar in both the species-ordination and the dual sample-ordination (NOY-MEIR 1970, 1971, HINNERI 1972, PIGNATTI & PIGNATTI 1975). On each axis there is a group of samples and of species with high positive projections, while the rest of the samples and species have zero or low positive projections; negative projections are rare and generally small. It was suggested

(Noy-Meir 1971) that each axis (noncentered varimax component) could best be interpreted as defining and typifying a 'nodum' of vegetation. A nodum in this context is an entity defined by the coincidence of a group of species in a group of samples, where neither the two groups nor the nodum is necessarily discretely bounded from, nor hierarchically related to, other groups and noda. This definition combines elements contained in earlier usages of the term (Poore 1955, 1956, Lambert & Williams 1962).

The projections (loadings, scores) of species and sites on axes are post-normalized to obtain a measure of affinity of each species and site to each nodum (varying between 0 and 1). A nodum represents a cluster of related samples (and species); it is consequently a floristically defined community-type or coenon in the sense of Westhoff et al. (1959, article 20). In nodal component analysis (as in uses of RA for classification) the coena are doubly defined by a group of typical (high-affinity) samples and a group of typical (high-affinity) species. Different data treatments before PCA will produce different definitions of noda. If nonstandardized or sample-standardized data are used, dominant species will have the major role in defining noda, which may then correspond to dominance-types; if species-standardized (or presence-absence) data are used, over-all floristic composition and rare species will have major roles in defining noda, and these units may then correspond to those of the school of Braun-Blanquet.

The nodum differs, however, from the coenon of phytosociology in that the affinity of both samples and species to it is measured on a continuous scale (rather than being a belongs/not belongs decision); thus different noda are not always discretely bounded and may overlap through samples and species that have affinities with more than one nodum. The degree of overlap between pairs of noda can be expressed by a standardized 'coefficient of conjunction' (Noy-Meir 1971). Complete discontinuity between noda (zero overlap) occurs only if it occurs also in the data matrix.

Since each axis can be interpreted (and named) independently, any number of axes can be used and interpreted and the relations between them can then be examined either graphically or by conjunction coefficients. The rotated solution depends on the number of principal axes used to obtain it, thus a rule for deciding on the number of meaningful axes is required. The best test is repeatability in replicate samples (Noy-Meir 1971, Seligman 1973) but criteria based on magnitude of sum-of-squares (eigenvalues) can give reasonable approximate rules. It is often useful to obtain rotated solutions at several levels of dimensionality, giving noda at different levels of resolution, between which quasi-hierarchical relations may be found. For a final synthetic description of the vegetation, a more or less subjective selection of one or two levels may be preferred to a presentation of the complete structure.

Nodal component analysis has been applied to a wide variety of vegetation samples from arid, mediterranean, tropical, and marine environments (Noy-Meir 1970, 1971, and unpublished, Shmida 1972, Seligman 1973, Luria 1975). [The method of noncentered varimax components was originally named 'nodal ordination' (Noy-Meir 1971) in accordance with a broader concept of ordination. Given the more restricted usage of ordination here, the term 'nodal component analysis' seems more appropriate.] Various data transformations have been used. As a rule, most useful results have been obtained by analyzing in parallel the untransformed presence data (floristic noda) and sample-standardized cover data (dominance noda), or logarithms of cover data as a compromise between the two. In most cases, nodal component analysis (together with matrices of conjunction between noda) gave a clear and easily interpretable picture of the patterns of compositional variation. The results have been further used to obtain catenations of noda and then of samples and species (Noy-Meir 1974a, 1974b), to derive divisive polythetic classifications of samples and species (Noy-Meir 1973b), and to test the correlation of environmental variables with transitions between noda (Noy-Meir 1970). The main problem with the technique is that the nodal components structure is obtained by a rather long and elaborate sequence of computations (similarity matrix, eigenvector solution, varimax rotation). It would be useful to have a simpler and more direct, approximate method that would give a structure with similar properties.

11.6.3 Other Recent Approaches

Several other recent multivariate methods result in a structure of overlapping groups with continuously variable affinity as generated by nodal component analysis. At the second stage of group analysis (Crawford & Wishart 1968) the 'potential' (= affinity) of each quadrat to each group is defined (and mapped as a continuous variable). A basic difference between group analysis and nodal component analysis seems to be that in the former the group vectors (axes) are not required to be orthogonal. Thus all group potentials are positive by definition, while in noncentered varimax ordination the requirement of orthogonality sometimes causes negative affinities, which are difficult to interpret. On the other hand, orthogonality has an advantage in that it prevents the assignment of separate axes to groups (noda) that are highly overlapping in both distribution and composition. Methods which describe continuous overlapping group influences around group cores were also suggested by LANGE (1968) and HALL (1970). YARRANTON et al. (1972) developed a classification with overlapping groups (but not with variable affinity). Another approach to a synthesis of classification and ordination is the two-parameter inosculate classification of DALE & ANDERSON (1973) and DALE & WEBB (1975).

11.7 Conclusion

After twenty-five years of development of continuous multivariate techniques in ecology, some of the early optimistic promises, as well as some of the sceptical criticisms, seem to have been overstated. No single method has emerged as a solution to all problems of describing and explaining patterns of compositional variation in natural communities. However, several ordination techniques have proved to be helpful tools in the analysis of community compositional data. Some technical problems which seemed severe only a few years ago (axis selection, *R*- or *Q*-analysis, choice of similarity measure, discontinuity) have been largely resolved by recent developments. Systematic tests of similarity measures and ordination techniques have clarified limitations of these and suggested appropriate choices for field investigators. Significant progress has been made on problems of nonlinearity and axis interpretation, but some further work on these is needed.

There seems to be a trend of departure from the desire for an all-purpose technique of continuous multivariate analysis (polar ordination or centered PCA) towards acceptance of two groups of specialized techniques, each of which is more effective in achieving one of the two main objectives of such analyses; (a) methods of nonlinear indirect ordination or catenation, for the detection and definition of coenoclines and community patterns, and (b) methods of nodal analysis, for definition of systems of groups while allowing for continuous variation. Three periods in the use of multivariate techniques in ecology may be recognized; (a) the early period of direct gradient analysis and other informal techniques, (b) a time of testing of multivariate techniques from other fields, and (c) the current selection among available techniques and design of new ones for appropriateness to ecological data and purposes. A significant conclusion from the last period is that continuous multivariate techniques can be found that are appropriate in ecology and that may serve in both of the major approaches to ordering community data: ordination and classification.

11.8 **Summary**

The early development of indirect ordination techniques in ecology was along two main lines: the Wisconsin approach, which led to the development and wide use of polar ordination (PO), and the multivariate Euclidean approach, which led to wide use of principal component analysis (PCA). Several methodological problems then emerged for both techniques: the basis of axis definition, the R/Q dichotomy, the choice of similarity measure and data transformation, the treatment of discontinuities, and the distortion due to nonlinearity and its effects on axis interpretation.

Developments in the last few years have contributed significantly towards the solutions of these problems. The procedures for axis selection in PO have been modified to allow either for similarities in the assemblage of samples (as in PCA) or for known environmental extremes (as in direct gradient analysis). The duality between the R and Q similarity matrices in PCA, and the simultaneous sample and species ordination in reciprocal averaging, turn the decision between R and Q techniques into a merely technical point. Some systematic relations have been recognized within the large array of similarity measures used in ordination, and rational guidelines for choices among them can be suggested. Noncentered PCA and reciprocal averaging (correspondence analysis) have proved to be useful in describing compositional variation that includes discontinuities.

Divergent solutions have been developed to cope with the problem of nonlinearity and axis interpretation: (a) The range of beta-diversity within which linear ordination methods do not cause too severe coenocline distortion has been defined. (b) Transformations of data and similarity measures (and in particular double standardization) have been shown to reduce distortion due to nonlinearity in ordination techniques assuming linear structure. (c) Some new methods of nonlinear ordination (catenation) have been fairly successful in detecting coenoclines in simulated and real data. No single answer to the problems of indirect ordination has emerged, but these various solutions make the problems generally manageable.

Some continuous multivariate techniques have been developed as means of classifying community samples. In these techniques, individual axes are interpretable as noda which define a structure of overlapping groups of samples connected by continuous variation. These techniques provide a further link between ordination and classification as approaches to understanding relationships of samples from natural communities to one another and environment.

ACKNOWLEDGEMENTS

This paper results from independent research on ordination by the first author while he was a Research Student at the Australian National University in Canberra, where he benefited from the advice and help of W. T. WILLIAMS and Donald WALKER, and by the second author and H. G. GAUCH at Cornell University with support by the National Science Foundation. The structure of the paper is based on a seminar given by the first author at the Centre d'Études Phytosociologiques et Écologiques at Montpellier in October 1974, and a joint article in *Vegetatio* (1976). Discussions with M. P. AUSTIN, Eilif DAHL, M. B. DALE, and László ORLÓCI, and exchange of manuscripts and ideas with François ROMANE, M. O. HILL, M. P. AUSTIN, M. J. FASHAM, and H. G. GAUCH have helped to clarify some of the problems and to bring this review up to date.

REFERENCES

AART, P. J. M. VAN DER, – 1973 – Distribution analysis of wolfspiders (Araneae, Lycosidae) in a dune area by means of principal component analysis. *Neth. J. Zool.* 23: 266-329.

ANDERSON, A. J. B., – 1971 – Ordination methods in ecology. *J. Ecol.* 59: 713-726.

AUSTIN, M. P., – 1968 – An ordination study of a chalk grassland community. *J. Ecol.* 56: 739-758.

AUSTIN, M. P., – 1972 – Models and analysis of descriptive ecological data. In 'Mathematical Models in Ecology,' ed. J. N. R. JEFFERS. *Symp. Brit. Ecol. Soc.* 12: 61-86.

AUSTIN, M. P., – 1976a – On non-linear species response models in ordination. *Vegetatio* 33: 33-41.

AUSTIN, M. P., – 1976b – Performance of four ordination techniques assuming three different non-linear species response models. *Vegetatio* 33: 43-49.

AUSTIN, M. P. & P. GREIG-SMITH, – 1968 – The application of quantitative methods to vegetation survey. II. Some methodological problems of data from rain forest. *J. Ecol.* 56: 827-844.

AUSTIN, M. P. & I. NOY-MEIR, – 1971 – The problem of non-linearity in ordination: experiments with two-gradient models. *J. Ecol.* 59: 763-773.

AUSTIN, M. P. & L. ORLÓCI, – 1966 – Geometric models in ecology. II. An evaluation of some ordination techniques. *J. Ecol.* 54: 217-227.

BANNISTER, P., – 1968 – An evaluation of some procedures used in simple ordinations. *J. Ecol.* 56: 27-34.

BEALS, E., – 1960 – Forest bird communities in the Apostle Islands of Wisconsin. *Wilson Bull.* 72: 156-181.

BEALS, E. W., – 1973 – Ordination: mathematical elegance and ecological naïveté. *J. Ecol.* 61: 23-35.

BENZÉCRI, J.-P., – 1969 – Statistical analysis as a tool to make patterns emerge from data. In 'Methodologies of Pattern Recognition,' ed. S. WATANABE, pp. 35-74. Academic Press, New York.

BIRKS, H. J. B., – 1974 – Numerical zonations of Flandrian pollen data. *New Phytol.* 73: 351-358.

BRAY, J. R. & J. T. CURTIS, – 1957 – An ordination of the upland forest communities of northern Wisconsin. *Ecol. Monogr.* 27: 325-349.

BROWN, R. T. & J. T. CURTIS, – 1952 – The upland conifer-hardwood forests of northern Wisconsin. *Ecol. Monogr.* 22: 217-234.

BURR, E. J., – 1968 – Cluster sorting with mixed character types. I. Standardization of character values. *Aust. Comput. J.* 1: 97-99.

CASSIE, R. M., – 1963 – Multivariate analysis in the interpretation of numerical plankton data. *N.Z. J. Sci.* 6: 36-59.

ČEŠKA, A. & H. ROEMER, – 1971 – A computer program for identifying species-relevé groups in vegetation studies. (Germ. Summ.) *Vegetatio* 23: 255-277.

CHARDY, P., M. GLEMAREC & A. LAUREC, – 1976 – Application of inertia methods to benthic marine ecology: practical implications of the basic options. *Estuar. coast. mar. Sci.* 4: 179-205.

CRAWFORD, R. M. M. & D. WISHART, – 1967 – A rapid multivariate method for the detection and classification of groups of ecologically related species. *J. Ecol.* 55: 505-524.

CRAWFORD, R. M. M. & D. WISHART, – 1968 – A rapid classification and ordination method and its application to vegetation mapping. *J. Ecol.* 56: 385-404.

CURTIS, J. T., – 1955 – A prairie continuum in Wisconsin. *Ecology* 36: 558-566.

CURTIS, J. T. & R. P. MCINTOSH, – 1951 – An upland forest continuum in the prairie-forest border region of Wisconsin. *Ecology* 32: 476-496.

DAGNELIE, P., – 1960 – Contribution à l'étude des communautés végétales par l'analyse factorielle. (Engl. summ.) *Bull. Serv. Carte phytogéogr.*, Paris, Sér. B, 5: 7-71, 93-195.

DAHL, E., – 1957 – Rondane: mountain vegetation in South Norway and its relation to the environment. *Skr. norske Vidensk.-Akad.*, Mat.-naturv. Kl. 1956 (3): 1-374.

DALE, M. B., – 1964 – The application of multivariate methods to heterogeneous data. Ph.D. Thesis. Univ. of Southampton.

DALE, M. B., – 1975 – On objectives of methods of ordination. *Vegetatio* 30: 15-32.

DALE, M. B. & D. J. ANDERSON, – 1973 – Inosculate analysis of vegetation data. *Aust. J. Bot.* 21: 253-276.

DALE, M. B. & L. J. WEBB, – 1975 – Numerical methods for the establishment of associations. (Germ. summ.) *Vegetatio* 30: 77-87.

ELLENBERG, H., – 1950 – Landwirtschaftliche Pflanzensoziologie. I. Unkrautgemeinschaften als Zeiger für Klima und Boden. Ulmer, Stuttgart. 141 pp.

ELLENBERG, H., – 1952 – Landwirtschaftliche Pflanzensoziologie. II. Wiesen und Weiden und ihre standörtliche Bewertung. Ulmer, Stuttgart. 143 pp.

ELLENBERG, H., – 1956 – Grundlagen der Vegetationsgliederung. 1. Teil. Aufgaben und Methoden der Vegetationskunde. In 'Einführung in die Phytologie,' by H. WALTER, Vol. 4, pt. 1, 136 pp. Ulmer, Stuttgart.

FALINSKI, J. B., – 1960 – Zastosowanie taksonomii wrocławskiej do fitosocjologii. (Germ. summ.: Anwendung der sog. 'Breslauer Taxonomie' in der Pflanzensoziologie) *Acta Soc. Bot. Polon.* 29: 333-361.

FASHAM, M. J. R., – 1977 – A comparison of nonmetric multidimensional scaling, principal components and reciprocal averaging for the ordination of simulated coenoclines, and coenoplanes. *Ecology* 58: 551-561.

FISHER, R. A., – 1940 – The precision of discriminant functions. *Ann Eugen., Lond.* 10: 422-429.

FLENLEY, J. R., – 1969 – The vegetation of the Wabag region, New Guinea Highlands: a numerical study. *J. Ecol.* 57: 465-490.

FRYDMAN, I. & R. H. WHITTAKER, – 1968 – Forest associations of southeast Lublin Province, Poland. (Germ. summ.) *Ecology* 49: 896-908.

GAUCH, H. G., Jr. – 1973a – The relationship between sample similarity and ecological distance. *Ecology* 54: 618-622.

Gauch, H. G., Jr. – 1973b – A quantitative evaluation of the Bray-Curtis ordination. *Ecology* 54: 829-836.
Gauch, H. G., Jr. & G. B. Chase, – 1974 – Fitting the Gaussian curve to ecological data. *Ecology* 55: 1377-1381.
Gauch, H. G., Jr., G. B. Chase & R. H. Whittaker, – 1974 – Ordination of vegetation samples by Gaussian species distributions. *Ecology* 55: 1382-1390.
Gauch, H. G., Jr. & T. R. Wentworth, – 1976 – Canonical correlation analysis as an ordination technique. *Vegetatio* 33: 17-22.
Gauch, H. G., Jr. & R. H. Whittaker, – 1972a – Coenocline simulation. *Ecology* 53: 446-451.
Gauch, H. G., Jr. & R. H. Whittaker, – 1972b – Comparison of ordination techniques. *Ecology* 53: 868-875.
Gauch, H. G. Jr., R. H. Whittaker & T. R. Wentworth, – 1977 – A comparative study of reciprocal averaging and other ordination techniques. *J. Ecol.* 65: 157-174.
Gittins, R., – 1969 – The application of ordination techniques. In 'Ecological Aspects of the Mineral Nutrition of Plants,' ed. I. H. Rorison. *Symp. Brit. Ecol. Soc.*, 1968, 9: 37-66.
Gleason, H. A., – 1926 – The individualistic concept of the plant association. *Bull. Torrey bot. Club* 53: 7-26.
Goff, F. G. & G. Cottam, – 1967 – Gradient analysis: the use of species and synthetic indices. *Ecology* 48: 793-806.
Goodall, D. W., – 1954 – Objective methods for the classification of vegetation. III. An essay in the use of factor analysis. *Aust. J. Bot.* 2: 304-324.
Goodall, D. W., – 1963 – The continuum and the individualistic association. (French summ.) *Vegetatio* 11: 297-316.
Gordon, A. D. & H. J. B. Birks, – 1974 – Numerical methods in Quaternary palaeoecology. II. Comparison of pollen diagrams. *New Phytol.* 73: 221-249.
Gower, J. C., – 1966 – Some distance properties of latent root and vector methods used in multivariate analysis. *Biometrika* 53: 325:338.
Green, R. H., – 1974 – Multivariate niche analysis with temporally varying environmental factors. *Ecology* 55: 73-83.
Greig-Smith, P., – 1964 – Quantitative Plant Ecology, 2nd ed. Butterworths, London. 256 pp.
Greig-Smith, P., – 1971 – Analysis of vegetation data: the user viewpoint. In 'Statistical Ecology,' ed. G. P. Patil, E. C. Pielou & W. E. Waters, Vol. 3, pp. 149-166. Pennsylvania State Univ.
Greig-Smith, P., M. P. Austin & T. C. Whitmore, – 1967 – The application of quantitative methods to vegetation survey. I. Association-analysis and principal component ordination of rain forest. *J. Ecol.* 55: 483-503.
Groenewoud, H. van, – 1965 – Ordination and classification of Swiss and Canadian coniferous forests by various biometric and other methods. (Germ. summ.) *Ber. geobot. Inst. ETA, Stiftg. Rübel*, Zürich, 1964, 36: 28-102.
Groenewoud, H. van, – 1973 – Covariation of plant species along gradients. (manuscript).
Guttman, L., – 1959 – Metricizing rank-ordered or unordered data for a linear factor analysis. *Sankhyā* 21: 257-268.
Guttman, L., – 1968 – A general nonmetric technique for finding the smallest coordinate space for a configuration of points. *Psychometrika* 33: 469-506.
Guinochet, M., – 1973 – Phytosociologie. Masson, Paris. 227 pp.
Hall, A. V., – 1970 – A computer-based method for showing continua and communities in ecology. *J. Ecol.* 58: 591-602.
Hansen, H. Mølholm, – 1930 – Studies on the vegetation of Iceland. In 'The

Botany of Iceland,' ed. J. L. A. Kolderup Rosenvinge & E. Warming, 3 (pt. 1, no. 10): 1-186. Frimodt, Copenhagen.

Hansen, H. Mølholm, – 1932 – Nørholm Hede, en formationsstatistisk Vegetationsmonografi. (Engl. summ.) *K. danske Vidensk Selsk. Skr.*, Nat.-Math. Afd., Ser. 9, 3(3): 99-196.

Hill, M. O., – 1973 – Reciprocal averaging: an eigenvector method of ordination. *J. Ecol.* 61: 237-249.

Hill, M. O., – 1974 – Correspondence analysis: a neglected multivariate method. *J. R. statist. Soc.*, C, 23: 340-354.

Hill, M. O., R. G. H. Bunce & M. W. Shaw, – 1975 – Indicator species analysis, a divisive polythetic method of classification, and its application to a survey of native pinewoods in Scotland. *J. Ecol.* 63: 597-613.

Hinneri, S., – 1972 – An ecological monograph on eutrophic deciduous woods in the SW archipelago of Finland. *Annls. Univ åbo* (*turku.*), Ser. A, II, 50: 1-131.

Hirschfeld, H. O., – 1935 – A connection between correlation and contingency. *Proc. Camb. phil. Soc.* 31: 520-524.

Holloway, J. D. & N. Jardine, – 1968 – Two approaches to zoogeography: a study based on the distributions of butterflies, birds and bats in the Indo-Australian area. *Proc. linn. Soc. Lond.* 179: 153-188.

Ihm, P. & H. van Groenewoud, – 1975 – A multivariate ordering of vegetation data based on Gaussian type gradient response curves. *J. Ecol.* 63: 767-777.

Ivimey-Cook, R. B. & M. C. F. Proctor, – 1967 – Factor analysis of data from an East Devon heath: a comparison of principal component and rotated solutions. *J. Ecol.* 55: 405-413.

Ivimey-Cook, R. B., M. C. F. Proctor & D. L. Wigston, – 1969 – On the problem of 'R/Q' terminology in multivariate analyses of biological data. *J. Ecol.* 57: 673-675.

James, F. C., – 1971 – Ordinations of habitat relationships among breeding birds. *Wilson Bull.* 83: 215-236.

Jeglum, J. K., C. F. Wehrhahn & J. M. A. Swan, – 1971 – Comparisons of environmental ordinations with principal component vegetational ordinations for sets of data having different degrees of complexity. (French summ.) *Can. J. For. Res.* 1: 99-112.

Johnson, R., – 1973 – A study of some multivariate methods for the analysis of botanical data. Ph.D. Thesis. Utah State Univ., Logan.

Kaiser, H. F., – 1958 – The varimax criterion for analytic rotation in factor analysis. *Psychometrika* 23: 187-200.

Kessell, S. R. & R. H. Whittaker, – 1976 – Comparisons of three ordination techniques. *Vegetatio* 32: 21-29.

Kershaw, K. A., – 1968 – Classification and ordination of Nigerian savanna vegetation. *J. Ecol.* 56: 467-482.

Kruskal, J. B., – 1964a – Multidimensional scaling by optimizing goodness of fit to a nonmetric hypothesis. *Psychometrika* 29: 1-27.

Kruskal, J. B., – 1964b – Nonmetric multidimensional scaling: a numerical method. *Psychometrika* 29: 115-129.

Kruskal, J. B. & F. Carmone, – 1971 – How to use M-D-SCAL (Version 5M) and other useful information. Bell Telephone Laboratories, Murray Hill, New Jersey, U.S.A., and Univ. of Waterloo, Waterloo, Canada. (mimeographed).

Kruskal, J. B. & J. D. Carroll, – 1969 – Geometric models and badness-of-fit functions. In 'Multivariate Analysis,' ed. P. R. Krishnaiah 2: 639-671.

Lacoste, A. & M. Roux, – 1971 – L'analyse multidimensionnelle en phytosociologie et en écologie. Application à des données de l'étage subalpin des

Alpes maritimes. I. L'analyse des données floristiques. (Engl. & Germ. summs.) *Oecol. Plant.* 6: 353-369.

LaFrance, C. R., – 1972 – Sampling and ordination characteristics of computer-simulated individualistic communities. *Ecology* 53: 387-397.

Lambert, J. M., – 1972 – Theoretical models for large-scale vegetation survey. In 'Mathematical Models in Ecology.' ed. J. N. R. Jeffers, *Symp. brit. Ecol. Soc.* 1971, 12: 87-109. Blackwell, Oxford.

Lambert, J. M. & M. B. Dale, – 1964 – The use of statistics in phytosociology. *Adv. ecol. Res.* 2.:59-99

Lambert, J. M., S. E. Meacock, J. Barrs & P. F. M. Smartt, – 1973 – AXOR and MONIT: two new polythetic-divisive strategies for hierarchical classification. *Taxon* 22: 173-176.

Lambert, J. M. & W. T. Williams, – 1962 – Multivariate methods in plant ecology. IV. Nodal analysis. *J. Ecol.* 50: 775-802.

Lance, G. N. & W. T. Williams, – 1967 – Mixed-data classificatory programs. I. Agglomerative systems. *Aust. Comput. J.* 1: 15-20.

Lange, R. T., – 1968 – Influence analysis in vegetation. *Aust. J. Bot.* 16: 555-564.

Lieth, H. & G. W. Moore, – 1971 – Computerized clustering of species in phytosociological tables and its utilization for field work. In, 'Statistical Ecology,' ed. G. P. Patil, E. C. Pielou & W. E. Waters, Vol. 1, pp. 404-422. Pennsylvania State Univ.

Lloyd, M. & R. J. Ghelardi, – 1964 – A table for calculating the 'equitability' component of species diversity. *J. Anim. Ecol.* 33: 217-225.

Loucks, O. L., – 1962 – Ordinating forest communities by means of environmental scalars and phytosociological indices. *Ecol. Monogr.* 32: 137-166.

Luria, M., – 1975 – Annual plant communities of a loessial plain in the Negev. M.Sc. Thesis. Hebrew Univ., Jerusalem. (Hebrew with Engl. summ.).

Maarel, E. van der, – 1969 – On the use of ordination models in phytosociology. (Germ. summ.) *Vegetatio* 19: 21-46.

Maarel, E. van der, – 1972 – Ordination of plant communities on the basis of their plant genus, family and order relationships. (Germ. summ.) In 'Grundfragen und Methoden in der Pflanzensoziologie,' ed. E. van der Maarel & R. Tüxen. *Ber. Symp. int. Ver. Vegetationskunde*, Rinteln, 1970: 183-192.

matuszkiewicz, W., – 1948 – Roślinność lasów okolik Lwowa. (Engl. summ.: Vegetation of the forests of the environs of Lvov) *Annls Univ. Mariae Curie-Skłodowska*, Lublin, Sect. C, 3: 119-193.

McDonald, R. P., – 1962 – A general approach to nonlinear factor analysis. *Psychometrika* 27: 397-415.

McDonald, R. P., – 1966 – Applications of nonlinear factor analysis in ecology. *Symp. ecol. Soc. Aust.* (manuscript).

Moral, R. del, – 1975 – Vegetation clustering by means of Isodata: revision by multiple discriminant analysis. *Vegetatio* 29: 179-190.

Motyka, J., – 1947 – O zadaniach i metodach badán geobotanicznych. (French summ.: Sur les buts et les méthodes des recherches géobotaniques) *Annls Univ. Mariae Curie-Skłodowska*, Lublin, Sect. C (Suppl.), 1: 1-168.

Norris, J. M. & J. P. Barkham, – 1970 – A comparison of some Cotswold beechwoods using multiple-discriminant analysis. *J. Ecol.* 58: 603-619.

Noy-Meir, I., – 1970 – Component analysis of semi-arid vegetation in south-eastern Australia. Ph.D. Thesis. Australian National Univ.

Noy-Meir, I., – 1971 – Multivariate analysis of the semi-arid vegetation in south-eastern Australia: nodal ordination by component analysis. *Proc. ecol. Soc. Aust.* 6: 159-193.

Noy-Meir, I., – 1973a – Data transformations in ecological ordination. I. Some advantages of non-centering. *J. Ecol.* 61: 329-341.
Noy-Meir, I., – 1973b – Divisive polythetic classification of vegetation by optimized division on ordination components. *J. Ecol.* 61: 753-760.
Noy-Meir, I., – 1974a – Catenation: quantitative methods for the definition of coenoclines. *Vegetatio* 29: 89-99.
Noy-Meir, I., – 1974b – Multivariate analysis of the semiarid vegetation in south-eastern Australia. II. Vegetation catenae and environmental gradients. *Aust. J. Bot.* 22: 115-140.
Noy-Meir, I. & M. P. Austin, – 1970 – Principal component ordination and simulated vegetation data. *Ecology* 51: 551-552.
Noy-Meir, I., D. Walker & W. T. Williams, – 1975 – Data transformations in ecological ordination. II. On the meaning of data standardization. *J. Ecol.* 63: 779-800.
Noy-Meir, I. & R. H. Whittaker, – 1977 – Continuous multivariate methods in community analysis: some problems and developments. *Vegetatio* 33: 79-88.
Orloci, L., – 1966 – Geometric models in ecology. I. The theory and application of some ordination methods. *J. Ecol.* 54: 193-215.
Orloci, L., – 1967a – An agglomerative method for classification of plant communities. *J. Ecol.* 55: 193-206.
Orloci, L., 1967b – Data centering: a review and evaluation with reference to component analysis. *Syst. Zool.* 16: 208-212.
Orlóci, L., – 1972 – On objective functions of phytosociological resemblance. *Amer. Midl. Nat.* 88: 28-55.
Orloci, L., – 1973 – Ordination by resemblance matrices. (Germ. summ.) In 'Ordination and Classification of Communities,' ed. R. H. Whittaker. *Handb. Veget. Sci.* 5: 249-286. Junk, The Hague.
Orlóci, L., – 1974a – On information flow in ordination. *Vegetatio* 29: 11-16.
Orlóci, L., – 1974b – Revisions for the Bray and Curtis ordination. (French summ.) *Can. J. Bot.* 52: 1773-1776.
Orlóci, L., – 1975 – Multivariate Analysis in Vegetation Research. Junk, The Hague. 276 pp.
Pielou, E. C., – 1969 – An Introduction to Mathematical Ecology. Wiley, New York. 286 pp.
Pignatti, E. & S. Pignatti, – 1975 – Syntaxonomy of the *Sesleria varia*-grasslands of the calcareous Alps. (Germ. summ.) *Vegetatio* 30: 5-14.
Poore, M. E. D., – 1955 – The use of phytosociological methods in ecological investigations. II. Practical uses involved in an attempt to apply the Braun-Blanquet system. *J. Ecol.* 43: 245-269.
Poore, M. E. D., – 1956 – The use of phytosociological methods in ecological investigations. IV. General discussion of phytosociological problems. *J. Ecol.* 44: 28-50.
Ramensky, L. G., – 1926 – Die Grundgesetzmässigkeiten im Aufbau der Vegetationsdecke. *Bot. Zbl.* N.F. 7: 453-455.
Ramensky, L. G., – 1930 – Zur Methodik der vergleichenden Bearbeitung und Ordnung von Pflanzenlisten und andere Objecten, die durch mehrere, verschiedenartig wirkende Factoren bestimmt werden. *Beitr. Biol. Pfl.* 18: 269-304.
Romane, F., – –1972 – Utilisation de l'analyse multivariable en phytoécologie. *Investigación pesq.*, Barcelona 36: 131-139.
Seligman, N. G., – 1973 – A quantitative geobotanical analysis of the vegetation of the Golan. Ph.D. Thesis. Hebrew Univ., Jerusalem (Hebrew with English summary).
Shepard, R. N., – 1962 – The analysis of proximities: multidimensional scaling with an unknown distance function. I, II. *Psychometrika* 27: 125-140, 219-246.

SHEPARD, R. N. & J. D. CARROLL, – 1966 – Parametric representation of non-linear data structures. In 'Multivariate Analysis,' ed. P. R. KRISHNAIAH 1: 561-592. Academic Press, New York & London.
SHMIDA, A., – 1972 – The vegetation of Jebel Maghara, Sinai. M.Sc. Thesis. Hebrew Univ., Jerusalem (Hebrew with English summary).
SNEATH, P. H. A., – 1966 – A method of curve seeking from scattered points. *Comput. J.* 8: 383-391.
SØRENSEN, T., – 1948 – A method of establishing groups of equal amplitude in plant sociology based on similarity of species content and its application to analysis of the vegetation on Danish commons. *Biol. Skr.*, K. danske Vidensk. Selsk. 5 (4): 1-34.
SWAN, J. M. A., – 1970 – An examination of some ordination problems by use of simulated vegetational data. *Ecology* 51: 89-102.
SWAN, J. M. A. & R. L. DIX, – 1966 – The phytosociological structure of upland forest at Candle Lake, Saskatchewan. *J. Ecol.* 54: 13-40.
SWAN, J. M. A., R. L. DIX & C. F. WEHRHAHN, – 1969 – An ordination technique based on the best possible stand-defined axes and its application to vegetational analysis. *Ecology* 50: 206-212.
TORGERSON, W. S., – 1958 – Theory and Methods of Scaling. Wiley, New York.
TUOMIKOSKI, R., – 1942 – Untersuchungen über die Vegetation der Bruchmoore in Ostfinnland: I. Zur Methodik der pflanzensoziologischen Systematik. (Finn. summ.) *Ann. bot., Soc. Zool.-Bot. Fenn. 'Vanamo'* (*Suomal eläin-ja kasvit. Seur. van kasvit. Julk.*) 17(1): 1-203.
VRIES, D. M. DE, – 1953 – Objective combinations of species. *Acta bot. neerl.* 1: 497-499.
VRIES, D. M. DE, J. P. BARETTA & G. HAMMING, – 1954 – Constellation of frequent herbage plants, based on their correlation in occurrence. *Vegetatio* 5/6: 105-111.
WALKER, B. H., – 1974 – Some problems arising from the preliminary manipulation of plant ecological data for subsequent numerical analysis. *J. S. Afr. Bot.* 40: 1-13.
WEBB, D. A., – 1954 – Is the classification of plant communities either possible or desirable? *Bot. Tiddskr.* 51: 362-370.
WEBB, L. J., J. G. TRACEY, W. T. WILLIAMS & G. N. LANCE, – 1967 – Studies in the numerical analysis of complex rain-forest communities. I. A comparison of methods applicable to site/species data. *J. Ecol.* 55: 171-191.
WENTWORTH, T. R., – 1976 – The vegetation of limestone and granite soils in the mountains of southeastern Arizona. Ph.D. Thesis. Cornell Univ.
WESTHOFF, V., J. J. BARKMAN, H. DOING & C. G. VAN LEEUWEN, – 1959 – Enige opmerkingen over de terminologie in de vegetatiekunde. *Jaarb. Kon. ned. bot. Ver.* 1958: 44-46.
WESTMAN, W. E., – 1975 – Edaphic climax pattern of the pygmy forest region of California. *Ecol. Monogr.* 45: 109-135.
WHITTAKER, R. H., – 1951 – A criticism of the plant association and climatic climax concepts. *NW. Sci.* 25: 17-31.
WHITTAKER, R. H., – 1952 – A study of summer foliage insect communities in the Great Smoky Mountains. *Ecol. Monogr.* 22: 1-44.
WHITTAKER, R. H., – 1956 – The vegetation of the Great Smoky Mountains. *Ecol. Monogr.* 26: 1-80.
WHITTAKER, R. H., – 1960 – Vegetation of the Siskiyou Mountains, Oregon and California. *Ecol. Monogr.* 30: 279-338.
WHITTAKER, R. H., – 1962 – Classification of natural communities. *Bot. Rev.* 28: 1-239.

WHITTAKER, R. H., – 1967 – Gradient analysis of vegetation. *Biol. Rev.* 42: 207-264.

WHITTAKER, R. H., – 1972a – Convergences of ordination and classification. (Germ. summ.) In, 'Grundfragen und Methoden in der Pflanzensoziologie,' ed. E. VAN DER MAAREL & R. TÜXEN. *Ber. Symp. int. Ver. Vegetationskunde*, Rinteln 1970: 39-57.

WHITTAKER, R. H., – 1972b – Evolution and measurement of species diversity. *Taxon* 21: 213-251.

WHITTAKER, R. H. (ed.), – 1973 – Ordination and Classification of Communities. *Handb. Veget. Sci.* 5: 1-737. Junk, The Hague.

WHITTAKER, R. H., – 1975 – Communities and Ecosystems. 2nd ed. Macmillan, New York. 385 pp.

WHITTAKER, R. H. & C. W. FAIRBANKS, – 1958 – A study of plankton copepod communities in the Columbia Basin, southeastern Washington. *Ecology* 39: 46-65.

WHITTAKER, R. H. & H. G. GAUCH, Jr., – 1973 – Evaluation of ordination techniques. (Germ. summ.) In 'Ordination and Classification of Communities,' ed. R. H. WHITTAKER. *Handb. Veget. Sci.* 5: 287-321. Junk, The Hague.

WHITTAKER, R. H. & W. A. NIERING, – 1965 – Vegetation of the Santa Catalina Mountains, Arizona. (II.) A gradient analysis of the south slope. *Ecology* 46: 429-452.

WHITTINGTON, H. B. & C. P. HUGHES, – 1972 – Ordovician geography and faunal provinces deduced from trilobite distribution. *Phil. Trans. R. Soc.* London, Ser. B, 263: 235-278.

WILLIAMS, W. T., – 1971 – Principles of clustering. *Ann. Rev. Ecol. Syst.* 2: 303-326.

WILLIAMS, W. T., M. B. DALE & G. N. LANCE, – 1971 – Two outstanding ordination problems. *Aust. J. Bot.* 19: 251-258.

WILLIAMS, W. T. & J. M. LAMBERT, – 1961 – Multivariate methods in taxonomy. *Taxon* 10: 205-211.

WILLIAMS, W. T. & G. N. LANCE, – 1968 – Choice of strategy in the analysis of complex data. *Statistician* 18: 31-44.

WILLIAMSON, M., – 1972 – The relation of principal component analysis to the analysis of variance. *Int. J. Math. Educ. Sci. Technol.* 3: 35:42.

WOLD, H., – 1966 – Estimation of principal components and related models by iterative least squares. In 'Multivariate Analysis,' ed. P. R. KRISHNAIAH 1: 391-420. Academic Press, New York.

WOOD, K. R., R. L. MCCORMARCK & L. T. VILLONE, – 1964 – Nonlinear factor analysis program. A-78A. Tech. Memo 1764. Systems Development Corp., Santa Monica, California.

YARRANTON, G. A., – 1967 – Principal components analysis of data from saxicolous bryophyte vegetation at Steps Bridge, Devon. I. A quantitative assessment of variation in the vegetation. *Can. J. Bot.* 45: 93-115.

YARRANTON, G. A., W. J. BEASLEIGH, R. G. MORRISON & M. I. SHAFI, – 1972 – On the classification of phytosociological data into nonexclusive groups with a conjecture about determining the optimum number of groups in a classification. *Vegetatio* 24: 1-12.

INDEX

* (Articles 12–20 make up the second volume, 'Classification of Plant Communities')

O

P

Q

R

T

U

V